Complex Analysis with Applications in Science and Engineering

Harold Cohen

Complex Analysis with Applications in Science and Engineering

Second Edition

with 212 illustrations

 Springer

Harold Cohen
Department of Physics and Astronomy
California State University, Los Angeles
5151 State University Drive
Los Angeles, CA 90032
U.S.A.
hcohen@calstatela.edu

Cover illustrations of Riemann, Gauss, and Cauchy by Helen Cohen.

Mathematics Subject Classification (2000): 30-01, 30-xx

ISBN 978-1-4419-4456-6   ISBN 978-0-387-73058-5 (eBook)

Printed on acid-free paper.

© 2007 Springer Science+Business Media, LLC
Softcover reprint of the hardcover 2nd edition 2007

First published in 2003 by Kluwer Academic Publishers as *Fundamentals and Applications of Complex Analysis*, ISBN: 0-306-47748-3.

9 8 7 6 5 4 3 2 1

springer.com   (LAP/SB)

*This book is dedicated with love to my wife Helen, my daughter
Allyson, my son David, and my daughter Lisa.*

*As with its predecessor, "Fundamentals and Applications of Complex
Analysis," this book is also dedicated to the memory of Dr. Charles Pine,
who provided the academic nurturing I needed when I first discovered
the beauty of mathematics and its applications to scientific problems.*

I am, therefore I think.

Contents

Preface

This book is the second edition of a text, the first edition of which is titled *Fundamentals and Applications of Complex Analysis.* The title has been changed to more clearly reflect the fact that the book contains many applications in the sciences and engineering as well as applications to purely mathematical problems.

Like the first edition, it is intended to serve as a text for both beginning and second courses in complex analysis. The treatment begins at a very elementary level with complex numbers. Many students in a first course may have been introduced to complex numbers without ever having taken a formal course in complex analysis. For those students, chapters 1 and 2 will be a review. As such, in an introductory course, it may be possible to begin instruction with the material in chapter 3. Depending on the students' experiences, chapters 3, 4, and 5 might serve as review material for a second course and may be covered quickly.

The book contains material that I have not found presented in other popular texts. In particular,

- Chapter 6 contains an extensive treatment of multivalued functions with applications of the structure of such functions to the evaluation of certain types of integrals.
- Chapter 7 is devoted to the analysis of functions that are defined by integrals which cannot be evaluated in terms of elementary functions. Methods are developed for determining the singularities of these integral-defined functions.
- Chapter 9 is devoted to the development and application of dispersion relations. These provide mathematical tools for determining a complete function from knowledge of just its real part, or just its imaginary part.

- Chapter 10, in which analytic continuation of functions is discussed, did not appear in the first edition. It is presented in this second edition in a form that does not seem to be found in other texts in that the material contains many examples showing how one analytically continues representations of a function (such as series representations) from one domain to another.

In order to provide the reader with concrete applications of the material, the book relies heavily on examples and contains many of them. I have provided a table of examples (like a table of contents) with descriptive titles that can quickly direct the reader to these sections.

Some examples involve the types of problems the reader is expected to have encountered in other courses.

For example:

1. Students who have taken an introductory physics course will have encountered analysis of simple AC circuits. This text revisits such analysis using complex numbers.
2. Methods of conformal mapping are used to solve problems in electrostatics.
3. Cauchy's residue theorem is used to evaluate many different types of definite integrals, such as those that students encounter in a beginning calculus sequence.

Occasionally, the reader will encounter the notation in which an asterisk is used to denote multiplication, as is done in writing computer code. I have found it a convenient notation that avoids the ambiguity that is inherent in more common notations. For numeric or algebraic quantities, if needed, the notation $A*B$ is used to represent the product of A and B. The meaning of $A*B$ as the product is unambiguous, as is the meaning of the four commonly used notations, AB, $A \times B$, $A \cdot B$, and $(A)(B)$. There are instances however, when one of these common notations leads to confusion. For example, one never denotes x^2 by xx or by $x \times x$. (Particularly, when $x \times x$ is written by hand rather than typed, it can easily turn out to look like a string of three xs.) However, denoting x^2 by $x * x$ would never be interpreted as anything other than algebraic multiplication of x with itself. If $A = 2$ and $B = 3$, the meaning of 2×3 is clear as is the interpretation of $(2)(3)$, but 23 will always be interpreted as the number twenty three, and $2 \cdot 3$ will often be read as two and three tenths. But $2 * 3$ clearly indicates the multiplication of 2 and 3. As such, using the asterisk to denote multiplication provides a single notation that never leads to confusion. Therefore, whenever it is deemed necessary in this text, algebraic or numeric multiplication is denoted with an asterisk.

In material involving functions of two different complex variables, it is necessary to discuss two different complex planes. Examples of this include the evaluation of an integral of a function that depends on two different complex variables and the mapping of a region of one complex to some region of another complex plane. When it is necessary to indicate a particular complex plane in a figure (such as the z-plane or the w-plane), the complex variable defining that plane is placed on a small "flag" on the imaginary axis of the plane. An example of this is shown in figs. 8.1 on pp. 250, 251.

This manuscript was prepared by the author using Microsoft Word and MathType, an equation editor developed and marketed by Design Science Co. of Long Beach, California. Because of constraints encountered using these programs, it is sometimes necessary to position a mathematical expression in a sentence that does not fit on the same line as the text in that sentence. Such expressions have been placed on a separate line, centered on the page. They should be read as if they were text within the sentence. They are distinguished from equations in that they are in the center of the page, they do not contain an "equal", "not equal", or "inequality" symbol, and are not designated with an equation number. Equations are displayed starting close to the left margin, have an "equal", "not equal", or "inequality" symbol, and are identified by an equation number. An example of this "out of line" part of a sentence can be found in chapter 3 at the beginning of section 3.2 on p. 48.

I am very grateful to Professors Wallace Etterbeek (Mathematics Department, California State University, Sacramento) and Theodore Gamlin (Mathematics Department, University of California, Los Angeles) for their help in the development of the material about analytic continuation. In particular, I wish to express my gratitude to Professor Richard Katz (Mathematics Department, California State University, Los Angeles) for the time he spent reviewing the chapter on analytic continuation, and his patient explanations that helped me to understand this material more fully. I want to also acknowledge the contributions of Mr. Richard Simpson for his helpful reviews of, and comments about various parts of the text. This careful reading of the manuscript helped improve the presentation of the subject matter and eliminate errors that appeared in the first edition. Thanks are also expressed to those students in the mathematical methods of physics courses I have taught over the years whose curiosity and questions helped in the development of this material.

I am most indebted to my wife Helen Z. Cohen whose artistic talent, patience, and expertise with drawing software produced the cover of this book.

Harold Cohen
Los Angeles, 2006

Examples

9 DISPERSION RELATIONS

10 ANALYTIC CONTINUATION

Chapter 1

INTRODUCTION

1.1 A Brief History

Real Numbers

When the ratio of two integers is not an integer, it is called a *rational fraction*. Rational fractions fill in part of the intervals between the integers. The set of numbers consisting of integers and rational fractions is referred to as the set of *rational numbers*. A sample of these is shown graphically in fig. 1.1.

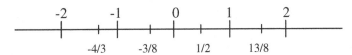

Figure 1.1

Graphical representation of a sample of rational numbers

Irrational numbers first arose in the study of solutions to nonlinear algebraic equations of integer order. For example, with $\alpha > 0$, the quadratic equation

$$x^2 - \alpha = 0 \tag{1.1a}$$

has solutions

$$x = \pm\sqrt{\alpha} \tag{1.1b}$$

When α is a perfect square (the square of an integer), the solutions to eq. 1.1a are integers. When α is the ratio of perfect squares (such as 1/4 and 9/16), the solutions to eq. 1.1a are rational fractions ($\pm 1/2$ and $\pm 3/4$ for this example). When α is not a perfect square or the ratio of perfect squares (numbers like 2, 5, 1/2, and 3/4), the values of x given in eq. 1.1b are irrational numbers. The set containing all integers, rational fractions and irrational numbers is referred to as the set of *real numbers*. This set consists of all the points on the axis of fig. 1.1.

Complex numbers

Girolamo (or Geronimo) Cardano first introduced *imaginary numbers* in the sixteenth century in the study of solutions to a quadratic equation of the form

$$x^2 + \beta = 0 \tag{1.2a}$$

with $\beta > 0$. The solutions to eq. 1.2a are

$$x = \pm\sqrt{-\beta} \tag{1.2b}$$

Cardano defined $\sqrt{-1}$ to be an "imagined" or imaginary number. In the eighteenth century Leonhard Euler introduced the symbol i to represent $\sqrt{-1}$. i has the property that

$$i^2 = \left(\sqrt{-1}\right)^2 = -1 \tag{1.3}$$

With this definition, eq. 1.2b can be written as

$$x = \pm i\sqrt{\beta} \tag{1.4}$$

Using $i = \sqrt{-1}$, a *complex number* is written as

$$z = x + iy \tag{1.5}$$

where x and y are real numbers. The part of z that is not multiplied by i (x in eq. 1.5) is called the *real part* of z and is denoted by

$$x = \text{Re}(z) \tag{1.6a}$$

The quantity that multiplies i (y in eq. 1.5) is referred to as the *imaginary part* of z and is denoted by

$$y = \text{Im}(z) \tag{1.6b}$$

The reader should be aware of the fact that $\text{Im}(z)$ is y, not iy. That is, $\text{Im}(z)$ does not contain a factor of i, and both $\text{Im}(z)$ and $\text{Re}(z)$ are real.

Argand diagram

The idea that a complex number could be used to represent a point in the two-dimensional coordinate plane had been developed by Karl Friedrich Gauss and proposed in his 1799 doctoral thesis at the University of Göttingen. However, Gauss did not publish this interpretation. It wasn't until 1806 that the interpretation appeared in print when an obscure mathematician named Jean Robert Argand published it in a privately printed book entitled *Essay on a Manner of Representing Imaginary Quantities in Geometric Constructions*. As such, the diagram used to describe points in the two-dimensional plane by a complex number is referred to as an *Argand diagram*. (See Abbott, 1985, pp. 11, 59 and Bell, 1937, pp. 232–234.)

Let a point in the plane have coordinates (x, y). It is evident from the Pythagorean theorem that the point is a distance r from the origin given by

$$r = \sqrt{x^2 + y^2} \tag{1.7a}$$

The line from the origin to the point makes an angle θ obtained from

$$\tan \theta = \frac{y}{x} \tag{1.7b}$$

Argand's idea was to represent the point graphically by the complex number

$$z = x + iy \tag{1.5}$$

by renaming the x and y axes as the *real* (Re) and *imaginary* (Im) *axes*, respectively. The graphical description of a complex number is the Argand diagram shown in fig. 1.2.

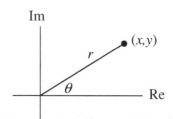

Figure 1.2
The Argand representation of a point in the complex plane

The idea of representing complex quantities as points in a two-dimensional plane paved the way for great progress and significant advances in the development of complex analysis, making it the elegant mathematical structure it has become.

Chapter 2

COMPLEX NUMBERS

2.1 Conjugation, Modulus, Argument, and Arithmetic

Complex conjugation

The *complex conjugate* of a complex number z is denoted by z^*. It is obtained by replacing i by $-i$ everywhere it appears in the complex number. The complex conjugate of

$$z = x + iy \tag{1.5}$$

is defined as

$$z^* = x - iy \tag{2.1}$$

The Argand diagrams of a complex number and its conjugate are shown in fig. 2.1.

Referring to eqs. 1.5 and 2.1, we see that

$$\text{Re}(z^*) = \text{Re}(z) \tag{2.2a}$$

and

$$\text{Im}(z^*) = -\text{Im}(z) \tag{2.2b}$$

Therefore, conjugation of a real number leaves the number unchanged and the complex conjugate of an imaginary number is its negative.

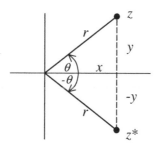

Figure 2.1
Argand diagram of z and $z*$

Example 2.1: Complex conjugation of selected complex numbers

The complex conjugate of

$$z = \ell n(1+i) \tag{2.3a}$$

is

$$z* = \ell n(1-i) \tag{2.3b}$$

Conjugation of

$$z = (1+i)^i \tag{2.4a}$$

yields

$$z* = (1-i)^{-i} \tag{2.4b}$$

Conjugating

$$z = i^{(3+i)} \tag{2.5a}$$

we have

$$z* = (-i)^{(3-i)} \tag{2.5b}$$

□

Modulus

The *modulus* or *magnitude* of a complex number z is denoted by $|z|$. It has a meaning similar to that of the modulus of a real number in that it represents the "size" of the number, that is, its distance from the origin. As discussed in chapter 1, the length of the radius line from the origin to a point with coordinates (x, y) is given by

$$r = |z| = \sqrt{x^2 + y^2} = \sqrt{(\text{Re}(z))^2 + (\text{Im}(z))^2} \tag{2.6}$$

For this reason, $|z|$ is sometimes referred to as the *length* of the complex number.

Argument

The angle θ in fig. 1.2 is called the *argument* of z. Denoted by

$$\theta = \arg(z) \tag{2.7a}$$

it is the angle, measured in the counterclockwise direction, from the positive real axis to the radius line and is found from

$$\tan \theta = \frac{\text{Im}(z)}{\text{Re}(z)} \tag{2.7b}$$

Referring to fig. 2.1, we see that

$$\arg(z*) = -\arg(z) \tag{2.8}$$

Addition and subtraction

Addition and subtraction of complex numbers are identical to addition and subtraction of real numbers. Thus,

$$z_1 \pm z_2 = (x_1 + iy_1) \pm (x_2 + iy_2) = (x_1 \pm x_2) + i(y_1 \pm y_2) \tag{2.9}$$

The Argand representations of two complex numbers and their sum are shown in figs. 2.2a and 2.2b.

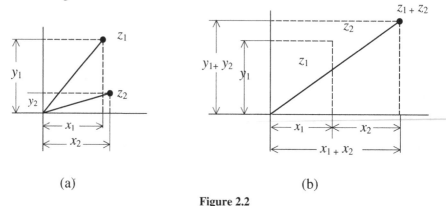

(a)   (b)

Figure 2.2

Argand diagram of (a) two complex numbers (b) the sum of two complex numbers

Comparing figs. 2.2b and 2.2c, we see that the sum of complex numbers results in the same line in the complex plane as the sum of two vectors in the x–y plane.

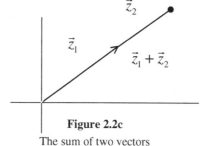

Figure 2.2c

The sum of two vectors

There are two possible ways to subtract two vectors, as shown in fig. 2.3. The direction associated with each difference vector makes vector subtraction unambiguous.

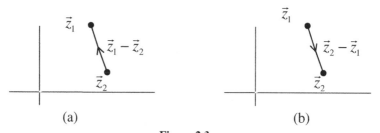

(a)   (b)

Figure 2.3

The difference between two vectors

However, the analogous differences between two complex numbers are not uniquely defined because complex numbers are not defined by directions. If the line between z_1 and z_2 in fig. 2.3c is $z_1 - z_2$, then the complex number represented by that line, when added to z_2 yields z_1.

$$z_1 = z_2 + (z_1 - z_2) \tag{2.10a}$$

This is analogous to the vector difference of fig. 2.3a. If that line represents $z_1 - z_2$, then the associated complex number, when added to z_1 yields z_2,

$$z_2 = z_1 + (z_2 - z_1) \tag{2.10b}$$

and is the analogue of the vector sum of fig. 2.3b.

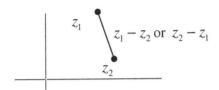

Figure 2.3c

The difference between two complex numbers

Multiplication

The product of two complex numbers is given by

$$z_1 z_2 = (x_1 + iy_1)(x_2 + iy_2) = (x_1 x_2 - y_1 y_2) + i(x_1 y_2 + x_2 y_1) \tag{2.11}$$

Thus,

$$\mathrm{Re}(z_1 z_2) = x_1 x_2 - y_1 y_2 \tag{2.12a}$$

and

$$\mathrm{Im}(z_1 z_2) = x_1 y_2 + x_2 y_1 \tag{2.12b}$$

The modulus of this product is

$$|z_1 z_2| = \sqrt{(x_1 x_2 - y_1 y_2)^2 + (x_1 y_2 + x_2 y_1)^2}$$
$$= \sqrt{(x_1^2 + y_1^2)}\sqrt{x_2^2 + y_2^2} = |z_1||z_2|$$

(2.13)

With $z_1 = z$ and $z_2 = z^*$, eq. 2.13 becomes

$$zz^* = x^2 + y^2 = r^2$$

(2.14)

Division

In order to determine the real and imaginary parts of

$$\frac{z_1}{z_2} = \frac{(x_1 + iy_1)}{(x_2 + iy_2)}$$

(2.15)

we *rationalize the denominator* of this complex number, which is the process of multiplying z_1/z_2 by 1 in the form z_2^*/z_2^*. Then

$$\frac{z_1 z_2^*}{z_2 z_2^*} = \frac{(x_1 + iy_1)}{(x_2 + iy_2)}\frac{(x_2 - iy_2)}{(x_2 - iy_2)} = \frac{(x_1 x_2 + y_1 y_2) + i(x_2 y_1 - x_1 y_2)}{(x_2^2 + y_2^2)}$$

(2.16)

Therefore,

$$\mathrm{Re}\left(\frac{z_1}{z_2}\right) = \frac{(x_1 x_2 + y_1 y_2)}{(x_2^2 + y_2^2)}$$

(2.17a)

and

$$\mathrm{Im}\left(\frac{z_1}{z_2}\right) = \frac{(x_2 y_1 - x_1 y_2)}{(x_2^2 + y_2^2)}$$

(2.17b)

Equality

The point $z = 0$ has the property that $\mathrm{Re}(z) = \mathrm{Im}(z) = 0$. If

$$z = z_1 - z_2 = (x_1 - x_2) + i(y_1 - y_2) = 0$$

(2.18)

then

$$\mathrm{Re}(z_1 - z_2) = x_1 - x_2 = 0 \quad \text{(2.19a)}$$

and

$$\text{Im}(z_1 - z_2) = y_1 - y_2 = 0 \tag{2.19b}$$

Therefore, if z_1 and z_2 are equal, then

$$\text{Re}(z_1) = \text{Re}(z_2) \tag{2.20a}$$

and

$$\text{Im}(z_1) = \text{Im}(z_2) \tag{2.20b}$$

2.2 Cartesian, Trigonometric, and Polar Forms

A complex number can be expressed in terms of x and y, its real and imaginary parts. The form

$$z = \text{Re}(z) + i\,\text{Im}(z) = x + iy \tag{1.5}$$

is called the *Cartesian* representation of z.

Referring to fig. 1.2, z can also be expressed in terms of its magnitude r and its argument θ. With

$$x = r\cos\theta \tag{2.21a}$$

and

$$y = r\sin\theta \tag{2.21b}$$

eq. 1.5 can be written

$$z = r(\cos\theta + i\sin\theta) \tag{2.22}$$

This is referred to as the *trigonometric* form of z. Conjugating eq. 2.22 and expressing the Pythagorean theorem in trigonometric form,

$$\cos^2\theta + \sin^2\theta = 1 \tag{2.23}$$

we again obtain the expected result

$$zz^* = r^2 \tag{2.24}$$

When several complex numbers have the same magnitude, the points they describe lie on the circumference of a circle centered at the origin. If the radius of the circle is 1, such a circle is called the *unit circle* and, from eq. 2.22, points on its circumference are of the form

$$z = \cos\theta + i\sin\theta \tag{2.25}$$

The variation of z along the circumference of the unit circle is given by

$$\frac{dz}{d\theta} = -\sin\theta + i\cos\theta = i^2\sin\theta + i\cos\theta = i\left(\cos\theta + i\sin\theta\right) = iz \tag{2.26a}$$

from which

$$\frac{dz}{z} = id\theta \tag{2.26b}$$

Integrating this, with the constraint $|z| = 1$, leads straightforwardly to

$$z = \cos\theta + i\sin\theta = e^{i\theta} \tag{2.27}$$

Therefore, from eq. 2.22, a complex number of modulus r and argument θ can be expressed as

$$z = re^{i\theta} \tag{2.28}$$

which is referred to as the *polar* or *exponential* form.
   The conjugate of z in polar form is

$$z^* = re^{-i\theta} \tag{2.29}$$

from which we easily obtain the expected result
$$zz^* = r^2 \tag{2.24}$$

Equating the polar and trigonometric forms of a complex number we obtain

$$re^{i\theta} = r\left(\cos\theta + i\sin\theta\right) \tag{2.30a}$$

and

$$re^{-i\theta} = r\left(\cos\theta - i\sin\theta\right) \tag{2.30b}$$

Example 2.2: Cartesian, trigonometric, and polar forms of selected complex numbers

(a) To express

$$z = \ell n(1+i) \tag{2.3a}$$

in the three standard forms, we first write $1+i$ in polar form. With modulus

$$\left|1+i\right| = \sqrt{2} \tag{2.31}$$

and argument given by

$$\tan\theta = 1 \Rightarrow \theta = \pi/4 \tag{2.32}$$

we obtain

$$1+i = \sqrt{2}e^{i\pi/4} \tag{2.33}$$

Therefore, the Cartesian form of $\ell n(1+i)$ is

$$\ell n(1+i) = \tfrac{1}{2}\ell n(2) + i\,\pi/4 \tag{2.34}$$

That is,

$$x = \operatorname{Re}\left[\ell n(1+i)\right] = \tfrac{1}{2}\ell n(2) \tag{2.35a}$$

and

$$y = \operatorname{Im}\left[\ell n(1+i)\right] = \pi/4 \tag{2.35b}$$

The modulus of $\ell n(1+i)$ is given by

$$r = \sqrt{\left(\tfrac{1}{2}\ell n(2)\right)^2 + \left(\pi/4\right)^2} \tag{2.36a}$$

and the argument is found from

$$\tan\theta = \frac{\pi/4}{\frac{1}{2}\ell n(2)} \tag{2.36b}$$

Then, in trigonometric and polar form,

$$\ell n(1+i) = \sqrt{\left(\tfrac{1}{2}\ell n(2)\right)^2 + \left(\pi/4\right)^2}$$

$$*\left\{\cos\left[\tan^{-1}\left(\frac{\pi/4}{\frac{1}{2}\ell n(2)}\right)\right] + i\sin\left[\tan^{-1}\left(\frac{\pi/4}{\frac{1}{2}\ell n(2)}\right)\right]\right\} \tag{2.37}$$

$$= \sqrt{\left(\tfrac{1}{2}\ell n(2)\right)^2 + \left(\pi/4\right)^2}\, e^{\,i\tan^{-1}\left(\frac{\pi/4}{\frac{1}{2}\ell n(2)}\right)}$$

(b) To express

$$z = (1+i)^i \tag{2.4a}$$

in Cartesian, trigonometric, and polar forms, we use the identity

$$s^k = e^{k\ln(s)} \tag{2.38}$$

to write

$$(1+i)^i = e^{i\ln(1+i)} \tag{2.39}$$

With

$$\ell n(1+i) = \tfrac{1}{2}\ell n(2) + i\,\pi/4 \tag{2.34}$$

the polar and trigonometric forms of $(1+i)^i$ are

$$(1+i)^i = e^{i\left[\frac{1}{2}\ln(2)+i\pi/4\right]} = e^{-\pi/4}e^{i\left[\frac{1}{2}\ln(2)\right]}$$

$$= e^{-\pi/4}\left[\cos\left(\tfrac{1}{2}\ell n(2)\right) + i\sin\left(\tfrac{1}{2}\ell n(2)\right)\right] \tag{2.40}$$

The Cartesian form of this complex number is obtained from eq. 2.40 by identifying

$$x = \text{Re}\left[(1+i)^i\right] = e^{-\pi/4} \cos\left(\tfrac{1}{2}\ell n(2)\right) \tag{2.41a}$$

and

$$y = \text{Im}\left[(1+i)^i\right] = e^{-\pi/4} \sin\left(\tfrac{1}{2}\ell n(2)\right) \tag{2.41b}$$

(c) To express

$$z = i^{(3+i)} \tag{2.5a}$$

in the three standard forms, we write

$$i^{(3+i)} = e^{(3+i)\ell n(i)} \tag{2.42}$$

Using the identity

$$i = \cos(\pi/2) + i\sin(\pi/2) = e^{i\pi/2} \tag{2.43}$$

we see that

$$\ell n(i) = i\pi/2 \tag{2.44}$$

Therefore, the polar, trigonometric, and Cartesian forms of $i^{(3+i)}$ are

$$i^{(3+i)} = e^{-\pi/2}e^{3i\pi/2} = e^{-\pi/2}\left[\cos\left(\frac{3\pi}{2}\right) + i\sin\left(\frac{3\pi}{2}\right)\right] = -ie^{-\pi/2} \tag{2.45}$$
$$\square$$

deMoivre's theorem

Writing the identity

$$\left(e^{i\theta}\right)^n = e^{in\theta} \tag{2.46}$$

in trigonometric form, we obtain

$$\left(\cos\theta + i\sin\theta\right)^n = \cos(n\theta) + i\sin(n\theta) \tag{2.47}$$

Expanding the left side of eq. 2.47, we obtain *deMoivre's theorem*

$$\sum_{m=0}^{n}(i)^m \frac{n!}{m!(n-m)!}\left(\cos^{n-m}\theta\right)\left(\sin^m\theta\right) = \cos(n\theta) + i\sin(n\theta) \tag{2.48}$$

Then

$$\cos(n\theta) = \mathrm{Re}\left[\sum_{m=0}^{n}(i)^m \frac{n!}{m!(n-m)!}\left(\cos^{n-m}\theta\right)\left(\sin^m\theta\right)\right] \tag{2.49a}$$

and

$$\sin(n\theta) = \mathrm{Im}\left[\sum_{m=0}^{n}(i)^m \frac{n!}{m!(n-m)!}\left(\cos^{n-m}\theta\right)\left(\sin^m\theta\right)\right] \tag{2.49b}$$

When m is even,

$$(i)^m = (-1)^{m/2} \tag{2.50a}$$

and when m is odd,

$$(i)^m = i(-1)^{(m-1)/2} \tag{2.50b}$$

Therefore,

$$\cos(n\theta) = \sum_{\substack{m=0 \\ m\ even}}^{n}(-1)^{m/2}\frac{n!}{m!(n-m)!}(\cos^{n-m}\theta)(\sin^m\theta) \tag{2.51a}$$

and

$$\sin(n\theta) = \sum_{\substack{m=1 \\ m\ odd}}^{n}(-1)^{(m-1)/2}\frac{n!}{m!(n-m)!}(\cos^{n-m}\theta)(\sin^m\theta) \tag{2.51b}$$

Example 2.3: deMoivre's theorem for sin(2θ) and cos(2θ)

For $n = 2$, eq. 2.47 becomes

$$\left(\cos\theta + i\sin\theta\right)^2 = \cos^2\theta - \sin^2\theta + 2i\sin\theta\cos\theta$$
$$= \cos(2\theta) + i\sin(2\theta) \tag{2.52}$$

Therefore,

$$\cos(2\theta) = \cos^2\theta - \sin^2\theta \tag{2.53a}$$

and

$$\sin(2\theta) = 2\sin\theta\cos\theta \tag{2.53b}$$
$\square$

Exponential representations of trigonometric functions

Setting $r = 1$, eqs. 2.30 become

$$e^{i\theta} = \cos\theta + i\sin\theta \tag{2.54a}$$

and

$$e^{-i\theta} = \cos\theta - i\sin\theta \tag{2.54b}$$

Adding these, we obtain

$$\cos\theta = \frac{e^{i\theta} + e^{-i\theta}}{2} \tag{2.55a}$$

Subtraction yields

$$\sin\theta = \frac{e^{i\theta} - e^{-i\theta}}{2i} \tag{2.55b}$$

Equations 2.55 are referred to as the *exponential representations* of $\cos\theta$ and $\sin\theta$. The exponential representations of $\tan\theta$, $\cot\theta$, $\sec\theta$, and $\csc\theta$ are obtained straightforwardly from these expressions. For example,

$$\tan\theta = -i\frac{e^{i\theta} - e^{-i\theta}}{e^{i\theta} + e^{-i\theta}} \tag{2.56}$$

Hyperbolic functions

The *hyperbolic functions* are defined from the trigonometric functions for imaginary angles. Let

$$\theta \equiv iw \tag{2.57}$$

with w real. Then eqs. 2.55 become

$$\cos(iw) = \frac{e^{i(iw)} + e^{-i(iw)}}{2} = \frac{e^w + e^{-w}}{2} \equiv \cosh w \tag{2.58}$$

and

$$\sin(iw) = \frac{e^{i(iw)} - e^{-i(iw)}}{2i} = i\frac{e^w - e^{-w}}{2} \equiv i\sinh w \tag{2.59a}$$

from which

$$\sinh w = \frac{e^w - e^{-w}}{2} \tag{2.59b}$$

The real functions, $\cosh w$ and $\sinh w$, are the *hyperbolic cosine* and *hyperbolic sine* of w. Other hyperbolic functions are defined from $\cosh w$ and $\sinh w$ analogous to the way other trigonometric functions are defined from $\cos\theta$ and $\sin\theta$. For example,

$$\tan(iw) = \frac{\sin(iw)}{\cos(iw)} = i\frac{\sinh w}{\cosh w} \equiv i\tanh w \tag{2.60a}$$

so that

$$\tanh w \equiv \frac{\sinh w}{\cosh w} = \frac{e^w - e^{-w}}{e^w + e^{-w}} \qquad (2.60b)$$

Referring to eqs. 2.58 and 2.59a, the Pythagorean theorem in terms of coshw and sinhw becomes

$$\cos^2(iw) + \sin^2(iw) = \cosh^2 w - \sinh^2 w = 1 \qquad (2.61)$$

2.3 Roots of Unity

Roots of 1

For any integer $k \geq 0$

$$e^{\pm 2\pi i k} = \cos(2\pi k) \pm i \sin(2\pi k) = 1 \qquad (2.62)$$

For an integer $N \geq 1$, the Nth root of 1 is given by

$$(1)^{1/N} = \left(e^{\pm 2\pi i k} \right)^{1/N} = e^{\pm 2\pi i k / N} = \cos\left(\frac{2\pi k}{N} \right) \pm i \sin\left(\frac{2\pi k}{N} \right) . \qquad (2.63)$$

Setting $k = 0$,

$$(1)^{1/N} = e^0 = 1 \qquad (2.64a)$$

The root for $k = 1$ is

$$(1)^{1/N} = e^{\pm 2\pi i / N} = \cos\left(\frac{2\pi}{N} \right) \pm i \sin\left(\frac{2\pi}{N} \right) \qquad (2.64b)$$

With $k = 2$,

$$(1)^{1/N} = e^{\pm 4\pi i / N} = \cos\left(\frac{4\pi}{N} \right) \pm i \sin\left(\frac{4\pi}{N} \right) \qquad (2.64c)$$

$$\vdots$$

When $k = N$, the root is

$$(1)^{1/N} = e^{\pm 2\pi i N / N} = 1 \tag{2.64d}$$

which is the root obtained for $k = 0$. Setting $k = N + 1$ we obtain

$$(1)^{1/N} = e^{\pm 2\pi i / N} \tag{2.64e}$$

which is the root found for $k = 1$. Thus, the range of k can be taken to be $0 \le k \le (N - 1)$, $1 \le k \le N$ or any equivalent range.

For example, when k is in the range $1 \le k \le N$, the integer $(N - k)$ is in the range $0 \le (N - k) \le (N - 1)$. Then,

$$(1)^{1/N} = e^{2\pi i k / N} \tag{2.65a}$$

and

$$(1)^{1/N} = e^{2\pi i (N-k)/N} = e^{-2\pi i k / N} \tag{2.65b}$$

Thus, the root obtained for the negative integer $-k$ with $1 \le k \le N$ is identical to the root obtained for the positive integer $N - k$ with $0 \le (N - k) \le (N - 1)$. Therefore, the sign of the exponent in eq. 2.62 can be taken to be positive and all the roots of 1 can be obtained from

$$(1)^{1/N} = e^{2\pi i k / N} = \cos\left(\frac{2\pi k}{N}\right) + i \sin\left(\frac{2\pi k}{N}\right) \tag{2.66}$$

with $0 \le k \le (N - 1)$, $1 \le k \le N$ or any equivalent range. Because any such range contains N different values of k, there are N distinct values of the Nth root of 1.

From the trigonometric form of the Pythagorean theorem, we see from eq. 2.66, that

$$\left|(1)^{1/N}\right| = 1 \tag{2.67}$$

That is, the roots of 1 represent points on the circumference of the unit circle.

Example 2.4: Two square roots and three cube roots of 1

(a) The square roots of 1 are given by

$$(1)^{1/2} = \left(e^{2\pi i k}\right)^{1/2} = e^{i\pi k} \tag{2.68}$$

For $k = 0$ the root is

$$(1)^{1/2} = e^0 = 1 \tag{2.69a}$$

and the root for $k = 1$ is

$$(1)^{1/2} = e^{i\pi} = -1 \tag{2.69b}$$

As expected, for $k = 2$, we obtain

$$(1)^{1/2} = e^{2\pi i} = 1 \tag{2.69c}$$

which is the root obtained with $k = 0$. Therefore, there are two distinct roots of 1. These are the two points on the unit circle shown in fig. 2.4a.

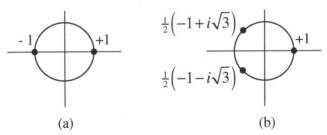

(a) (b)

Figure 2.4

(a) Two square roots of 1 and (b) three cube roots of 1 on the unit circle

(b) The cube roots of 1 are obtained from

$$(1)^{1/3} = e^{2\pi i k/3} = \cos\left(\frac{2\pi k}{3}\right) + i\sin\left(\frac{2\pi k}{3}\right) \tag{2.70}$$

The roots for $k = 0$, 1 and 2 are

$$(1)^{1/3} = e^0 = 1 \qquad (2.71a)$$

$$(1)^{1/3} = e^{2\pi i/3} = \cos\left(\frac{2\pi}{3}\right) + i\sin\left(\frac{2\pi}{3}\right) = -\frac{1}{2} + i\frac{\sqrt{3}}{2} \qquad (2.71b)$$

and

$$(1)^{1/3} = e^{4\pi i/3} = \cos\left(\frac{4\pi}{3}\right) + i\sin\left(\frac{4\pi}{3}\right) = -\frac{1}{2} - i\frac{\sqrt{3}}{2} \qquad (2.71c)$$

respectively. Of course, the root for $k = 3$,

$$(1)^{1/3} = e^{6\pi i/3} = 1 \qquad (2.71d)$$

is the same as the root obtained for $k = 0$. The three distinct roots of 1 are the points on the circumference of the unit circle shown in fig. 2.4b. □

Roots of −1

The Nth roots of −1 are found by writing

$$(-1)^{1/N} = e^{(i\pi + 2\pi i k)/N} = e^{i\pi(2k+1)/N} = \cos\left(\frac{\pi(2k+1)}{N}\right) + i\sin\left(\frac{\pi(2k+1)}{N}\right)$$

$$(2.72)$$

with $0 \le k \le (N-1)$ or some equivalent range of k. Just as with the roots of +1, the roots of −1 for $k \ge N$ are the same as those for $0 \le k \le (N-1)$. We see from eq. 2.72 that the roots of −1 are also of modulus 1 and so represent points on the unit circle.

Example 2.5: Two square roots and three cube roots of −1

(a) The two square roots of −1 are given by

$$(-1)^{1/2} = \left(e^{i\pi(2k+1)}\right)^{1/2} = e^{i\pi k}e^{i\pi/2} = ie^{i\pi k} \qquad (2.73)$$

For $k = 0$, and 1, the roots are

$$(-1)^{1/2} = ie^0 = i \tag{2.74a}$$

and

$$(-1)^{1/2} = ie^{i\pi} = -i \tag{2.74b}$$

As expected, the root obtained for $k = 2$ is

$$(-1)^{1/2} = ie^{2\pi i} = i \tag{2.74c}$$

which is the same as the root for $k = 0$.
    (b) The cube roots of -1 are given by

$$(-1)^{1/3} = e^{i\pi(2k+1)/3} = \cos\left(\frac{(2k+1)\pi}{3}\right) + i\sin\left(\frac{(2k+1)\pi}{3}\right) \tag{2.75}$$

Taking, for example, $k = 0$, 1, and 2, the roots are

$$(-1)^{1/3} = \cos\frac{\pi}{3} + i\sin\frac{\pi}{3} = \frac{1}{2} + i\frac{\sqrt{3}}{2} \tag{2.76a}$$

$$(-1)^{1/3} = \cos\pi + i\sin\pi = -1 \tag{2.76b}$$

and

$$(-1)^{1/3} = \cos\left(\frac{5\pi}{3}\right) + i\sin\left(\frac{5\pi}{3}\right) = \frac{1}{2} - i\frac{\sqrt{3}}{2} \tag{2.76c}$$

    Because $7\pi/3 = 2\pi + \pi/3$, it is straightforward that the root for $k = 3$ is identical to the root for $k = 0$. □

2.4 Complex Numbers and AC Circuits

**(Optional; requires the reader to have some knowledge of DC circuit
         analysis at the level of an introductory physics or electrical**

engineering course. See, for example, Serway and Jewett, 2004, pp. 835–836, 862–865).

When one or more electrical devices are connected to a battery, the circuit is a *Direct Current* (*DC*) circuit. This designation comes from the fact that the current delivered by the battery always flows in one direction. By convention, that direction is from the positive to the negative terminal of the battery.

In a circuit driven by an AC generator (such as the generator that delivers power to the household wall socket), the current reverses direction regularly, commonly many times per second. A circuit powered by this type of generator is an *Alternating Current* (*AC*) circuit. The number of reversals of the direction of the current in one second is called the *frequency*. For example, the frequency of the AC current delivered by electric utility companies in the United States is 60 oscillations per second, or 60 hertz (Hz).

DC circuits with resistors

Two resistors can be connected to a battery in either of two ways. The circuit diagram of fig. 2.5a illustrates resistors connected in *series*. Figure 2.5b is the circuit diagram for two resistors in *parallel*.

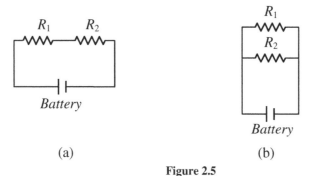

(a)   (b)

Figure 2.5

Two resistors in a DC circuit in (a) series (b) parallel

Each of the circuits of figs. 2.5 can be replaced by a circuit with a single *equivalent resistor* which draws the same current from the battery as the original circuit does. For N resistors in series, the resistance of the equivalent resistor is given by

$$R_{eq} = R_1 + R_2 + \cdots + R_N \quad \text{(series)} \tag{2.77a}$$

The resistance of the resistor that is equivalent to N resistors in parallel is found from

$$\frac{1}{R_{eq}} = \frac{1}{R_1} + \frac{1}{R_2} + \cdots + \frac{1}{R_N} \quad \text{(parallel)} \qquad (2.77b)$$

Using these expressions, one can reduce a complicated circuit comprised of subcircuits with series and parallel combinations of resistors, to a circuit containing a single equivalent resistor.

Example 2.6: Finding the equivalent resistance of a DC circuit

Referring to the circuit of fig. 2.6, subcircuit $\underline{A}$ contains an $8\ \Omega$ resistor in series with a $10\ \Omega$ resistor. From eq. 2.77a, this series combination can be replaced by a single resistor of resistance

$$R_{8,10} = 8 + 10 = 18\,\Omega \qquad (2.78)$$

This $18\,\Omega$ of resistance is in parallel with a $9\,\Omega$ resistor. Therefore, the entire subcircuit $\underline{A}$ can be replaced by one $6\,\Omega$ resistor as found from eq. 2.77b;

$$\frac{1}{R_A} = \frac{1}{18} + \frac{1}{9} = \frac{1}{6} \qquad (2.79)$$

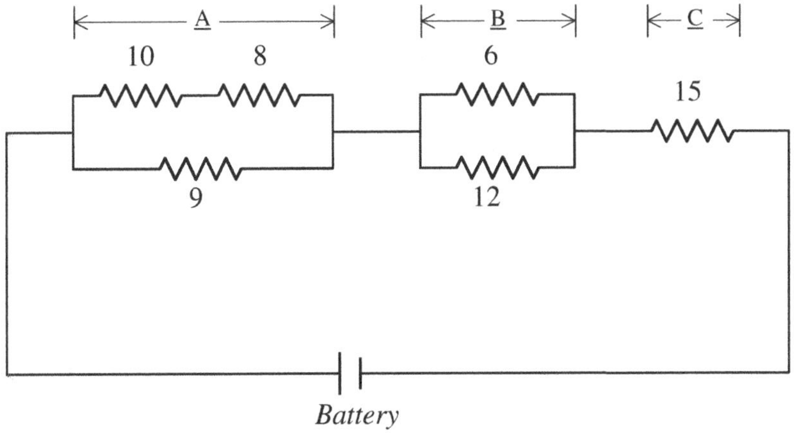

Figure 2.6
Multiple resistor DC circuit (resistances given in ohms)

The resistors in parallel in subcircuit $\underline{B}$ can be replaced by a single resistor, the resistance of which is obtained by

$$\frac{1}{R_B} = \frac{1}{12} + \frac{1}{6} = \frac{1}{4} \tag{2.80}$$

The circuit of fig. 2.6 is therefore equivalent to one containing resistors of resistances 6 Ω, 4 Ω, and 15 Ω in series. This circuit is equivalent to one with a single resistor of resistance given by

$$R_{eq} = 6 + 4 + 15 = 25\,\Omega \tag{2.81}$$
□

AC circuits with resistors, capacitors, and inductors

As shown in appendix 1, an AC circuit can be analyzed in the same way as a DC circuit containing resistors by using complex numbers. Resistors, capacitors, and inductance coils in an AC circuit are treated as resistors with complex impedences given by

$$Z_R = R \tag{A1.11a}$$

$$Z_C = -iX_C \tag{A1.11b}$$

and

$$Z_L = iX_L \tag{A1.11c}$$

where R is the resistance of the resistor and the capacitive and inductive reactances are given by

$$X_C = \frac{1}{2\pi fC} \tag{A1.5}$$

and

$$X_L = 2\pi fL \tag{A1.7}$$

In these expressions, f is the frequency of the AC generator, C is the capacitance of the capacitor measured in farads, and L is the inductance of the inductor measured in henrys.

With these assignments, each subcircuit can be reduced to an equivalent circuit containing two devices in series. This is achieved by replacing several devices in a series subcircuit by a two-device series combination with equivalent impedance given by

$$Z_{eq} = Z_1 + Z_2 + \cdots Z_N \tag{2.82a}$$

and replacing several devices in a parallel subcircuit by a two device series subcircuit with equivalent impedence given by

$$\frac{1}{Z_{eq}} = \frac{1}{Z_1} + \frac{1}{Z_2} + \cdots + \frac{1}{Z_N} \tag{2.82b}$$

Each resulting two-device series subcircuit is combined with other such subcircuits until the original AC circuit is reduced to two devices in series (or in parallel if one chooses) connected to the generator.

Example 2.7: Analysis of an AC circuit using complex numbers

Referring to eqs. A1.5 and A1.7, the reactances of the devices in the circuit of fig. 2.7 are

$$X_C = \frac{1}{2\pi fC} = 13.3\,\Omega \tag{2.83a}$$

and

$$X_L = 2\pi fL = 11.3\,\Omega \tag{2.83b}$$

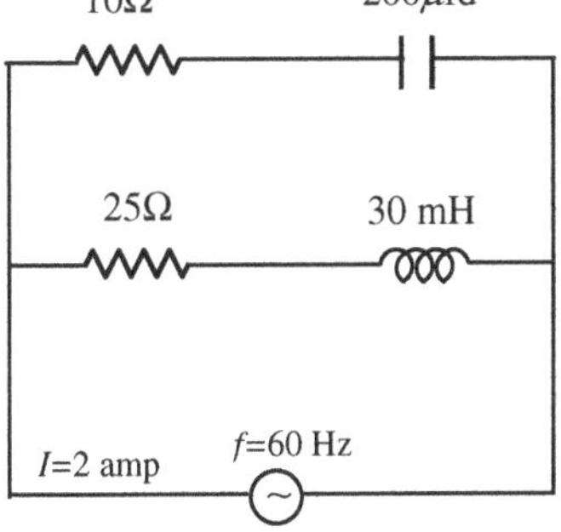

Figure 2.7
An AC circuit

Therefore, the net impedence of the upper branch of the circuit is

$$Z_1 = R_1 - iX_C = (10.0 - 13.3i)\,\Omega \tag{2.84a}$$

and the lower branch has impedence

$$Z_2 = R_2 + iX_L = (25.0 + 11.3i)\,\Omega \tag{2.84b}$$

Because these two branches are in parallel, the equivalent impedance of the circuit is given by

$$\frac{1}{Z_{eq}} = \frac{1}{Z_1} + \frac{1}{Z_2} = \frac{1}{10.0 - 13.3i} + \frac{1}{25.0 + 11.3i}$$

$$= \frac{10.0 + 13.3i}{10.0^2 + 13.3^2} + \frac{25.0 - 11.3i}{25.0^2 + 11.3^2} = .069 + .033i \tag{2.85}$$

from which

$$Z_{eq} = R_{eq} + iX_{eq} = \frac{1}{.069 + .033i} = (11.79 - 5.64i)\,\Omega \tag{2.86}$$

Then, because the circuit draws a real current of 2 amps, the net voltage is given by

$$V_{net} = IZ_{eq} = (23.58 - 11.28i)\ \text{volts} \tag{2.87}$$

Referring to eq. 2.7b and to fig. 2.8, the phase angle between voltage and current is given by

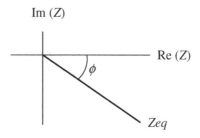

Figure 2.8
Phase diagram for the circuit of figure 2.7

$$\phi = \tan^{-1}\left(\frac{X_{eq}}{R_{eq}}\right) = \tan^{-1}\left(-\frac{5.64}{11.79}\right) = -25.57° = -0.45 \text{ rad} \qquad (2.88)$$

Because the current is real and the imaginary part of Z_{eq} (and therefore V_{eq}) is negative, the net voltage lags behind the current. This is the same phase relation (ICE) as that of a circuit containing a resistor and a capacitor in series. Such a circuit, called a *capacitive* series circuit, can be replaced by an equivalent one containing a resistor of resistance

$$R_{eq} = \text{Re}(Z_{eq}) \qquad (2.89a)$$

in series with a capacitor of reactance given by

$$X_{C_{eq}} = \left|\text{Im}(Z_{eq})\right| \qquad (2.89b)$$

For the circuit of this example, we see from eq. 2.86 that the equivalent resistance is 11.79 Ω and the equivalent capacitance is obtained from

$$X_{C_{eq}} = \frac{1}{2\pi f C_{eq}} = 5.64\,\Omega \qquad (2.90)$$

With $f = 60$ hz, we obtain

$$C_{eq} = 470.32\,\mu fd \qquad (2.91)$$

$\square$

In a circuit in which the imaginary part of Z_{eq} is positive, the voltage is ahead of the current (ELI). This is characteristic of an AC circuit containing a resistor and an inductor in series. The equivalent resistance of such an *inductive* series circuit is

$$R_{eq} = \text{Re}(Z_{eq}) \qquad (2.92a)$$

The equivalent inductance of the inductive series circuit is found from

$$X_{L_{eq}} = \text{Im}(Z_{eq}) = 2\pi f L_{eq} \qquad (2.92b)$$

Problems

1. Let $z = 3 + 2i$. Construct an Argand diagram of each of the following complex numbers.

   (a) $-z$ (b) $2z$ (c) iz (d) $-z^*$ (e) z^2 (f) $-\left(z^*\right)^2$ (g) zz^*

2. Prove that

   (a) $\arg(z_1 z_2) = \arg(z_1) + \arg(z_2)$ (b) $\arg(z_1/z_2) = \arg(z_1) - \arg(z_2)$

   Determine

   (c) $\arg(z_1 z_2^*)$ (d) $\arg(z_1^*/z_2)$ (e) $\arg(z_1^*/z_2^*)$

   in terms of $\arg(z_1)$ and $\arg(z_2)$.

3. Express each of the following complex numbers in its Cartesian, trigonometric, and polar representations.

   (a) $\dfrac{\left[(1+2i)^2\right]^*}{(7+i)}$ (b) $\left(\sqrt{3}-i\right)^{-i}$ (c) $\cos\left(\dfrac{\pi}{4} + i\dfrac{\pi}{3}\right)$ (d) $\ln\left(-\sqrt{3}\right)$

   (e) $e^{\sqrt{1+i}}$ (f) $e^{ie^{i\theta}}$ θ real   (g) $\left(i^i\right)^i$ (h) $\left[i^{\left(i^i\right)}\right]^*$

   (i) $\sinh(1+i)$ (j) $\ell n\left[\dfrac{1}{3+4i} - \dfrac{1}{3-4i}\right]$

4. z_1 and z_2 are complex numbers. Find all values of the real numbers A and B that satisfy $z_1 = z_2$ when

   (a) $z_1 = (A+3) + i(B+2)$ and $z_2 = (B-3) + i(2-A)$

   (b) $z_1 = \left(A^2 + 2AB\right) + i(B-1)$ and $z_2 = \left(A^2 - 2A + 2B\right) - i(A+B)$

5. If $z = A + iB$, what values of A and B satisfy

(a) $z = \dfrac{1}{2z}$ (b) $z^* = \dfrac{1}{2 - z^*}$

6. For $z \neq 1$,

$$S = \sum_{\ell=0}^{N} z^{\ell} = 1 + z + z^2 + \cdots + z^N = \frac{1 - z^N}{1 - z}$$

(see appendix 3.) Set $z = e^{i\theta}$, and with θ not an integer multiple of 2π, show that

$$S_s = \sum_{\ell=0}^{N} \sin(\ell\theta) = \frac{\sin\left[\frac{1}{2}(N-1)\theta\right]\sin\left[\frac{1}{2}N\theta\right]}{\sin\left[\frac{1}{2}\theta\right]}$$

and

$$S_c = \sum_{\ell=0}^{N} \cos(\ell\theta) = \frac{\cos\left[\frac{1}{2}(N-1)\theta\right]\sin\left[\frac{1}{2}N\theta\right]}{\sin\left[\frac{1}{2}\theta\right]}$$

7. (a) Use deMoivre's theorem to derive expressions for $\sin(3\theta)$ and $\cos(3\theta)$ in terms of $\sin\theta$ and $\cos\theta$.
   (b) Use the results of part (a) to derive expressions for $\sinh(3w)$ and $\cosh(3w)$ in terms of $\sinh w$ and $\cosh w$.

8. (a) Use deMoivre's theorem to prove

   (i) $\sin(\theta + \phi) = \sin\theta\cos\phi + \cos\theta\sin\phi$

   (ii) $\cos(\theta + \phi) = \cos\theta\cos\phi - \sin\theta\sin\phi$

   (b) Use the results of part (a) to derive expressions for $\cosh(v + w)$ and $\sinh(v + w)$ in terms of $\cosh v$, $\sinh v$, $\cosh w$, and $\sinh w$.

9. The equations $1 + \tan^2\theta = \sec^2\theta$ and $1 + \cot^2\theta = \csc^2\theta$ are two expressions of the Pythagorean theorem. What are the analogous expressions of the Pythagorean theorem for $\tanh w$, $\coth w$, $\operatorname{sech} w$, and $\operatorname{csch} w$?

10. Determine the exponential representations of coth w, sech w, and csch w.

11. (a) Use the exponential representation for $z = \cosh w$ given in eq. 2.58
to show that

$$w = \cosh^{-1}(z) = \ell n\left(z \pm \sqrt{z^2 - 1}\right)$$

(b) Derive an analogous expression for $w = \sinh^{-1}z$ using eq. 2.59b.
(c) Use eqs. 2.55 to derive analogous expressions for

(i) $\theta = \cos^{-1}(z)$ (ii) $\theta = \sin^{-1}(z)$

12. (a) Find the real and imaginary parts for each of the following sums.

$$S_{a1}^+ = \sum_{\ell=0}^{3} i^\ell \qquad S_{a2}^+ = \sum_{\ell=0}^{4} i^\ell \qquad S_{a1}^- = \sum_{\ell=0}^{3} (-i)^\ell \qquad S_{a2}^- = \sum_{\ell=0}^{4} (-i)^\ell$$

(b) $M \geq 2$ is an integer. Using the results of part (a), evaluate each of
the sums below.

$$S_{b1}^+ = \sum_{\ell=0}^{4M} i^\ell \qquad S_{b2}^+ = \sum_{\ell=0}^{4M-1} i^\ell \qquad S_{b3}^+ = \sum_{\ell=0}^{4M-2} i^\ell \qquad S_{b4}^+ = \sum_{\ell=0}^{4M-3} i^\ell$$

$$S_{b1}^- = \sum_{\ell=0}^{4M} (-i)^\ell \quad S_{b2}^- = \sum_{\ell=0}^{4M-1} (-i)^\ell \quad S_{b3}^- = \sum_{\ell=0}^{4M-2} (-i)^\ell \quad S_{b4}^- = \sum_{\ell=0}^{4M-3} (-i)^\ell$$

13. (a) Find all the distinct fourth roots of $+1$ and show their positions on the
circumference of the unit circle.
(b) Find all the distinct fourth roots of -1 and show their positions on
the circumference of the unit circle.

14. For what values of α does $z = 1/2 + i(\alpha - 1)$ lie on the unit circle?

15. Every point on the circumference of the unit circle satisfies $|z| = 1$.
    (a) Show that for each such point

$$\text{Re}\left(\frac{1}{1+z}\right) = \text{Re}\left(\frac{1}{1-z}\right) = \frac{1}{2}$$

   The reader should note that at the points where the unit circle crosses the real axis, this result is meant in the sense of a limit as $z \to \pm 1$.
    (b) Prove that the only points on the circumference of the unit circle that satisfy

$$\text{Im}\left(\frac{1}{1+z}\right) = \text{Im}\left(\frac{1}{1-z^*}\right)$$

   are the points where the unit circle crosses the imaginary axis.

16. If α and β are positive real numbers,
    (a) Show that

$$e^{i\alpha} - e^{i\beta} = 2ie^{i(\alpha+\beta)/2}\sin\left(\frac{\alpha-\beta}{2}\right)$$

    (b) Derive an expression equivalent to that of part (a) for $e^{i\alpha} + e^{i\beta}$.

17. If α is a real positive number, prove that the complex number

$$z = \alpha - \sqrt{(1+\alpha^2)}e^{i\,\tan^{-1}(1/\alpha)}$$

   is a point on the unit circle.

18. For α real and $\alpha^2 < 1$, prove that the two solutions to each quadratic equation below represent points on the circumference of the unit circle.

   (a) $z^2 + 2i\alpha z - 1 = 0$ (b) $z^2 - 2i\alpha z - 1 = 0$

   (c) $z^2 + 2\alpha z + 1 = 0$ (d) $z^2 - 2\alpha z + 1 = 0$

19. In the circuits of figs. P2.1, $R = 20\,\Omega$, $L = 50$ mH, and $C = 200$ μfd and the generator frequency is $f = 100$ Hz. Find the equivalent two-element series circuit for each circuit.

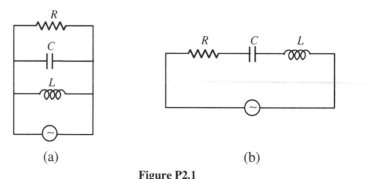

(a)   (b)

Figure P2.1

An RLC (a) parallel circuit (b) series circuit

20. (a) For each circuit shown in figs. P2.1, find the frequency of the generator in terms of R, L, and C so that the circuit is equivalent to a circuit with just a resistor. (A circuit that is equivalent to one with only a resistor is called a *resonant* circuit.) What is the resistance of the resistor of each of the resonant circuits?

   (b) If the circuit of ex. 2.7 were a resonant circuit, what would be the frequency of the generator? What would be the resistance of the resonant circuit?

21. Find the two-element parallel circuit equivalent to the circuit of ex. 2.7. That is, find the element that is in parallel with the equivalent resistor, find its device number (C or L) and find the resistance of the equivalent resistor.

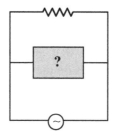

Figure P2.2

An equivalent 2-element parallel circuit

22. The parallelogram of fig. P2.3 is defined by sides z_1 and z_2. Prove that the area of this parallelogram is given by

$$A_p = \left| \mathrm{Im}\left(z_1^* z_2 \right) \right|$$

23. A vector A in the x-y plane can be expressed either in terms of unit basis vectors e_x and e_y or as a complex quantity A. That is,

$$A = A_x e_x + A_y e_y \Rightarrow A = A_x + i A_y$$

The x-y plane is defined by nonparallel vectors A and B. Let A and B be their equivalent complex representations. Show that the dot product of the two vectors can be written in terms of their complex equivalents as

$$A \bullet B = \mathrm{Re}\left(A^* B \right)$$

and that the cross-product of the two vectors (which perpendicular to the x-y plane) is given in terms of their complex equivalents by

$$(A \times B) = \mathrm{Im}\left(A^* B \right)$$

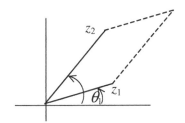

Figure P2.3
Parallelogram with sides z_1 and z_2

Chapter 3

COMPLEX VARIABLES

When the real and/or imaginary parts of z are variables instead of constants, z is called a *complex variable* instead of a complex number. Then a function $F(x, y)$ can be viewed as a function of z, and calculus operations can be performed on $F(z)$.

3.1 Derivatives, Cauchy–Riemann Conditions, and Analyticity

Derivatives and the Cauchy–Riemann conditions

The derivative of a function of one variable is defined by

$$\frac{dF}{dx} \equiv \lim_{\Delta x \to 0} \frac{F(x + \Delta x) - F(x)}{\Delta x} \tag{3.1a}$$

Because z is a linear combination of two variables, the meaning of dF/dz is ambiguous. Referring to eq. 3.1a, the most reasonable way to define dF/dz is

$$\frac{dF}{dz} \equiv \lim_{\Delta z \to 0} \frac{F(z + \Delta z) - F(z)}{\Delta z} = \lim_{\substack{\Delta x \to 0 \\ \Delta y \to 0}} \frac{F(x + \Delta x, y + \Delta y) - F(x, y)}{\Delta x + i\Delta y}$$

$$\tag{3.1b}$$

If dF/dz is unique, the limit must be independent of how Δz approaches zero. That is, dF/dz has meaning only if

$$\lim_{\Delta z \to 0} \frac{F(x + \Delta x, y + \Delta y) - F(x, y)}{\Delta x + i \Delta y}$$

$$= \lim_{\Delta x \to 0} \left[\lim_{\Delta y \to 0} \frac{F(x + \Delta x, y + \Delta y) - F(x, y)}{\Delta x + i \Delta y} \right] \qquad (3.2)$$

$$= \lim_{\Delta y \to 0} \left[\lim_{\Delta x \to 0} \frac{F(x + \Delta x, y + \Delta y) - F(x, y)}{\Delta x + i \Delta y} \right]$$

As indicated in the second line of eq. 3.2, when we first set $\Delta y = 0$, then take the limit $\Delta x \to 0$, we obtain

$$\frac{dF}{dz} = \lim_{\Delta x \to 0} \frac{F(x + \Delta x, y) - F(x, y)}{\Delta x} = \frac{\partial F}{\partial x} \qquad (3.3a)$$

When we set $\Delta x = 0$ first, then take the limit $\Delta y \to 0$, as indicated in the third line of eq. 3.2, we obtain

$$\frac{dF}{dz} = \lim_{\Delta y \to 0} \frac{F(x, y + \Delta y) - F(x, y)}{i \Delta y} = -i \frac{\partial F}{\partial y} \qquad (3.3b)$$

Because dF/dz has meaning only if these limits are the same, $F(z)$ must satisfy

$$\frac{\partial F}{\partial x} = -i \frac{\partial F}{\partial y} \qquad (3.4)$$

Equation 3.4 is referred to as the *Cauchy–Riemann* (abbreviated *CR*) *condition* expressed in terms of $F(x, y)$.

It is also convenient to express this CR condition in terms of the real and imaginary parts of $F(z)$. Let $U(x, y)$ and $V(x, y)$ be two real functions defined by

$$F(z) \equiv U(x, y) + i V(x, y) \qquad (3.5)$$

Substituting this into eq. 3.4, we have

$$\frac{\partial U}{\partial x} + i\frac{\partial V}{\partial x} = -i\frac{\partial U}{\partial y} + \frac{\partial V}{\partial y} \qquad (3.6)$$

From the equality of the real and imaginary parts of eq. 3.6, we obtain

$$\frac{\partial U}{\partial x} = \frac{\partial V}{\partial y} \qquad (3.7a)$$

and

$$\frac{\partial U}{\partial y} = -\frac{\partial V}{\partial x} \qquad (3.7b)$$

These are the CR conditions expressed in terms of the real and imaginary parts of $F(z)$. Both of these equations must be satisfied in a region R containing z in order for dF/dz to be defined at the point z. If either one of these conditions is not satisfied, then dF/dz is not defined at z.

The CR conditions can also be expressed in polar coordinates r and θ. A a continuous, differentiable function $F(x, y)$ can be expressed as $F(r,\theta)$ using

$$x = r\cos\theta \qquad (2.21a)$$

and

$$y = r\sin\theta \qquad (2.21b)$$

Then, with

$$r = \sqrt{x^2 + y^2} \qquad (1.7a)$$

and

$$\tan\theta = \frac{y}{x} \qquad (1.7b)$$

we have

$$\frac{\partial F(r,\theta)}{\partial x} = \frac{\partial F}{\partial r}\frac{\partial r}{\partial x} + \frac{\partial F}{\partial \theta}\frac{\partial \theta}{\partial x} = \cos\theta\frac{\partial F}{\partial r} - \frac{\sin\theta}{r}\frac{\partial F}{\partial \theta} \qquad (3.8a)$$

and

$$\frac{\partial F(r,\theta)}{\partial y} = \frac{\partial F}{\partial r}\frac{\partial r}{\partial y} + \frac{\partial F}{\partial \theta}\frac{\partial \theta}{\partial y} = \sin\theta\frac{\partial F}{\partial r} + \frac{\cos\theta}{r}\frac{\partial F}{\partial \theta} \qquad (3.8b)$$

Referring to eq. 3.4, the CR condition in terms of $F(r,\theta)$ becomes

$$\frac{\partial F}{\partial r} = -i\frac{1}{r}\frac{\partial F}{\partial \theta} \qquad (3.9)$$

and the CR conditions in terms of $U(r,\theta)$ and $V(r,\theta)$ are

$$\frac{\partial U}{\partial r} = \frac{1}{r}\frac{\partial V}{\partial \theta} \qquad (3.10a)$$

and

$$\frac{\partial V}{\partial r} = -\frac{1}{r}\frac{\partial U}{\partial \theta} \qquad (3.10b)$$

Analyticity

Let $F(z)$ satisfy the CR condition(s) at all points in a region containing z. Referring to eqs. 3.3 and 3.5, the derivative of $F(z)$ at z can be expressed either as

$$\frac{dF}{dz} = \frac{\partial F}{\partial x} = \frac{\partial U}{\partial x} + i\frac{\partial V}{\partial x} \qquad (3.11a)$$

or as

$$\frac{dF}{dz} = -i\frac{\partial F}{\partial y} = -i\frac{\partial U}{\partial y} + \frac{\partial V}{\partial y} \qquad (3.11b)$$

Referring to eqs. 3.3a and 3.8a or to eqs. 3.3b and 3.8b, dF/dz can be expressed in terms of r and θ as

$$\frac{dF}{dz} = \frac{\partial F}{\partial x} = \cos\theta\frac{\partial F}{\partial r} - \frac{\sin\theta}{r}\frac{\partial F}{\partial \theta} = e^{-i\theta}\frac{\partial F}{\partial r} = -\frac{ie^{-i\theta}}{r}\frac{\partial F}{\partial \theta} \qquad (3.12)$$

dF/dz can also be expressed in terms of $U(r,\theta)$ and $V(r,\theta)$ as

$$\frac{dF}{dz} = e^{-i\theta}\frac{\partial F}{\partial r} = e^{-i\theta}\left[\frac{\partial U}{\partial r} + i\frac{\partial V}{\partial r}\right] \qquad (3.13a)$$

or equivalently

$$\frac{dF}{dz} = -i\frac{e^{-i\theta}}{r}\frac{\partial F}{\partial\theta} = -i\frac{e^{-i\theta}}{r}\left[\frac{\partial U}{\partial\theta} + i\frac{\partial V}{\partial\theta}\right] \qquad (3.13b)$$

If $F(z)$ satisfies the CR condition(s), then dF/dz is uniquely defined. If dF/dz satisfies the CR condition(s), then d^2F/dz^2 is defined, and so on. If $F(z)$ and all its derivatives satisfy the CR conditions at z, then $F(z)$ is *infinitely differentiable* at z, and is said to be *analytic* at z.

Example 3.1: Testing the analyticity of a function

(a) Consider

$$F(z) = e^z = e^x e^{iy} = e^x\left(\cos y + i\sin y\right) \qquad (3.14)$$

Testing the CR condition in terms of $F(z)$, we have

$$\frac{\partial F}{\partial x} = \frac{\partial}{\partial x}\left(e^x e^{iy}\right) = e^x e^{iy} = e^z \qquad (3.15a)$$

and

$$-i\frac{\partial F}{\partial y} = -i\frac{\partial}{\partial y}\left(e^x e^{iy}\right) = e^x e^{iy} = e^z \qquad (3.15b)$$

so e^z satisfies the CR condition.

Thus, from either of eqs. 3.11, we see that

$$\frac{d}{dz}e^z = e^z \qquad (3.16)$$

Therefore, de^z/dz also satisfies the CR condition and so d^2e^z/dz^2 is defined. Continuing this process, we see that e^z is infinitely differentiable and so is analytic.

(b) Let N be a positive integer. Then, to apply the CR conditions in terms of $U(r,\theta)$ and $V(r,\theta)$, we write

$$F(z) = z^N = r^N e^{iN\theta} = r^N \left[\cos(N\theta) + i\sin(N\theta)\right] \tag{3.17}$$

Therefore,

$$U(r,\theta) = r^N \cos(N\theta) \tag{3.18a}$$

and

$$V(r,\theta) = r^N \sin(N\theta) \tag{3.18b}$$

The CR conditions of eqs. 3.10 yield

$$\frac{\partial U}{\partial r} = Nr^{N-1}\cos(N\theta) = \frac{1}{r}\frac{\partial V}{\partial \theta} \tag{3.19a}$$

and

$$\frac{\partial V}{\partial r} = Nr^{N-1}\sin(N\theta) = -\frac{1}{r}\frac{\partial U}{\partial \theta} \tag{3.19b}$$

So, for positive integer N, z^N *satisfies* the CR condition. Using either form of eqs. 3.13, we see, for example, that

$$\frac{d}{dz}z^N = e^{-i\theta}\left[\frac{\partial}{\partial r}\left(r^N\cos(N\theta)\right) + i\frac{\partial}{\partial r}\left(r^N\sin(N\theta)\right)\right] = Nz^{N-1} \tag{3.20}$$

It is straightforward to show that z^{N-1}, z^{N-2}, ... satisfy the CR conditions. Thus, z^N is infinitely differentiable and therefore is analytic. $\square$

We note that in ex. 3.1, it is unnecessary to specify the values of x and y (or r and θ) at which the CR conditions are applied to e^z and z^N and their derivatives. For these functions, the CR conditions are valid at all finite z and therefore, they are analytic at all finite z. A function that is analytic at all finite z is called an *entire* function.

Referring to eqs. 3.16 and 3.20, we see that the derivative of each of these functions is the result one obtains by treating z as a single variable, even though it is the sum of two independent variables. These are examples of the general result that when a function is analytic everywhere in a region containing the point z, its derivative at z is that obtained by treating z as if it were a single variable.

To see this in general, consider the differential of $F(x, y)$, given by

$$dF = \frac{\partial F}{\partial x}dx + \frac{\partial F}{\partial y}dy \tag{3.21}$$

When $F(x, y)$, described as $F(z)$, is analytic in a region containing z, then, from eqs. 3.11,

$$\frac{\partial F}{\partial x} = \frac{dF}{dz} \tag{3.22a}$$

and

$$\frac{\partial F}{\partial y} = i\frac{dF}{dz} \tag{3.22b}$$

Thus, eq. 3.21 can be expressed as

$$dF = \frac{dF}{dz}(dx + idy) = \frac{dF}{dz}dz \tag{3.23}$$

which is the differential of $F(z)$ if z were a single variable.

Example 3.2: Demonstrating that a fractional root function is not analytic

In ex. 3.1b, it was shown that for integer $N \geq 0$, z^N is infinitely differentiable at all z, and therefore is an entire function. To illustrate the analysis of z^N when N is not an integer, we apply the CR condition to

$$F(z) = z^{3/2} = (x + iy)^{3/2} \tag{3.24}$$

We have

$$\frac{\partial F}{\partial x} = \tfrac{3}{2}(x+iy)^{1/2} = -i\frac{\partial F}{\partial y} \tag{3.25}$$

Applying the CR condition to dF/dz yields

$$\frac{\partial}{\partial x}\left[\tfrac{3}{2}(x+iy)^{1/2}\right] = \tfrac{3}{2}\tfrac{1}{2}(x+iy)^{-1/2} = -i\frac{\partial}{\partial y}\left[\tfrac{3}{2}(x+iy)^{1/2}\right] \tag{3.26}$$

which is not defined at $z = 0$. Therefore, $z^{3/2}$ is not analytic at $z = 0$. □

We see from eq. 3.25 that $z^{3/2}$ seems to satisfy the CR condition(s) even though it is not analytic. In chapter 6, we show that as with the fractional roots of unity discussed in chapter 2, the function z^p with p a rational fraction or irrational number, has more than one value at every point z and therefore is not uniquely defined anywhere. Therefore, the derivatives of such functions are not uniquely defined at any z and thus do not satisfy the CR condition(s) anywhere.

Examples 3.1 and 3.2 illustrate the general result that only a function which satisfies the CR condition(s) in a region containing a point z is uniquely defined and is infinitely differentiable. Therefore, a function is analytic at z if it satisfies the CR condition(s) at z.

Laplace's equation for an analytic function

Let $F(z)$ be analytic at all points in a region R containing z. Then the real and imaginary parts of $F(z)$ satisfy

$$\frac{\partial U}{\partial x} = \frac{\partial V}{\partial y} \tag{3.7a}$$

and

$$\frac{\partial U}{\partial y} = -\frac{\partial V}{\partial x} \tag{3.7b}$$

at all x and y in R. Adding $\partial^2 U/\partial x^2$ obtained from eq. 3.7a to $\partial^2 U/\partial y^2$ obtained from eq. 3.7b, we obtain

$$\frac{\partial^2 U}{\partial x^2} + \frac{\partial^2 U}{\partial y^2} = \frac{\partial^2 V}{\partial x\partial y} - \frac{\partial^2 V}{\partial y\partial x} = 0 \tag{3.27a}$$

Similarly, adding $\partial^2 V/\partial x^2$ obtained from eq. 3.7b to $\partial^2 V/\partial y^2$ from eq. 3.7a, we obtain

$$\frac{\partial^2 V}{\partial x^2} + \frac{\partial^2 V}{\partial y^2} = \frac{\partial^2 U}{\partial y\partial x} - \frac{\partial^2 U}{\partial x\partial y} = 0 \tag{3.27b}$$

The differential operator

$$\nabla^2 \equiv \frac{\partial^2}{\partial x^2} + \frac{\partial^2}{\partial y^2} \tag{3.28}$$

is called the *Laplacian* or *"del squared"* operator in two dimensions. With this definition, eqs. 3.27 can be expressed in compact notation as

$$\nabla^2 U = 0 \tag{3.29a}$$

and

$$\nabla^2 V = 0 \tag{3.29b}$$

Equations 3.27 and equivalently eqs. 3.29 are the two-dimensional *Laplace's equations* for U and V. Thus, we see that a pair of functions that satisfy the CR conditions at all points in a region R also satisfy Laplace's equation in R. Such functions are called *harmonic functions*.

If $U(x, y)$ and $V(x, y)$, the real and imaginary parts of $F(z)$, are harmonic in a region R, then $F(z)$ also satisfies Laplace's equation. That is,

$$\nabla^2 F = \nabla^2 U + i\nabla^2 V = 0 \tag{3.30}$$

at all z in R.

Determination of an analytic function

If only $U(x, y)$, the real part (or only $V(x, y)$, the imaginary part) of $F(z)$ is known in a region R containing z, the CR conditions can be used to determine if an imaginary part (or a real part) can be found that makes $F(z)$ analytic in R.

For example, let the real part of $F(z)$, $U(x, y)$, be a known function. If $F(z)$ is analytic in R, its imaginary part, $V(x, y)$ satisfies

$$\frac{\partial V}{\partial y} = \frac{\partial U}{\partial x} \tag{3.7a}$$

Then $V(x, y)$ is given by

$$V(x, y) = \int \frac{\partial U}{\partial x} dy + A \tag{3.31}$$

The partial derivative of a function of x and y with respect to y, for example, means differentiating only the explicit y dependence of the function, treating x as a constant. In the same way, the *partial integral* of a function with respect to y means integrating only the explicit y dependence of the function, treating x as a constant. Therefore, the "constant of integration" A, must satisfy

$$\frac{\partial A}{\partial y} = 0 \tag{3.32}$$

This means that A cannot contain any explicit y dependence, and so, can at most depend on x. As such, from

$$\frac{\partial V}{\partial x} = -\frac{\partial U}{\partial y} \tag{3.7b}$$

eq. 3.31 requires $A(x)$ to satisfy

$$\frac{dA}{dx} = -\frac{\partial U}{\partial y} - \int \frac{\partial^2 U}{\partial x^2} dy \tag{3.33a}$$

Then

$$A(x) = -\int \left[\frac{\partial U}{\partial y} + \int \frac{\partial^2 U}{\partial x^2} dy \right] dx + C \tag{3.33b}$$

where C is a true constant, independent of x and y.

We see from eq. 3.33b that in order for a function A that depends on x to exist, the integrand

$$\frac{\partial U}{\partial y} + \int \frac{\partial^2 U}{\partial x^2} dy$$

must only depend on x. If this integrand contains any explicit y dependence, then a function $A(x)$ does not exist, and therefore an analytic function with the specified real part $U(x, y)$ does not exist. If this integrand does contain only explicit x dependence, the imaginary part of $F(z)$ is given by

$$V(x, y) = \int \frac{\partial U}{\partial x} dy - \int \left[\frac{\partial U}{\partial y} + \int \frac{\partial^2 U}{\partial x^2} dy \right] dx + C \tag{3.34}$$

We note that if

$$\frac{\partial U}{\partial y} + \int \frac{\partial^2 U}{\partial x^2} dy$$

is independent of y, then

$$\frac{\partial}{\partial y} \left[\frac{\partial U}{\partial y} + \int \frac{\partial^2 U}{\partial x^2} dy \right] = 0 = \frac{\partial^2 U}{\partial x^2} + \frac{\partial^2 U}{\partial y^2} \tag{3.35}$$

This is consistent with the fact that when $F(z)$ is analytic in a region containing z, its real (and its imaginary) part satisfies Laplace's equation.

In problem 6 of this chapter the reader will derive an analogous process for the case where $V(x, y)$, the imaginary part of a function is, specified. The CR conditions are used to determine whether $U(x, y)$ exists, and if it does, the expression for $U(x, y)$ will be derived.

Example 3.3: Determination of an analytic function

(a) Let us determine if an analytic function exists with a real part given by

$$U(x, y) = x^2 y^2 \tag{3.36}$$

Substituting this into eq. 3.33b, we obtain

$$A(x) = \int \left[\frac{\partial U}{\partial y} + \int \frac{\partial^2 U}{\partial x^2} dy \right] dx = \int \left[2x^2 y + \tfrac{1}{3} y^3 \right] dx \tag{3.37}$$

Because the integrand explicitly depends on y, an analytic function with a real part $x^2 y^2$ does not exist.

(b) Let the real part of a function be defined by

$$U(x, y) = \sin x \cosh y \tag{3.38}$$

Then

$$A(x) = \int \left[\frac{\partial U}{\partial y} + \int \frac{\partial^2 U}{\partial x^2} dy \right] dx = 0 \tag{3.39}$$

and, from eq. 3.34,

$$V(x, y) = \int \frac{\partial U}{\partial x} dy = \cos x \sinh y + C \tag{3.40}$$

Therefore, the analytic function with real part given by eq. 3.38 is

$$F(z) = \sin x \cosh y + i \cos x \sinh y + C = \sinh z + C \tag{3.41}$$

□

3.2   Integrals of Analytic Functions

Consider the integral of a function $F(z)$ along some path or *contour* C in the complex plane from z_1 to z_2. We denote such an integral by

$$\int_{C}^{z_2} {}_{z_1} F(z)dz$$

There are an infinite number of possible paths that can be taken between these endpoints. Two such contours, C_1 and C_2, are shown in fig. 3.1.

In general, we expect the value of the integral to depend on which contour is taken. But if the integral is independent of the contour, then

$$\int_{C_1}^{z_2} {}_{z_1} F(z)dz - \int_{C_2}^{z_2} {}_{z_1} F(z)dz = 0 \tag{3.42a}$$

Inverting the limits on the integral over C_2 changes the sign of the integral and this becomes

$$\int_{C_1}^{z_2} {}_{z_1} F(z)dz + \int_{C_2}^{z_2} {}_{z_1} F(z)dz = 0 \tag{3.42b}$$

The path from z_1 to z_2 along a contour C_1 then from z_2 back to z_1 along a different contour C_2 forms a *closed contour* C. The *closed contour integral* around C is denoted by

$$\oint_{C} F(z)dz$$

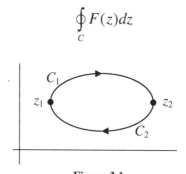

Figure 3.1

Two contours between z_1 and z_2

Referring to eqs. 3.42, we see that when $F(z)$ is independent of the path taken, the integral around a closed contour C satisfies

$$\oint_C F(z)dz = 0 \qquad\qquad (3.43)$$

To see what conditions must be placed on $F(z)$ at points enclosed by C, and thus between the segments C_1 and C_2, we consider *Green's theorem* (which is the two-dimensional form of *Stokes' theorem*) for a closed contour integral in the x–y plane. For completeness, a derivation of Green's theorem is presented in appendix 2. The reader is also referred to textbooks on vector analysis (for example, Marion, 1965).

Let C be a closed contour in the x–y plane, and let all points inside and on C lie within a region R. Let $G(x, y)$ and $H(x, y)$ be two functions, the derivatives of which exist at all points in R. Green's theorem states that

$$\oint_C (Gdx + Hdy) = \iint_R \left(\frac{\partial H}{\partial x} - \frac{\partial G}{\partial y} \right) dxdy \qquad\qquad (3.44)$$

Expressing $F(z)$ and dz in terms of their real and imaginary parts, eq. 3.43 becomes

$$\oint_C F(z)dz = \oint_C (Udx - Vdy) + i\oint_C (Vdx + Udy) = 0 \qquad\qquad (3.45)$$

Comparing the real part of eq. 3.45 to eq. 3.44, we set $G = U$ and $H = -V$. Then, Green's theorem yields

$$\oint_C (Udx - Vdy) = \iint_R \left(-\frac{\partial V}{\partial x} - \frac{\partial U}{\partial y} \right) dxdy = 0 \qquad\qquad (3.46a)$$

at all points in R. Thus,

$$\frac{\partial V}{\partial x} = -\frac{\partial U}{\partial y} \qquad\qquad (3.7b)$$

must hold at all points in R. Comparing the imaginary part of eq. 3.45 to eq. 3.44, we take $G = V$ and $H = U$. Then Green's theorem requires that

$$\oint_C (Vdx + Udy) = \iint_R \left(\frac{\partial U}{\partial x} - \frac{\partial V}{\partial y} \right) dxdy = 0 \qquad\qquad (3.46b)$$

from which we deduce that

$$\frac{\partial U}{\partial x} = \frac{\partial V}{\partial y} \tag{3.7a}$$

everywhere in R. Therefore, we see that if the closed contour integral of a function is zero, $F(z)$ must be analytic everywhere inside the contour.

This result can be interpreted in another important way. Let C_1 and C_2 be two different contours connecting endpoints z_1 and z_2 as shown in fig. 3.1. The integral along C_1 is unchanged when C_1 is deformed (reshaped) into C_2 as long as all the points between C_1 and C_2, the region over which the deformation occurs, are points of analyticity of the integrand.

Example 3.4: Independence of contour for the integral of an analytic integrand

As shown in ex. 3.1, z^N is an entire function for integer $N > 0$. Thus, it is expected that the integral of z^N between any two points should be independent of the contour between those points.

Consider evaluating the integral

$$I = \int_0^{1+i} z^2 dz \tag{3.47}$$

over each of the contours shown in figs. 3.2.

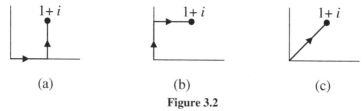

(a)   (b)   (c)

Figure 3.2

Three contours from $z = 0$ to $z = 1 + i$

Referring to the contour of fig. 3.2a, we see that over the horizontal segment, $z = x$ with $0 \leq x \leq 1$. Along the vertical segment, $z = 1 + iy$ with $0 \leq y \leq 1$. Therefore,

$$I_a = \int_0^1 x^2 dx + \int_0^1 (1+iy)^2 i dy = \tfrac{1}{3}(1+i)^3 \tag{3.48a}$$

For the contour shown in fig. 3.2b, $z = iy$ along the vertical segment, and $z = x + i$ along the horizontal piece. As with the contour of fig. 3.2a, $0 \leq x \leq 1$ and $0 \leq y \leq 1$. Thus,

$$I_b = \int_0^1 (iy)^2 \, idy + \int_0^1 (x+i)^2 \, dx = \tfrac{1}{3}(1+i)^3 \qquad (3.48b)$$

Points along the contour of fig. 3.2c satisfy $x = y$ so z can be expressed as $z = (1 + i)x$ or equivalently $z = (1 + i)y$. Then, as for the contours of figs. 3.2a and 3.2b,

$$I_c = \int_0^{1+i} (1+i)^2 x^2 (1+i) \, dx = \tfrac{1}{3}(1+i)^3 \qquad (3.48c)$$

Thus, as expected, the integral of eq. 3.47 is independent of which contour is taken between $z = 0$ and $z = 1 + i$.

Equivalently, we can interpret this result in terms of the idea that the integral is unchanged when one of the contours is reshaped into one of the others.

We also note that integration over each contour yields

$$\int_0^{1+i} z^2 \, dz = \tfrac{1}{3} z^3 \Big|_0^{1+i} = \tfrac{1}{3}(1+i)^3 \qquad (3.49)$$

which is the result one obtains by evaluating this integral treating z as a single variable rather than the sum of two independent variables. □

As with derivatives, this is an example of the general result that when integrating between two points along a contour that traverses a region of analyticity of a function, then the function can be integrated treating z as a single variable.

To see this, let $F(z)$ be analytic at all points in a region R. Then, as shown earlier,

$$dF = \frac{dF}{dz} \, dz \qquad (3.23)$$

Because $F(z)$ is analytic in R, all derivatives of $F(z)$ are analytic in R. In particular

$$G(z) = \frac{dF}{dz} \qquad (3.50)$$

is analytic in R. Integrating eq. 3.23,

$$\int_{z_1}^{z_2} G(z)dz = \int_{z_1}^{z_2} dF = F(z_2) - F(z_1) \tag{3.51}$$

the result obtained by treating z as a single variable in the integral of $G(z)$.

3.3 Pole Singularities

If $F(z)$ does not satisfy the CR conditions in an infinitesimal region containing a point z, and so is not analytic at z, it is said to have a *singularity* at z.

Example 3.5: A function with a singularity at the origin

Let

$$F(z) = \frac{1}{z} = \frac{x - iy}{x^2 + y^2} \tag{3.52}$$

Applying the CR condition in terms of $F(z)$, we obtain

$$\frac{\partial F}{\partial x} = -\frac{1}{(x + iy)^2} = -\frac{1}{z^2} \tag{3.53a}$$

and

$$-i\frac{\partial F}{\partial y} = -i\left[\frac{-i}{(x + iy)^2}\right] = -\frac{1}{z^2} \tag{3.53b}$$

Note that the CR condition is satisfied at all points except $z = 0$. Therefore, $1/z$ is singular at $z = 0$, and is analytic at all other values of z. □

Removable and nonremovable singularities

If $F(z)$ is singular at a point z_0, then for many functions, there is a smallest nonnegative integer k for which

$$\lim_{z \to z_0} \frac{d^k F}{dz^k} = \infty \tag{3.54}$$

As shown in ex. 3.2, $z^{3/2}$ and its first derivative are finite at $z = 0$, but the second derivative is infinite at the origin. Similarly, in ex. 3.5, it was shown that $1/z$ and all of its derivatives are infinite at the origin.

If there exists a positive real number p such that

$$\lim_{z \to z_0} (z - z_0)^p \frac{d^k F}{dz^k} \neq \infty \tag{3.55}$$

the singularity of $F(z)$ at z_0 is called a *removable* or *nonessential* singularity. If there is no positive real number p for which eq. 3.55 is satisfied, the singularity at z_0 is said to be *nonremovable* or *essential*.

Example 3.6: Removable and nonremovable singularities

(a) Referring to ex. 3.2, $z^{3/2}$ and its first derivative are zero at $z = 0$, and its second derivative satisfies

$$\lim_{z \to 0} \frac{d^2}{dz^2} z^{3/2} = \tfrac{3}{4} \lim_{z \to 0} z^{-1/2} = \infty \tag{3.56}$$

We see that

$$\lim_{z \to 0} z^{1/2} \left(\frac{d^2}{dz^2} z^{3/2} \right) \neq \infty \tag{3.57a}$$

$$\lim_{z \to 0} z^{3/2} \left(\frac{d^3}{dz^3} z^{3/2} \right) \neq \infty \tag{3.57b}$$

$$\lim_{z \to 0} z^{5/2} \left(\frac{d^4}{dz^4} z^{3/2} \right) \neq \infty \tag{3.57c}$$

and so on. Thus, the singularity of $z^{3/2}$ at $z = 0$ is removable.

(b) The exponential function

$$F(z) = e^{1/z} \tag{3.58}$$

is also singular at $z = 0$. This can be seen by noting that when $z \to 0$ from points along the positive real axis,

$$\lim_{z \to 0} e^{1/z} = \infty \tag{3.59a}$$

when $z \to 0$ along the negative real axis,

$$\lim_{z \to 0} e^{1/z} = 0 \tag{3.59b}$$

and when $z \to 0$ along the positive or negative imaginary axis,

$$\lim_{z \to 0} e^{1/z} = e^{\pm i\infty} \tag{3.59c}$$

which are complex numbers of magnitude 1. That is, $e^{1/z}$ is not uniquely defined, and is therefore singular at the origin. There is no power of z that will remove that indeterminacy of $e^{1/z}$ at $z = 0$, therefore the singularity of $e^{1/z}$ at the origin is nonremovable. $\square$

Only functions with removable singularities are integrable. There are two types of removable singularities. One is called a *pole*, the other, a *branch point*. A detailed discussion of branch points is presented in chapter 6. We focus the present discussion on pole singularities.

Pole singularities

Let

$$\lim_{z \to z_0} F(z) = \infty \tag{3.60}$$

Let there be some integer $N > 0$ such that

$$R(z) \equiv (z - z_0)^N F(z) \tag{3.61a}$$

is analytic everywhere in a region containing z_0 and

$$R(z_0) = \lim_{z \to z_0} (z - z_0)^N F(z) \tag{3.61b}$$

is a finite, nonzero complex number. Then $F(z)$ is said to have a *pole of order* N at z_0 and the factor $(z - z_0)^N$ removes the pole.

When $N = 1$, the pole is a *first-order* or *simple pole*. The function

$$R(z) = (z - z_0)F(z) \tag{3.62a}$$

is analytic in a region containing z_0 and

$$R(z_0) = \lim_{z \to z_0}(z - z_0)F(z) \neq 0 \tag{3.62b}$$

is a finite complex number. For a simple pole, $R(z_0)$ is called the *residue of the pole*.

Example 3.7: Selected functions with poles

(a) The function $1/z$ of ex. 3.5 has a pole at $z = 0$. This is easily seen by noting that

$$R(0) = \lim_{z \to 0} z\frac{1}{z} = 1 \tag{3.63}$$

Because $R(0)$ is finite and nonzero, the pole at $z = 0$ is removed by multiplying $1/z$ by z. Therefore, the pole is a simple pole.

(b) Let

$$F(z) = \frac{\sin(z+1)}{z^3} \tag{3.64}$$

Clearly,

$$\lim_{z \to 0} F(z) = \infty \tag{3.65}$$

and

$$\lim_{z \to 0} z^3 F(z) = \sin(1) \neq 0 \tag{3.66}$$

Therefore, $[\sin(z + 1)]/z^3$ has a third-order pole at $z = 0$ arising from the factor $1/z^3$.

    (c) Let

$$F(z) = \frac{\sin z}{z^3} \tag{3.67}$$

As with $[\sin(1 + z)]/z^3$, $\sin z/z^3$ also has a pole at $z = 0$ due to the factor $1/z^3$. If we try to remove this pole by defining

$$R(z) = z^3 F(z) = \sin z \tag{3.68}$$

we see that

$$R(0) = \lim_{z \to 0} \sin z = 0 \tag{3.69}$$

Because $R(0)$ must be nonzero, z^3 is not the factor that removes the pole at $z = 0$. Thus, the pole of $\sin z/z^3$ is not a third order pole.

    To see what is needed to remove that pole, we refer to the graphs of $\sin z$ and z for real z shown in fig. 3.3.

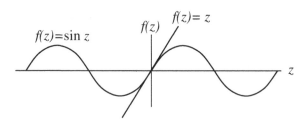

Figure 3.3

Graphs of $\sin z$ and z

At $z = 0$, $f(z) = \sin z$ and $f(z) = z$ have the same value, and in the region near the origin, they appear to have similar slopes. This indicates that for small z,

$$\sin z \simeq z \tag{3.70}$$

with z expressed in radians. A numerical analysis of z and $\sin z$, given in table 3.1, tends to confirm this result (which is proven rigorously for complex z in chapter 4).

Table 3.1 – Values of z and $\sin z$

z (degrees)	z (radians)	Sin z
0.5	$\pi/360 = 0.00873$	0.00873
1.0	$\pi/180 = 0.01745$	0.01745
2.0	$\pi/90 = 0.03491$	0.03490
5.0	$\pi/36 = 0.08727$	0.08716

Using eq. 3.70, we see that near zero,

$$\frac{\sin z}{z^3} \approx \frac{1}{z^2} \tag{3.71}$$

Therefore

$$R(0) = \lim_{z \to 0} z^2 \frac{\sin z}{z^3} = \lim_{z \to 0} \frac{z^2}{z^2} = 1 \tag{3.72}$$

so $\sin z/z^3$ has a second-order pole at $z = 0$. □

3.4 Cauchy's Residue Theorem

Cauchy's residue theorem provides a method for evaluating a closed contour integral of a function that has one or more poles inside the contour.

Cauchy's residue theorem for one simple pole

In fig. 3.4a, the contour C_a encloses a simple pole of $F(z)$ at z_0.
To develop Cauchy's residue theorem for evaluating the closed contour integral of $F(z)$ around C_a shown in fig. 3.4a, we consider the integral around C_b shown in fig. 3.4b. Because the pole at z_0 is excluded from C_b, this integral satisfies

$$\oint_{C_b} F(z)dz = 0 \tag{3.73}$$

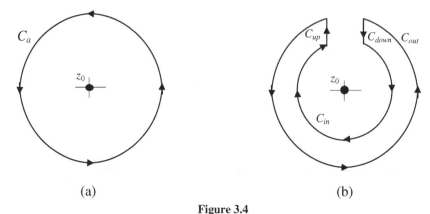

(a)   (b)

Figure 3.4

(a) C_a, a contour enclosing and   (b) C_b, a contour excluding a simple pole of $F(z)$ at z_0

Referring to fig. 3.4b for the notation used to describe the four segments that comprise C_b and the direction in which each is traversed, eq. 3.73 can be written as

$$\oint_{C_b} F(z)dz = \int_{C_{out}} F(z)dz + \int_{C_{down}} F(z)dz + \int_{C_{in}} F(z)dz + \int_{C_{up}} F(z)dz = 0$$

$$(3.74)$$

If there were no gap between C_{up} and C_{down}, the contour C_{out} would be identical to C_a, and C_{in} would become a closed contour. That is,

$$\lim_{gap \to 0} \int_{C_{out}} F(z)dz = \oint_{C_a} F(z)dz \qquad\qquad (3.75a)$$

and

$$\lim_{gap \to 0} \int_{C_{in}} F(z)dz = \oint_{C_{in}} F(z)dz \qquad\qquad (3.75b)$$

One way to eliminate this gap is to use the fact that there are no singularities between C_{up} and C_{down}. Therefore, as shown in fig. 3.5, the gap can be closed without changing the value of

$$\oint_{C_{out}} F(z)dz \quad \text{and} \quad \oint_{C_{in}} F(z)dz$$

When the width of the gap is zero, C_{up} and C_{down} lie along the same line and are traversed in opposite directions. Therefore,

$$\lim_{gap \to 0} \left[\int_{C_{down}} F(z)dz + \int_{C_{up}} F(z)dz \right] = 0 \qquad (3.76)$$

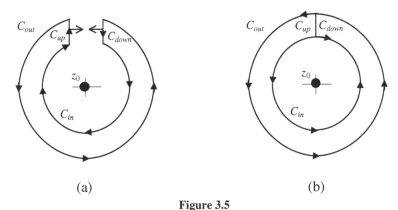

(a)   (b)

Figure 3.5

(a)   The contour can be deformed to close the gap and (b) gap is closed

As shown in fig. 3.5, the integral around C_{out} is traversed in the counterclockwise direction and the integral around C_{in} is traversed in the clockwise direction. Therefore, with eqs. 3.75 and. 3.76, eq. 3.74 becomes

$$\lim_{gap \to 0} \oint_{C_b} F(z)dz = \oint_{C_a} F(z)dz + \oint_{C_{in}} F(z)dz = 0 \qquad (3.77)$$

Referring to fig. 3.6, this result can also be obtained by considering the integral around the contour C_i enclosing a region of analyticity of the integrand. Because there are no singularities within C_i,

$$\oint_{C_i} F(z)dz = 0 \qquad (3.78)$$

so that

$$\oint_{C_b} F(z)dz = \oint_{C_b} F(z)dz + \oint_{C_i} F(z)dz$$

$$= \int_{C_{out}} F(z)dz + \int_{C_{down}} F(z)dz + \int_{C_{in}} F(z)dz + \int_{C_{up}} F(z)dz = 0$$

$$(3.79)$$

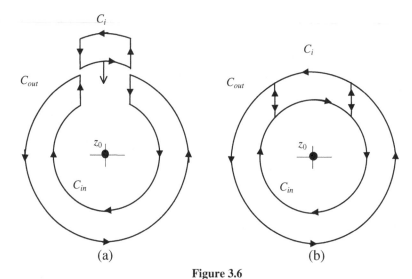

Figure 3.6

(a) Contour with an open gap and (b) section inserted that closes the gap

We note from fig. 3.6b that C_i is constructed so that when it is inserted into the gap, the curved segments of C_i close C_{out} and C_{in} and the sides of C_i are along the same paths as C_{up} and C_{down}. Because the integrals along these sides are oppositely directed to C_{up} and C_{down}, these integrals cancel and

$$\lim_{\substack{\text{section} \\ \text{inserted } C_b}} \oint F(z)dz = \oint_{C_a} F(z)dz + \oint_{C_{in}} F(z)dz = 0 \qquad (3.80)$$

which is equivalent to the result we obtained in eq. 3.76.

The positive real axis is defined by $0 \le r \le \infty$, $\theta = 0$. An increase in θ is achieved by a rotation in the counterclockwise direction. Rotation in the clockwise direction results in a decrease in θ. Therefore, $d\theta > 0$ implies an infinitesimal counterclockwise rotation, and $d\theta < 0$ indicates an infinitesimal clockwise rotation. For this reason, an integral around a contour in the clockwise direction is the negative of an integral around that contour in the counterclockwise direction.

Referring to fig. 3.5b or 3.6b, the integral around the contour C_a is in the counterclockwise direction, and the integral around C_{in} is in the clockwise direction. When needed, we use superscripts "*cc*" and "*c*" to designate counterclockwise and clockwise traversals along a contour. Thus, the results stated in eqs. 3.77 and 3.80 become

$$\oint_{C_a}^{cc} F(z)dz + \oint_{C_{in}}^{c} F(z)dz = \oint_{C_a}^{cc} F(z)dz - \oint_{C_{in}}^{cc} F(z)dz = 0 \qquad (3.81)$$

from which we obtain

$$\oint_{C_a}^{cc} F(z)dz = \oint_{C_{in}}^{cc} F(z)dz \qquad (3.82)$$

This result can be obtained more easily without forming and then eliminating the gap. The only singularity of $F(z)$ inside C_a is at z_0, therefore there are no singularities between C_a and C_{in}. Thus, C_a can be shrunk to C_{in} without changing the value of the integral and eq. 3.82 is obvious.

Because z_0 is the only singularity of $F(z)$ enclosed by C_{in}, this contour can be shaped into a circle of radius ρ centered at z_0, without affecting this integral around C_{in}.

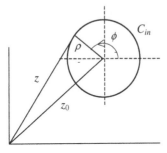

Figure 3.7
Circular contour centered on z_0

We see in fig. 3.7 that any point z on the circle C_{in} can be written as

$$z = z_0 + \rho e^{i\phi} \qquad (3.83a)$$

from which

$$dz = i\rho e^{i\phi} d\phi = i(z - z_0)d\phi \qquad (3.83b)$$

The integral around the closed contour implies that ϕ ranges from an arbitrary ϕ_0 to $\phi_0 + 2\pi$, so

$$\oint_{C_{in}}^{cc} F(z)dz = i\int_{\phi_0}^{\phi_0+2\pi}(z-z_0)F(z)d\phi \qquad (3.84)$$

Because z_0 is the only point of singularity enclosed by C_{in}, this inner contour can be shrunk to a circle of infinitesimal radius without encountering any singularities of $F(z)$. With the definition of the residue given in eq. 3.62b, this becomes

$$\lim_{\rho\to 0}\oint_{C_{in}}^{cc} F(z)dz = \int_{\phi_0}^{\phi_0+2\pi}\lim_{z\to z_0}(z-z_0)F(z)id\phi = 2\pi iR(z_0) \qquad (3.85a)$$

Therefore, from eq. 3.82,

$$\oint_{C_a}^{cc} F(z)dz = 2\pi iR(z_0) \qquad (3.85b)$$

This is Cauchy's residue theorem for the counterclockwise integral around a contour enclosing one simple pole of $F(z)$.

It is straightforward to see that if the integral is taken in the clockwise direction, Cauchy's residue theorem becomes

$$\oint_{C_a}^{c} F(z)dz = -2\pi iR(z_0) \qquad (3.85c)$$

With Cauchy's residue theorem, the problem of evaluating the closed contour integral of a function with one simple pole inside the contour is reduced to the algebraic problem of determining the residue of that pole using the definition given in eq. 3.62b.

Example 3.8: Cauchy's residue theorem for an integrand with one simple pole

Let us evaluate

$$I \equiv \oint \frac{e^z \sin z}{(z-2i)}dz \qquad (3.86)$$

integrating in the counterclockwise direction around the contours shown in fig. 3.8.

(a) For the circle of radius 5, the pole is inside the contour. Therefore,

$$\oint_{C_a} \frac{e^z \sin z}{(z-2i)}\,dz = 2\pi i R(2i) = -2\pi e^{2i}\sinh(2) \tag{3.87}$$

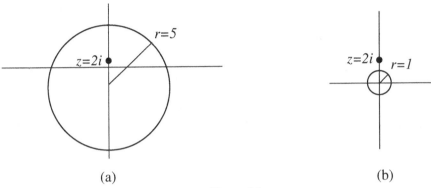

(a)   (b)

Figure 3.8

Contours for evaluating the integral of eq. 3.86

(b) The pole at $2i$ is outside the circle of radius 1. Thus,

$$\oint_{C_b} \frac{e^z \sin z}{(z-2i)}\,dz = 0 \tag{3.88}$$

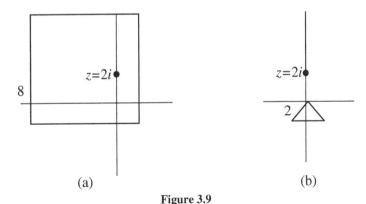

(a)   (b)

Figure 3.9

Contours equivalent to those of fig. 3.8 for evaluating the integral of eq. 3.86

Referring to fig. 3.8, we see that the integrand is analytic everywhere in the region between the circle of radius 5 centered at the origin of fig. 3.8a, and the square of side 8 centered at $(-2,1)$ shown in fig. 3.9a. Therefore, the circle can be reshaped into the square without changing the value of the integral. As such, the integral of eq. 3.86 around this square contour is also $-2\pi e^{2i}\sinh 2$.

Similarly, because the circle of fig. 3.8b can be reshaped into the equilateral triangle of fig. 3.9b without crossing over a singularity of the integrand, the integral of 3.86 around the triangle is zero. □

Cauchy's residue theorem for more than one simple pole

To extend Cauchy's theorem to integrands that have more than one simple pole enclosed by the contour, we apply the same approach used for one simple pole.

Let $F(z)$ have two simple poles at z_0 and $z_0{}'$ enclosed by a contour shown in fig. 3.10a. We first construct a contour that excludes each of the poles as shown in fig. 3.10b, and then close the gaps (or equivalently insert sections that contribute nothing to the closed contour integral).

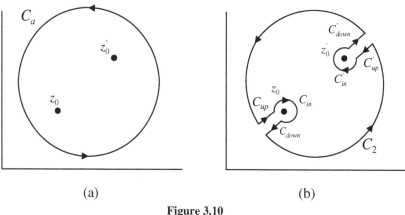

(a)   (b)

Figure 3.10

Contour (a) enclosing and (b) excluding two simple poles

Writing the integral over the contour of fig. 3.10b as a sum of integrals over each of its segments, we have

$$\int_{C_1}^{cc} F(z)dz + \int_{C_{down}} F(z)dz + \int_{C_{in}}^{c} F(z)dz + \int_{C_{up}} F(z)dz$$

$$+\int_{C_2}^{cc} F(z)dz + \int_{C_{down}'} F(z)dz + \int_{C_{in}'}^{c} F(z)dz + \int_{C_{up}'} F(z)dz = 0 \tag{3.89a}$$

When the gaps between C_{up} and C_{down} and between C'_{up} and C'_{down} are closed (or equivalently, when regions of analyticity are inserted to fill the gaps), the integrals along both pairs of "up" and "down" segments add to zero and the contours C_1 and C_2 combine to become the closed contour C_a. Then, reversing the directions on the two "in" contours, we obtain

$$\lim_{2\,gaps\to 0}\left[\int_{C_1}^{cc} F(z)dz + \int_{C_{in}}^{c} F(z)dz + \int_{C_2}^{cc} F(z)dz + \int_{C'_{in}}^{c} F(z)dz\right]$$

$$= \oint_{C_a}^{cc} F(z)dz - \oint_{C_{in}}^{cc} F(z)dz - \oint_{C'_{in}}^{cc} F(z)dz = 0$$

$$(3.89b)$$

from which

$$\oint_{C_a}^{cc} F(z)dz = \oint_{C_{in}}^{cc} F(z)dz + \oint_{C'_{in}}^{cc} F(z)dz \qquad (3.90)$$

Again we describe points on the "in" contours as

$$z = z_0 + \rho e^{i\phi} \qquad (3.91a)$$

and

$$z' = z'_0 + \rho' e^{i\phi'} \qquad (3.91b)$$

Following the analysis for one simple pole (eqs. 3.83, through 3.85), it is straightforward to show that the integral around each "in" contour yields $2\pi i$ multiplied by the residue at each pole. Then, eq. 3.90 becomes

$$\oint_{C_a}^{cc} F(z)dz = 2\pi i R(z_0) + 2\pi i R(z'_0) \qquad (3.92)$$

Extending this analysis to an integrand that has N simple poles at points $z_1, z_2, \ldots, z_N$ enclosed by a contour C, Cauchy's residue theorem becomes

$$\oint_C^{cc} F(z)dz = 2\pi i \sum_{k=1}^{N} R(z_k) \qquad (3.93a)$$

and

$$\oint_C^c F(z)dz = -2\pi i \sum_{k=1}^{N} R(z_k) \tag{3.93b}$$

From now on, it is to be understood that a closed contour is traversed in the counterclockwise direction unless otherwise specified. The notation "*cc*" or "*c*" is only displayed when it is necessary to do so.

Example 3.9: Cauchy's residue theorem for an integrand with two simple poles

Consider

$$I \equiv \oint \frac{e^z}{\left(z^2+1\right)}dz \tag{3.94}$$

around each of the contours shown in fig. 3.11. Clearly, the integrand has simple poles at $z = \pm i$.

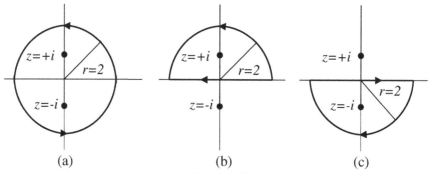

(a)   (b)   (c)

Figure 3.11

Three contours for the integral of eq. 3.94

(a) Because both poles are enclosed by the contour of fig. 3.11a,

$$\oint_{C_a} \frac{e^z}{\left(z^2+1\right)}dz = 2\pi i\left[R(+i)+R(-i)\right] \tag{3.95}$$

Writing the integrand as $e^z/[(z+i)(z-i)]$ the residues are found to be

$$R(+i) = \lim_{z\to+i}(z-i)\frac{e^z}{\left[(z-i)(z+i)\right]} = \frac{e^i}{2i} \tag{3.96a}$$

and

$$R(-i) = \lim_{z \to -i}(z+i)\frac{e^z}{\left[(z-i)(z+i)\right]} = \frac{e^{-i}}{-2i} \tag{3.96b}$$

Therefore,

$$\oint_{C_a} \frac{e^z}{\left(z^2+1\right)}\,dz = 2\pi i \left[\frac{e^i - e^{-i}}{2i}\right] = 2\pi i \sin(1) \tag{3.97}$$

(b) For the contour of fig. 3.11b, the pole at $z = +i$ is enclosed in the contour but not the pole at $z = -i$. As can be seen, this contour is traversed in the counterclockwise direction. Therefore,

$$\oint_{C_b} \frac{e^z}{\left(z^2+1\right)}\,dz = 2\pi i R(+i) = \pi e^i = \pi\left(\cos(1) + i\sin(1)\right) \tag{3.98}$$

(c) The contour of fig. 3.11c encloses only the pole at $z = -i$ and is traversed in the clockwise direction. Thus,

$$\oint_{C_c} \frac{e^z}{\left(z^2+1\right)}\,dz = -2\pi i R(-i) = \pi e^{-i} = \pi\left(\cos(1) - i\sin(1)\right) \tag{3.99}$$

$\square$

Cauchy's residue theorem for high order poles

A pole of order greater than 1 is referred to as a *high order* pole. To evaluate the closed contour integral of a function with a high order pole, let C be a contour in the w-plane and let z be a point inside the contour. If $f(w)$ is analytic at all points inside and on a contour, then

$$F(w) \equiv \frac{f(w)}{(w-z)} \tag{3.100}$$

has one simple pole at $w = z$. Therefore, by Cauchy's residue theorem,

$$\oint_C F(w)\,dw = \oint_C \frac{f(w)}{(w-z)}\,dw = 2\pi i \lim_{w \to z}(w-z)\frac{f(w)}{(w-z)} = 2\pi i f(z) \tag{3.101a}$$

from which,

$$f(z) = \frac{1}{2\pi i} \oint_C \frac{f(w)}{(w-z)} dw \qquad (3.101b)$$

This is *Cauchy's integral representation* for the analytic function $f(z)$.

Let R be a region in the z-plane, and let $f(z)$ be analytic at all z in R. Then df/dz exists everywhere in R. Because all the z dependence of $f(z)$ is in the denominator of the integrand in eq. 3.101b, df/dz is obtained by taking the derivative of $1/(w-z)$ with respect to z. Thus,

$$\frac{df}{dz} = \frac{1}{2\pi i} \oint_C \frac{f(w)}{(w-z)^2} dw \qquad (3.102a)$$

$$\frac{d^2 f}{dz^2} = \frac{2}{2\pi i} \oint_C \frac{f(w)}{(w-z)^3} dw \qquad (3.102b)$$

$$\frac{d^3 f}{dz^3} = \frac{3*2}{2\pi i} \oint_C \frac{f(w)}{(w-z)^4} dw = \frac{3!}{2\pi i} \oint_C \frac{f(w)}{(w-z)^4} dw \qquad (3.102c)$$

and so on. From these results it is straightforward to see that

$$\frac{d^n f}{dz^n} = \frac{n!}{2\pi i} \oint_C \frac{f(w)}{(w-z)^{n+1}} dw \qquad (3.102d)$$

Because the $n = 0$ derivative is $f(z)$, eq. 3.102d includes Cauchy's theorem for an integrand with a simple pole, which is also Cauchy's integral representation of an analytic function.

This Cauchy integral representation of $d^n f/dz^n$ provides us with a prescription for evaluating an integral in which the integrand has a pole of any order. By identifying $f(z)$, the analytic part of the integrand, the contour integral of a function with a pole at z, of order $(n + 1)$ is given by eq. 3.102d. If z is a fixed point z_0, then eq. 3.102d yields

$$\oint_C \frac{f(w)}{(w-z_0)^{n+1}} dw = \frac{2\pi i}{n!} \lim_{w \to z_0} \frac{d^n f(w)}{dw^n} \equiv \frac{2\pi i}{n!} f^{(n)}(z_0) \qquad (3.103)$$

**Example 3.10: Cauchy's residue theorem for an integrand with a third
order pole**

We consider

$$I \equiv \oint \frac{ze^z}{(z-1)^3} dz \qquad (3.104)$$

around a circle of radius 2, centered at the origin.

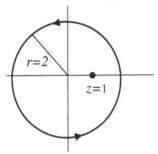

Figure 3.12
Contour and pole for the integral of eq. 3.104

It is evident that the integrand has a pole of order 3 at $z = 1$, and that the
part of the integrand that is analytic at $z = 1$ is

$$f(z) = ze^z \qquad (3.105)$$

Therefore, from eq. 3.103b,

$$\oint_C \frac{ze^z}{(z-1)^3} dz = \frac{2\pi i}{2!} \lim_{z \to 1} \frac{d^2}{dz^2}\left(ze^z\right) = 3\pi ie \qquad (3.106)$$

□

Integrands with more than one high order pole

The closed contour integral of a function with two or more poles, each of
arbitrary order, is obtained straightforwardly from eq. 3.103b. To illustrate,
we consider

$$I \equiv \oint_C \frac{g(z)}{(z-z_1)^{m+1}(z-z_2)^{n+1}} dz \qquad (3.107)$$

where $g(z)$ is analytic at all points inside and on C, and z_1 and z_2 are inside the contour.

The part of the integrand that is analytic at z_1 is

$$f_1(z) = \frac{g(z)}{\left(z - z_2\right)^{n+1}} \tag{3.108a}$$

and the part that is analytic at z_2 is

$$f_2(z) = \frac{g(z)}{\left(z - z_1\right)^{m+1}} \tag{3.108b}$$

Therefore, the contribution to the integral from the pole at z_1 is given by

$$\frac{1}{m!} \lim_{z \to z_1} \frac{d^m f_1}{dz^m} = \frac{1}{m!} \lim_{z \to z_1} \frac{d^m}{dz^m} \left[\frac{g(z)}{\left(z - z_2\right)^{n+1}} \right] \tag{3.109a}$$

and the contribution from the pole at z_2 is

$$\frac{1}{n!} \lim_{z \to z_2} \frac{d^n f_2}{dz^n} = \frac{1}{n!} \lim_{z \to z_2} \frac{d^n}{dz^n} \left[\frac{g(z)}{\left(z - z_1\right)^{m+1}} \right] \tag{3.109b}$$

so that

$$\oint_C \frac{g(z)}{\left(z - z_1\right)^{m+1} \left(z - z_2\right)^{n+1}} \, dz$$

$$= 2\pi i \left[\frac{1}{m!} \frac{d^m}{dz^m} \left(\frac{g(z)}{\left(z - z_2\right)^{n+1}} \right)_{z \to z_1} + \frac{1}{n!} \frac{d^n}{dz^n} \left(\frac{g(z)}{\left(z - z_1\right)^{m+1}} \right)_{z \to z_2} \right] \tag{3.110}$$

This result can be generalized to any number of poles in a straightforward way. When there are more than two poles, the expression equivalent to eq. 3.110 is straightforward but becomes unwieldy and so is not presented here. It is left to the interested reader to deduce the expression for the closed contour integral of a function containing a general number of poles within the contour.

Example 3.11: Cauchy's residue theorem for an integrand with several poles

We consider

$$I \equiv \oint \frac{e^z}{(z-1)^2 z^3 (z-2)} dz \qquad\qquad (3.111)$$

around a circle of radius 3/2, centered at the origin. In asmuch as the pole at $z = 2$ is outside the contour, according to eq. 3.110, the result is

$$\oint \frac{e^z}{(z-1)^2 z^3 (z-2)} dz$$

$$= 2\pi i \left[\frac{1}{1!} \frac{d}{dz} \left(\frac{e^z}{z^3 (z-2)} \right)_{z=1} + \frac{1}{2!} \frac{d^2}{dz^2} \left(\frac{e^z}{(z-1)^2 (z-2)} \right)_{z=0} \right] \qquad (3.112)$$

$$= 2\pi i \left(e - \frac{29}{8} \right) \qquad\qquad \Box$$

Problems

1. For each function below, determine the values or range of values of x and y (if any) for which the function is analytic.

   (a) $F(z) = z^3$ (b) $F(z) = x^2 + y^2$ (c) $F(z) = x^2 + iy^2$

   (d) $F(z) = y^2 - ix^2$ (e) $F(z) = e^{y+ix}$ (f) $F(z) = e^{-y+ix}$

   (g) $F(z) = e^{x-iy}$ (h) $F(z) = z*$

2. Prove that each function below is an entire function.

   (a) $F(z) = \sin^2 z$ (b) $F(z) = \cosh^3(z)$

   (c) $F(z) = z^n$ n a positive integer.   (d) $F(z) = e^{z^2}$

3. Show that each function below satisfies Laplace's equation.

(a) $F(z) = e^{z^2}$ (b) $F(z) = \sinh(z)$

(c) $F(z) = \sin(z^2)$ (d) $F(z) = \tan(z)$ $|z| < \dfrac{\pi}{2}$

4. $F(z) = U(x, y) + iV(x, y)$ is analytic everywhere in a region R. Therefore, dF/dz is also analytic everywhere in R. Referring to eqs. 3.11, apply the CR conditions to $\mathrm{Re}(dF/dz)$ and $\mathrm{Im}(dF/dz)$ to show that $U(x, y)$ and $V(x, y)$ each satisfy Laplace's equation.

5. (a) Express Laplace's equation $\nabla^2 F(x, y) = 0$ in polar coordinates.
    (b) Express each function of problem 2 in polar coordinates, then show that each function satisfies Laplace's equation in polar coordinates.
    (c) Let $F(z)$ be analytic everywhere in a region R that contains z. Expressing z in exponential form, use the identities

$$\frac{\partial F}{\partial r} = \frac{\partial F}{\partial z}\frac{\partial z}{\partial r} = e^{i\theta}\frac{\partial F}{\partial z} \quad \text{and} \quad \frac{\partial F}{\partial \theta} = \frac{\partial F}{\partial z}\frac{\partial z}{\partial \theta} = iz\frac{\partial F}{\partial z}$$

to show that

$$\frac{1}{r}\frac{\partial}{\partial r}\left(r\frac{\partial F}{\partial r}\right) + \frac{1}{r^2}\frac{\partial^2 F}{\partial \theta^2} = 0$$

everywhere in R.

6. $V(x, y)$ is the imaginary part of a function $F(z)$.

    (a) Find the condition that determines whether a real part can be found that makes $F(z)$ analytic in a region R containing the point (x, y).
    (b) When that condition holds so that $F(z)$ is analytic, derive an expression for the real part, $U(x, y)$.

7. The real and imaginary parts of $F(z)$ are designated $U(x, y)$ and $V(x, y)$, respectively. For each function below, determine if the unknown part of $F(z)$ can be determined that makes it analytic. If $F(z)$ can be made analytic, find its unknown part.

    (a) $V(x, y) = 2x + y$ (b) $U(x, y) = 2x^2 + y^2$

(c) $U(x, y) = \cos x \cosh y$ (d) $V(x, y) = y^2 - x^2$

(e) $U(x, y) = \sin y \cosh x$ (f) $V(x, y) = e^x \cosh y$

(g) $V(x, y) = e^y \sinh x$ (h) $U(x, y) = e^y \cos x$

8. Let $F(z) = U(x, y) + iV(x, y)$ be analytic everywhere in a region R. Show that if $U(x, y) = P(x) + Q(y)$, and $P(x)$ and $Q(y)$ satisfy $P''(x) = p_1$ and $Q'(y) = q_1 - y$ (where p_1 and q_1 are constants) then $V(x, y)$ exists.

9. Let $F(z) = U(x, y) + iV(x, y)$ be analytic everywhere in a region R. Let $U(x, y) = P(x)Q(y)$.
   (a) Show that $V(x, y)$ can be determined if $P''(x) = 0$ and $Q'(y) = 0$.

   (b) Show that $V(x, y)$ can be determined if $P(x)$ and $Q(x)$ satisfy
   $P''(x) - \alpha^2 P(x) = 0$ and $Q''(x) + \alpha^2 Q(x) = 0$ where α is a constant.

10. (a) Show that $\int_2^{2i} \cos z\, dz$ along the linear path shown in fig. P3.1a
    is unchanged when that linear path is deformed into the segment of the perimeter of the square shown in fig. P3.1b.

    (b) Show that $\int_2^{2i} z^2\, dz$ along the linear path shown in fig. P3.1a is
    unchanged when that linear path is deformed into the perimeter of the quarter circle shown in fig. P3.1c.

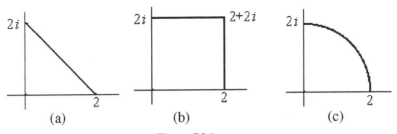

Figure P3.1

(a) A straight line, (b) a section of a square, and (c) a quarter circle of radius 2
from $z = 2$ to $z = 2i$

11. Evaluate $\int_0^{1+i\pi} e^z dz$ along each of the paths shown in fig. P3.2.

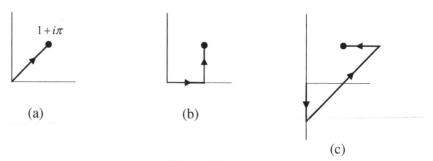

(a)   (b)

(c)

Figure P3.2

Three paths from the origin to the point $z = 1 + i\pi$

12. Verify that $\int_0^{1+i} z dz$ is the same for the two paths shown in fig. P3.3 and is the result obtained by treating z as a single variable even though it is the sum of two independent variables.

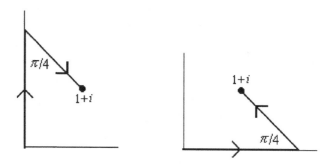

Figure P3.3

Two paths from the origin to the point $z = 1 + i$

13. Verify that $\int_0^{1+i} z\, dz$ is the same for the two paths shown in fig. P3.4 and is the result obtained by treating z as a single variable even though it is the sum of two independent variables.

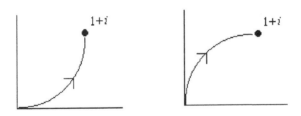

Figure P3.4

Two quarter circular paths from the origin to the point $z = 1 + i$

14. $F(z)$ is an analytic function everywhere inside and on the unit circle. Evaluate the integral below for integer N with $N > 0$, $N = 0$, and $N < 0$.

$$\int_0^{2\pi} e^{iN\phi} F\left(e^{i\phi}\right) d\phi$$

15. Each function below has one or more simple poles. Identify the value(s) of z at which each function has the simple pole(s) and determine the value of each pole.

(a) $F(z) = \dfrac{1}{z + 3i}$ (b) $F(z) = \dfrac{z}{z^2 - 4}$ (c)

$$F(z) = \dfrac{e^z}{z^2 + 4\pi^2}$$

(d) $F(z) = \cot z$ (e) $F(z) = \tanh z$ (f) $F(z) = \dfrac{\sin z}{4z - \pi}$

(g) $F(z) = \dfrac{e^z}{4z^2 + 9\pi^2}$ (h) $F(z) = \dfrac{e^{-z}}{z + 3\pi i}$ (j) $F(z) = \dfrac{\sinh z}{2z - i\pi}$

The contours in problems 16 through 20 are to be traversed in the counterclockwise direction.

16. Evaluate $\displaystyle\oint \dfrac{z \cos z}{(z + i - 1)} dz$ around the following contours:

(a) A circle centered at the origin of radius $1/2$
(b) A circle centered at the origin of radius $3/2$

(c) A square centered at the origin of side 5

17. Evaluate $\displaystyle\oint \frac{z \cos z}{(z+i-1)^3} dz$ around the following contours:

(a) A circle centered at the origin of radius 1/2
(b) A circle centered at the origin of radius 3/2
(c) A square centered at the origin of side 5

18. Evaluate $\displaystyle\oint \frac{\cos z}{(z-1)(z+2i)} dz$ around the following contours:

(a) A circle centered at the origin of radius 1/2
(b) A circle centered at the origin of radius 3/2
(c) A square centered at the origin of side 5

19. Evaluate $\displaystyle\oint \frac{\cos z}{(z-1)^3(z+2i)} dz$ around the following contours:

(a) A circle centered at the origin of radius 1/2
(b) A circle centered at the origin of radius 3/2
(c) A circle centered at the origin of radius 5/2
(d) A square centered at the origin of side 5

20. Evaluate each integral below around a circle centered at the origin of radius 2. In parts (h) and (i), $n \geq 1$ is an integer.

(a) $\displaystyle\oint \frac{e^{i\pi z}}{(z-1)} dz$
 (b) $\displaystyle\oint \frac{e^{i\pi z}}{(z-1)^2} dz$
 (c) $\displaystyle\oint \frac{e^{i\pi z}}{(z-1)^3} dz$

(d) $\displaystyle\oint \frac{e^{i\pi z}}{(z-i)^3} dz$
 (e) $\displaystyle\oint \frac{e^{i\pi z}}{(z-i)^4} dz$
 (f) $\displaystyle\oint \frac{e^{i\pi z}}{(z^2+1)} dz$

(g) $\displaystyle\oint \frac{e^{i\pi z}}{(z^2+1)^2} dz$
 (h) $\displaystyle\oint \frac{e^{i\pi z}}{(z+1)^n} dz$
 (i) $\displaystyle\oint \frac{e^{i\pi z}}{(z+i)^n} dz$

Chapter 4

SERIES, LIMITS, AND RESIDUES

4.1 Taylor Series for an Analytic Function

Let a function $F(w)$ be analytic at all points in a region R of the w-plane. Let R contain the point z_0 and let C be a circular contour in R centered at z_0.

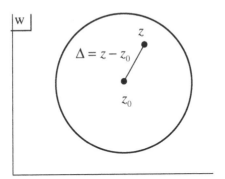

Figure 4.1

Circular contour in the w-plane centered at z_0

The Cauchy integral representation of $F(z)$ at some point z inside the contour is

$$F(z) = \frac{1}{2\pi i} \oint_C \frac{F(w)}{(w-z)} dw \qquad (3.101b)$$

Defining

$$\Delta \equiv z - z_0 \tag{4.1}$$

eq. 3.101b can be written

$$F(z) = F(z_0 + \Delta) = \frac{1}{2\pi i} \oint_C \frac{F(w)}{(w - z_0 - \Delta)} dw \tag{4.2}$$

Because z is inside the contour, the distance from z_0 to z is smaller than the distance from z_0 to any point on the contour. Therefore, for all points w on the contour,

$$\left| \frac{\Delta}{(w - z_0)} \right| < 1 \tag{4.3}$$

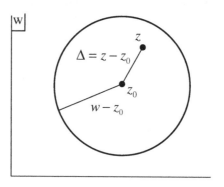

Figure 4.2
$$\Delta = |z - z_0| < |w - z_0|$$

As derived in appendix 3, for a complex quantity t, the function $1/(1 - t)$ can be expanded in the *geometric series* as

$$\frac{1}{(1-t)} = 1 + t + t^2 + \cdots = \sum_{n=0}^{\infty} t^n \qquad |t| < 1 \tag{A3.4}$$

Letting

$$t = \frac{\Delta}{(w - z_0)} \tag{4.4}$$

we write

$$\frac{1}{(w - z_0 - \Delta)} = \frac{1}{(w - z_0)} \frac{1}{\left(1 - \dfrac{\Delta}{(w - z_0)}\right)} = \sum_{n=0}^{\infty} \frac{\Delta^n}{(w - z_0)^{n+1}} \tag{4.5}$$

Then, referring to eq. 3.103b, eq. 4.2 can be expressed as

$$F(z) = \sum_{n=0}^{\infty} \frac{\Delta^n}{2\pi i} \oint_C \frac{F(w)}{(w - z_0)^{n+1}} \, dw = \sum_{n=0}^{\infty} \Delta^n \frac{F^{(n)}(z_0)}{n!} \tag{4.6}$$

With $\Delta = (z - z_0)$, eq. 4.6 becomes

$$F(z) = \sum_{n=0}^{\infty} \frac{F^{(n)}(z_0)}{n!} (z - z_0)^n \tag{4.7}$$

This is called the *Taylor series representation* or the *Taylor series expansion* of $F(z)$ around z_0. When $z_0 = 0$, the Taylor series is called the *MacLaurin series*,

$$F(z) = \sum_{n=0}^{\infty} \frac{F^{(n)}(0)}{n!} z^n \tag{4.8}$$

Convergence of the Taylor series

To say that the Taylor series is a valid representation of $F(z)$ means that the value of the series at the point z is the same as the value of the function at that z. The terminology is that the series (on the right side of eq. 4.7 or 4.8) *converges* to the function (on the left side of eq. 4.7 or 4.8) at z.

It was shown in chapter 3 that as long as we do not cross over a singularity of the integrand, a contour can be reshaped without affecting the integral. Let z_1 denote the singularity of $F(z)$ outside the contour that is nearest to z_0. Then, without affecting the integral representation of $F(z)$, the circular contour of fig. 4.1 can be enlarged so that it is infinitesimally close to, but does not touch z_1. The Taylor series converges to $F(z)$ at all points inside this circle centered at z_0, of radius $|z_1 - z_0|$ which is called the *circle of convergence*. Its radius is the *radius of convergence* of the series. Because an entire function is analytic at all z with $|z| < \infty$, the Taylor series for an entire function has an infinite radius of convergence.

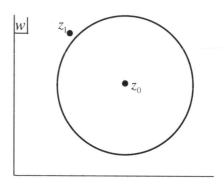

Figure 4.3

Circle of convergence of the series of a function expanded around z_0 with its nearest singularity at z_1

Example 4.1: The Taylor series for selected analytic functions

(a) Let us determine the Taylor series for

$$F(z) = \frac{1}{(1-z)} \tag{4.9}$$

around the point $z_0 = 1/4$. With

$$F^{(n)}\left(\tfrac{1}{4}\right) = \frac{n!}{(1-z)^{n+1}}\bigg|_{z=1/4} = n!\left(\tfrac{4}{3}\right)^{n+1} \tag{4.10}$$

the Taylor series for $1/(1 - z)$, expanded around $z_0 = 1/4$, is

$$\frac{1}{(1-z)} = \sum_{n=0}^{\infty} \left(\tfrac{4}{3}\right)^{n+1} \left(z - \tfrac{1}{4}\right)^n \tag{4.11}$$

Because $1/(1 - z)$ has a singularity at $z = 1$, the circular contour, centered at $z = 1/4$, can be enlarged to a circle that is infinitesimally close to 1. Thus, the radius of convergence of the series of eq. 4.11 is 3/4. By the identical argument, the MacLaurin series for $1/(1 - z)$ (the geometric series) converges to this function at all z within a circle centered at $z = 0$, of radius 1.

(b) Because all the derivatives of

$$F(z) = e^z \qquad\qquad (4.12)$$

are

$$F^{(n)}(z_0) = e^{z_0} \qquad\qquad (4.13)$$

the Taylor series for e^z around z_0 is given by

$$e^z = e^{z_0}\left[1 + (z - z_0) + \frac{(z - z_0)^2}{2!} + \cdots\right] = e^{z_0}\sum_{n=0}^{\infty} \frac{(z - z_0)^n}{n!} \qquad (4.14a)$$

or

$$e^{(z - z_0)} = \sum_{n=0}^{\infty} \frac{(z - z_0)^n}{n!} \qquad\qquad (4.14b)$$

Because e^z is an entire function, the radius of convergence of this Taylor series is infinite.

The MacLaurin series for e^z is easily obtained by setting $z_0 = 0$. The result is

$$e^z = \sum_{n=0}^{\infty} \frac{z^n}{n!} \qquad\qquad (4.15)$$

(c) The coefficients of the MacLaurin series for

$$F(z) = \sin z \qquad\qquad (4.16)$$

are found to be

$$F(0) = \sin z\big|_{z=0} = 0 \qquad\qquad (4.17a)$$

$$F'(0) = \cos z\big|_{z=0} = 1 \qquad\qquad (4.17b)$$

$$F''(0) = -\sin z\big|_{z=0} = 0 \qquad\qquad (4.17c)$$

and so on. Then the MacLaurin series for $\sin z$ is

$$\sin z = z - \frac{z^3}{3!} + \frac{z^5}{5!} - \cdots = \sum_{n=0}^{\infty} (-1)^n \frac{z^{2n+1}}{(2n+1)!} \qquad (4.18)$$

As with the exponential series, $\sin z$ is an entire function, and so its Taylor series around any finite z_0 has an infinite radius of convergence.

We note that when z is small, the terms in z^2, z^3, ... are negligibly small compared to z and can be ignored. Therefore, as indicated in ex. 3.7c, when z is small,

$$\sin z \approx z \qquad\qquad \begin{matrix}(3.70)\\[4pt]\square\end{matrix}$$

4.2 Laurent Series for a Singular Function

If a function $F(z)$ has a singularity at z_0, it cannot be represented by a Taylor series. To develop a series for such a function around z_0, we again start with the Cauchy integral representation of $F(z)$ given in eq. 3.101b, with a contour that excludes z_0 as shown in fig. 4.4.

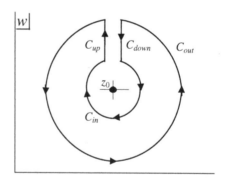

Figure 4.4

Contour that excludes a singularity of $F(z)$ at z_0

With $\Delta = (z - z_0)$, we write the Cauchy integral representation as

$$F(z) = \frac{1}{2\pi i} \oint_C \frac{F(w)}{(w - z_0 - \Delta)}\, dw \qquad (4.19)$$

around the contour of fig. 4.4. Following the arguments used in chapter 3 to develop Cauchy's residue theorem, we shrink the gap between C_{up} and C_{down}

to zero, set the sum of integrals along C_{up} and C_{down} to zero and reverse the direction of traversal around the inner contour changing the sign on that integral. Then, with the outer and inner contours traversed in the counterclockwise direction, eq. 4.19 becomes

$$F(z) = \frac{1}{2\pi i} \left[\oint_{C_{out}} \frac{F(w)}{(w - z_0 - \Delta)} dw - \oint_{C_{in}} \frac{F(w)}{(w - z_0 - \Delta)} dw \right] \qquad (4.20)$$

We then expand $1/(w - z_0 - \Delta)$ in a geometric series for each contour in terms of a quantity that has a magnitude less than 1.

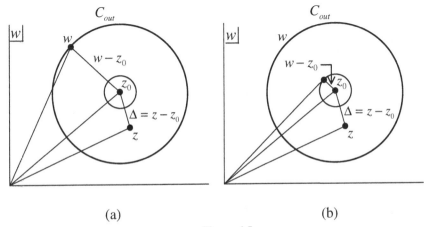

(a)   (b)

Figure 4.5

Comparison of $|w - z_0|$ and $|z - z_0|$ for the (a) outer and (b) inner contours

As shown in fig. 4.5a, points on the outer contour satisfy $|\Delta| < |w - z_0|$. Therefore, for the integral around the outer contour,

$$\frac{1}{(w - z_0 - \Delta)} = \frac{1}{(w - z_0)} \frac{1}{\left(1 - \dfrac{\Delta}{(w - z_0)}\right)} = \sum_{n=0}^{\infty} \frac{\Delta^n}{(w - z_0)^{n+1}} \qquad (4.21a)$$

Points on the inner contour satisfy $|\Delta| > |w - z_0|$. Therefore, for the integral around this contour,

$$\frac{1}{(w - z_0 - \Delta)} = -\frac{1}{\Delta} \frac{1}{\left(1 - \dfrac{(w - z_0)}{\Delta}\right)} = -\sum_{n=0}^{\infty} \frac{(w - z_0)^n}{\Delta^{n+1}} \qquad (4.21b)$$

Then, the integral representation of $F(z)$ becomes

$$F(z) = \frac{1}{2\pi i} \left[\sum_{n=0}^{\infty} \Delta^n \oint_{C_{out}} \frac{F(w)}{(w-z_0)^{n+1}} dw + \sum_{n=0}^{\infty} \frac{1}{\Delta^{n+1}} \oint_{C_{in}} F(w)(w-z_0)^n dw \right]$$

(4.22a)

The letter used for the summation index does not affect the sum, therefore we substitute $n' = -n - 1$ in the second sum, then drop the prime. Then eq. 4.22a becomes

$$F(z) = \frac{1}{2\pi i} \left[\sum_{n=0}^{\infty} \Delta^n \oint_{C_{out}} \frac{F(w)}{(w-z_0)^{n+1}} dw + \sum_{n=-\infty}^{-1} \Delta^n \oint_{C_{in}} \frac{\tilde{F}(w)}{(w-z_0)^{n+1}} dw \right]$$

(4.22b)

Recall that both $F(w)$ and $1/(w-z_0)^{n+1}$ are singular at $w = z_0$. Therefore, the integrands in both sums are singular at z_0. The singularities of the integrands in the first sum arise from poles of order $n + 1$ at z_0 and the singularity of $F(w)$ at z_0. Because $-\infty \leq n \leq -1$, there is no pole singularity in the integrands in the second sums arising from $1/(w-z_0)^{n+1}$. But those integrands are singular because of the singularity of $F(w)$ at z_0.

We note that there are no other singularities of the integrands inside the contours. Thus, the region between C_{out} and C_{in} is a region of analyticity for both integrands. Therefore, one of these contours can be deformed into the other without changing the value of the integral and we can write

$$\oint_{C_{out}} \frac{F(w)}{(w-z_0)^{n+1}} dw = \oint_{C_{in}} \frac{F(w)}{(w-z_0)^{n+1}} dw$$

(4.23)

Then eq. 4.22b becomes

$$F(z) = \frac{1}{2\pi i} \left[\sum_{n=0}^{\infty} \Delta^n \oint_{C_{out}} \frac{F(w)}{(w-z_0)^{n+1}} dw + \sum_{n=-\infty}^{-1} \Delta^n \oint_{C_{out}} \frac{F(w)}{(w-z_0)^{n+1}} dw \right]$$

$$= \frac{1}{2\pi i} \sum_{n=-\infty}^{\infty} \Delta^n \oint_{C_{out}} \frac{F(w)}{(w-z_0)^{n+1}} dw$$

(4.24)

We now drop the subscript "out" on the contour, recognizing that C is any contour that encloses only the singularity at z_0.

Let

$$a_n(z_0) \equiv \frac{1}{2\pi i} \oint_C \frac{F(w)}{(w-z_0)^{n+1}} dw \qquad (4.25)$$

Referring to eqs. 3.102, we point out that because $F(w)$ is singular at z_0, this integral is not related to $F^{(n)}(z_0)$. That is,

$$a_n(z_0) \ne \frac{1}{n!} F^{(n)}(z_0) \qquad (4.26)$$

Then, with $\Delta = (z - z_0)$, eq. 4.24 becomes

$$F(z) = \sum_{n=-\infty}^{\infty} a_n(z_0)(z-z_0)^n \qquad (4.27)$$

This is the *Laurent series representation* of a function that is singular at z_0.

Because the integrand of the integral defining $a_n(z_0)$ is singular only at z_0, the contour can be taken to be a circle, centered at z_0, that can be enlarged out to the next singularity of $F(z)$. Denoting this next singularity as z_1 the circle of convergence for the Laurent series, like that for the Taylor series, is centered at z_0 with a radius $|z_1 - z_0|$ as shown in fig. 4.3. The difference between the circle of convergence for the Taylor series and the circle of convergence for the Laurent series is that the Taylor series converges to $F(z)$ for all z within the circle, including z_0. The Laurent series converges to $F(z)$ for all z within the circle except z_0.

Laurent series for an analytic function

Referring to eq. 4.25, the Laurent series coefficients for $n \le -1$ can be written

$$a_{-n}(z_0) = \frac{1}{2\pi i} \oint_C F(w)(w-z_0)^{n-1} dw \qquad (4.28a)$$

and for $n \ge 0$,

$$a_n(z_0) = \frac{1}{2\pi i} \oint_C \frac{F(w)}{(w-z_0)^{n+1}} dw \qquad (4.28b)$$

If $F(w)$ is analytic everywhere inside and on the contour, the integrand of the integral of eq. 4.28a is analytic everywhere inside and on C. Then

$$a_{-n}(z_0) = \frac{1}{2\pi i} \oint_C F(w)(w-z_0)^{n-1} dw = 0 \qquad (4.29a)$$

and from eq. 4.28b,

$$a_n(z_0) = \frac{1}{2\pi i} \oint_C \frac{F(w)}{(w-z_0)^{n+1}} dw = \frac{F^{(n)}(z_0)}{n!} \qquad (4.29b)$$

Therefore, when $F(w)$ is analytic at all points inside and on C, the Laurent series for $F(z)$ becomes

$$F(z) = \sum_{n=0}^{\infty} \frac{F^{(n)}(z_0)}{n!} (z-z_0)^n \qquad (4.7)$$

which is the Taylor series for an analytic function. That is, the Taylor series is the Laurent series for a function that is analytic everywhere inside the contour.

Laurent series for a function with a pole of order M

Let the singularity of $F(w)$ be a pole of order M. For an integer $L \geq 1$, we see from eq. 4.25 that

$$a_{-(M+L)}(z_0) = \frac{1}{2\pi i} \oint_C F(w)(w-z_0)^{(M+L-1)} dw \qquad (4.30)$$

Because the $M + L - 1 \geq M$, the factor $(w - z_0)^{(M+L-1)}$ removes the singularity from $F(w)$ and the integrand of the integral of eq. 4.30 is analytic everywhere inside and on C. Therefore,

$$a_{-(M+1)}(z_0) = a_{-(M+2)}(z_0) = \cdots = 0 \qquad (4.31)$$

and the Laurent series for a function with a Mth order pole at z_0 becomes

$$F(z) = \frac{a_{-M}(z_0)}{(w-z_0)^M} + \frac{a_{-(M-1)}(z_0)}{(w-z_0)^{(M-1)}} + \cdots + \frac{a_{-1}(z_0)}{(w-z_0)} \tag{4.32}$$
$$+ a_0(z_0) + a_1(z_0)(z-z_0) + a_2(z_0)(z-z_0)^2 + \cdots$$

We note that each term in the second line of eq. 4.32 contains a positive power of $(z - z_0)$. Therefore, this sum of terms is a function that is analytic at all points inside the contour. Henceforth, when necessary, this analytic part is denoted by

$$\Phi_A(z) \equiv a_0(z_0) + a_1(z_0)(z-z_0) + a_2(z_0)(z-z_0)^2 + \cdots \tag{4.33}$$

and the Laurent series for a function with an Mth order pole at z_0 is expressed as

$$F(z) = \frac{a_{-M}(z_0)}{(w-z_0)^M} + \frac{a_{-(M-1)}(z_0)}{(w-z_0)^{(M-1)}} + \cdots + \frac{a_{-1}(z_0)}{(w-z_0)} + \Phi_A(z) \tag{4.34}$$

Example 4.2: The Laurent series for selected functions with singularities

(a) Let $N \geq 0$ be an integer. Referring to ex. 3.6b,

$$F(z) = z^N e^{1/z} \tag{4.35}$$

has an essential singularity at $z = 0$. From the Taylor series for the exponential function given in ex. 4.1b, the Laurent series for $F(z)$ around $z = 0$ is

$$z^N e^{1/z} = \sum_{k=0}^{\infty} \frac{z^{N=k}}{k!} \tag{4.36a}$$

Replacing the summation index by $\ell = N - k$, the Laurent series can be expressed as

$$z^N e^{1/z} = \sum_{\ell=-\infty}^{N} \frac{z^\ell}{(N-\ell)!}$$
$$= \cdots + \frac{z^{-2}}{(N+2)!} + \frac{z^{-1}}{(N+1)!} + \cdots + \frac{z^{N-2}}{2!} + \frac{z^{N-1}}{1!} + z^N \tag{4.36b}$$

(b) Let $N > 0$ be an integer. The Laurent series representation of

$$F(z) = \frac{\sin z}{z^N} \tag{4.37}$$

can be determined from the MacLaurin series for $\sin z$ given in eq. 4.18. The result is

$$\frac{\sin z}{z^N} = \sum_{k=0}^{\infty} (-1)^k \frac{z^{2k+1-N}}{(2k+1)!} \tag{4.38}$$

Because $2k + 1$ is an odd integer, when N is even, the Laurent series is an expansion in odd powers of z. For odd N, the powers of z are even. We define a new summation index m by

$$2k + 1 - N = 2m + 1 \quad (N \text{ even}) \tag{4.39a}$$

and

$$2k + 1 - N = 2m \quad (N \text{ odd}) \tag{4.39b}$$

Then the Laurent series can be more transparently expressed as an odd or even series as

$$\frac{\sin z}{z^N} = \sum_{m=-N/2}^{\infty} (-1)^{m+\frac{1}{2}N} \frac{z^{2m+1}}{(2m+1+N)!} \quad (N \text{ even}) \tag{4.40a}$$

and

$$\frac{\sin z}{z^N} = \sum_{m=-(N-1)/2}^{\infty} (-1)^{m+\frac{1}{2}(N-1)} \frac{z^{2m}}{(2m+N)!} \quad (N \text{ odd}) \tag{4.40b}$$

$\square$

4.3 Radius of Convergence and the Cauchy Ratio

The radius of convergence of a series is determined from the singularity structure of the function that the series represents. It may be possible to determine the radius using the Cauchy ratio test derived in appendix 4.

For a series

$$F(z) = \sum_{n=0}^{\infty} \sigma_n(z) \tag{4.41}$$

to converge, the Cauchy ratio must satisfy

$$\rho = \lim_{n \to \infty} \left| \frac{\sigma_{n+1}(z)}{\sigma_n(z)} \right| < 1 \tag{A4.10}$$

where $\sigma_n(z)$ and $\sigma_{n+1}(z)$ are two consecutive terms in the series.
If this inequality can be solved in the form

$$|z - z_0| < r_0 = constant \tag{4.42}$$

then r_0 is the radius of convergence of the series. This is particularly straightforward for a power series in $(z - z_0)$.

Example 4.3: Radius of convergence of the Taylor series for an entire function

As shown above, the Taylor series representation of the exponential function is given by

$$e^{(z-z_0)} = \sum_{n=0}^{\infty} \frac{(z - z_0)^n}{n!} \tag{4.14b}$$

By replacing z by $z - z_0$ in eq. 4.18, the series for $\sin z$ expanded about z_0 is

$$\sin z = (z - z_0) - \frac{(z - z_0)^3}{3!} + \frac{(z - z_0)^5}{5!} - \cdots = \sum_{n=0}^{\infty} (-1)^n \frac{(z - z_0)^{2n+1}}{(2n+1)!} \tag{4.43}$$

These (entire) functions are analytic at all finite z and thus have an infinite radius of convergence.
A general expression for such a series is of the form

$$F(z) = \sum_{n=0}^{\infty} c_n \frac{(z - z_0)^{Mn+M_0}}{(Mn + M_0)!} \tag{4.44}$$

with

$$\lim_{n \to \infty} \left| \frac{c_{n+1}}{c_n} \right| \equiv C \tag{4.45}$$

where C is a finite nonzero constant. Obviously, for the exponential series, $c_n = 1$, $M = 1$, and $M_0 = 0$. For the sine series, $c_n = (-1)^n$, $M = 2$, and $M_0 = 1$. For both series, $C = 1$.

The Cauchy ratio for this series is

$$\begin{aligned}
\rho &= \lim_{n \to \infty} \left| \frac{c_{n+1}}{c_n} \right| \left| \frac{(z-z_0)^{[M(n+1)+M_0]}}{(z-z_0)^{[Mn+M_0]}} \right| \frac{[Mn+M_0]!}{[M(n+1)+M_0]!} \\
&= C \left| (z-z_0) \right|^M \lim_{n \to \infty} \frac{1}{[M(n+1)+M_0][M(n+1)+M_0-1]\cdots[Mn+M_0+1]} \\
&= C \left| (z-z_0) \right|^M {}_0^* 0
\end{aligned} \tag{4.46}$$

For the series to converge, this ratio must be less than 1. Therefore, the condition for the series to converge is

$$|z - z_0| < \infty \tag{4.47}$$

That is, the radius of convergence is infinite. □

Example 4.4: Radius of convergence of the Laurent series for a function with a singularity

The generator of the geometric series, and the series it generates are

$$F(z) = \frac{1}{[1-(z-z_0)]} = \sum_{n=0}^{\infty} (z-z_0)^n \tag{4.48}$$

Because $F(z)$ has a pole at $z - z_0 = 1$, the circle of convergence is centered at z_0 with a radius of 1.

The Taylor series for $ln\,[1-(z-z_0)]$ is given by

$$F(z) = -ln\left[1-(z-z_0)\right] = \sum_{n=0}^{\infty} \frac{(z-z_0)^{n+1}}{(n+1)} \qquad (4.49)$$

Because the logarithm function has a singularity at $z - z_0 = 1$, the circle of convergence of this series is also the circle centered at z_0 of radius 1.

A general form of such series of this type is

$$F(z) = \sum_{n=0}^{\infty} c_n \frac{(z-z_0)^{Mn+M_0}}{(Mn+M_0)} \qquad (4.50)$$

where, for the geometric series, $c_n = 1$, $M = 0$, and $M_0 = 1$. For the logarithm series, $c_n = 1$, $M = 1$, and $M_0 = 1$. Therefore, $C = 1$ for both series.

The Cauchy ratio for this series is

$$\rho = \lim_{n\to\infty} \left|\frac{c_{n+1}}{c_n}\right| \left|\frac{(z-z_0)^{[M(n+1)+M_0]}}{(z-z_0)^{[Mn+M_0]}}\right| \frac{Mn+M_0}{M(n+1)+M_0}$$

$$= C\left|(z-z_0)\right|^M \lim_{n\to\infty} \frac{M + \dfrac{M_0}{n}}{\left[M\left(1+\dfrac{1}{n}\right)+\dfrac{M_0}{n}\right]} = C\left|(z-z_0)\right|^M \qquad (4.51)$$

By requiring $\rho < 1$, we see that the series converges for

$$\left|z-z_0\right| < \frac{1}{C^{1/M}} \qquad (4.52)$$

From this it is straightforward to see that both the geometric and logarithm series of eqs. 4.48 and 4.49 have a radius of convergence of 1. □

From this discussion and the examples above, we see that the region of convergence of a series is obtained by requiring the Cauchy ratio to be strictly less than 1. This defines a set of points within a region that does not have a definite boundary. Such a region is called an *open region*. That is, a series converges in an open circle, at points of analyticity of the function (which excludes the center of the circle for a Laurent series).

Referring to fig. 4.3, the boundary of the circle of convergence of a series contains z_1, the singularity closest to z_0, the center of the circle. For any point on the boundary, part of the neighborhood of that point would be outside the region (the shaded area of fig. 4.6) where the series is not defined.

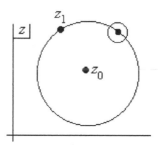

Figure 4.6
A point on the boundary of a closed region

Because the series is not defined outside its circle of convergence, the derivatives of the series do not exist at points on the boundary, and the series is not analytic at points on the boundary. The function that is represented by the series at points within the circle of convergence can be analytic at all points on the boundary except z_1, but the series is not. Therefore, the function is not represented by the series at points on the boundary. For example, $1/(1 - z)$ is analytic at $z = -1$ and $\pm i$ but the geometric series

$$\sum_{n=0}^{\infty} z^n$$

which represents $1/(1 - z)$ for $|z| < 1$, is not.

4.4 Limits and Series

L'Hopital's[†] rule provides a prescription for evaluating the limit of a function when that limit is in the form $0/0$ or ∞/∞. In some cases, it is possible to derive the rule using series representations of functions.

[†] The name of the Marquis de l'Hopital (1661–1704) appears in the current literature spelled in two different ways. During his lifetime, the name was spelled in the old French form "l'Hospital". The "s" was not pronounced. The modern French spelling of his name is "l'Hopital". The pronunciation has not changed.

Limit in the form 0/0

If

$$F(z) \equiv \frac{P(z)}{Q(z)} \tag{4.53}$$

such that

$$\lim_{z \to z_0} P(z) = 0 \tag{4.54a}$$

and

$$\lim_{z \to z_0} Q(z) = 0 \tag{4.54b}$$

then

$$\lim_{z \to z_0} F(z) = \frac{0}{0} \tag{4.55}$$

Let $P(z)$ and $Q(z)$ be analytic everywhere in a region R containing z_0. Then each can be expanded in a Taylor series around z_0. The form of the limit of eq. 4.56 arises because $P(z)$ and $Q(z)$ are zero at z_0. This occurs because the functions and possibly several of their lowest-order derivatives are zero at z_0.

If a function $A(z)$ is analytic at z_0, and satisfies

$$A(z_0) = A'(z_0) = A''(z_0) = \cdots = A^{(n-1)}(z_0) = 0 \tag{4.56a}$$

with

$$A^{(n)}(z_0) \neq 0 \tag{4.56b}$$

its Taylor series around z_0 is

$$A(z) = \frac{A^{(n)}(z_0)}{n!}(z - z_0)^n + \frac{A^{(n+1)}(z_0)}{(n+1)!}(z - z_0)^{(n+1)} + \cdots$$

$$= \sum_{m=n}^{\infty} \frac{A^{(m)}(z_0)}{m!}(z - z_0)^m \tag{4.57}$$

As $z \to z_0$, terms of order higher than $(z - z_0)^n$ are much smaller than $(z - z_0)^n$ and can be ignored. Therefore, when

$$\lim_{z \to z_0} A(z) \to \lim_{z \to z_0} \frac{A^{(n)}(z_0)}{n!}(z - z_0)^n = 0 \qquad (4.58)$$

$A(z)$ is said to have an nth *order zero at* z_0.
   For example, if $A(z)$ has a first-order zero at z_0, then

$$A(z_0) = 0 \qquad (4.59)$$

with

$$A'(z_0) \neq 0 \qquad (4.60)$$

Then the Taylor series for $A(z)$ is

$$A(z) = \frac{A'(z_0)}{1!}(z - z_0) + \frac{A''(z_0)}{2!}(z - z_0)^2 + \cdots \qquad (4.61)$$

If $P(z)$ has a zero of order M and $Q(z)$ has a zero of order N, then

$$\lim_{z \to z_0} F(z) =$$

$$\lim_{z \to z_0} \left[\frac{\frac{1}{M!}P^{(M)}(z_0)(z - z_0)^M + \frac{1}{(M+1)!}P^{(M+1)}(z_0)(z - z_0)^{M+1} + \cdots}{\frac{1}{N!}Q^{(N)}(z_0)(z - z_0)^N + \frac{1}{(N+1)!}Q^{(N+1)}(z_0)(z - z_0)^{N+1} + \cdots} \right] \qquad (4.62a)$$

As $z \to z_0$, terms containing $(z - z_0)^{M+1}$, $(z - z_0)^{M+2}$, ... are small compared to $(z - z_0)^M$ and can be ignored. Likewise, terms of order $(z - z_0)^{N+1}$ and higher can be ignored relative to $(z - z_0)^M$. Therefore,

$$\lim_{z \to z_0} F(z) = \lim_{z \to z_0} \left[\frac{\frac{1}{M!}P^{(M)}(z_0)(z - z_0)^M}{\frac{1}{N!}Q^{(N)}(z_0)(z - z_0)^N} \right] = \begin{cases} 0 & M > N \\[2mm] \dfrac{P^{(M)}(z_0)}{Q^{(M)}(z_0)} & M = N \\[2mm] \infty & M < N \end{cases} \qquad (4.62b)$$

Limit in the form ∞/∞

When

$$\lim_{z \to z_0} F(z) = \lim_{z \to z_0} \frac{P(z)}{Q(z)} = \frac{\infty}{\infty} \tag{4.63}$$

l'Hopital's rule can also be applied to evaluate this limit. The general proof of this is beyond the level of this text, but is presented in various advanced calculus texts. (Two such resources are Buck and Buck, 1978, p. 21 and Taylor, 1955, pp. 121–123.)

If $1/P(z)$ and $1/Q(z)$ are analytic at z_0, the limit of eq. 4.63 can be evaluated by writing it as

$$\lim_{z \to z_0} F(z) = \lim_{z \to z_0} \frac{1/Q(z)}{1/P(z)} \equiv \lim_{z \to z_0} \frac{R(z)}{S(z)} = \frac{0}{0} \tag{4.64}$$

and applying l'Hopital's rule to $R(z)$ and $S(z)$.

One reason $P(z_0)$ and $Q(z_0)$ could be infinite is that each has a pole at z_0. If $P(z)$ has a pole of order M and $Q(z)$ has an Nth-order pole at z_0, their Laurent series representations are

$$P(z) = \frac{p_{-M}}{(z - z_0)^M} + \frac{p_{-(M-1)}}{(z - z_0)^{M-1}} + \cdots \tag{4.65a}$$

and

$$Q(z) = \frac{q_{-N}}{(z - z_0)^N} + \frac{q_{-(N-1)}}{(z - z_0)^{N-1}} + \cdots \tag{4.65b}$$

Then

$$F(z) = \frac{\dfrac{p_{-M}}{(z - z_0)^M} + \dfrac{p_{-(M-1)}}{(z - z_0)^{M-1}} + \cdots}{\dfrac{q_{-N}}{(z - z_0)^N} + \dfrac{q_{-(N-1)}}{(z - z_0)^{N-1}} + \cdots} \tag{4.66a}$$

Multiplying numerator and denominator by $(z - z_0)^N$, this becomes

$$F(z) = \frac{p_{-M}(z - z_0)^{N-M} + p_{-(M-1)}(z - z_0)^{N-M+1} + \cdots}{q_{-N} + q_{-(N-1)}(z - z_0) + \cdots} \tag{4.66b}$$

From this we see that when $P(z)$ and $Q(z)$ have M^{th} and N^{th} order poles respectively, then

$$\lim_{z \to z_0} F(z) = \begin{cases} 0 & N > M \\ \dfrac{p_{-M}}{q_{-M}} & N = M \\ \infty & N < M \end{cases} \tag{4.67}$$

4.5 Arithmetic Combinations of Power Series

Let $S(z)$ and $T(z)$ be two functions for which the Taylor or Laurent series expansions around z_0 are known. Let $F(z)$ be an arithmetic combination of $S(z)$ and $T(z)$, for which we wish to find the Taylor or Laurent series representation. In some cases, in addition to determining the coefficients of the series for $F(z)$ directly from eq. 4.7 or 4.25, it may be possible to find these coefficients by arithmetically combining the series for $S(z)$ and $T(z)$.

We define

$$S(z) = \sum_{k=-\infty}^{\infty} \sigma_k (z - z_0)^k \tag{4.68a}$$

$$T(z) = \sum_{\ell=-\infty}^{\infty} \tau_\ell (z - z_0)^\ell \tag{4.68b}$$

and

$$F(z) = \sum_{m=-\infty}^{\infty} a_m(z_0)(z - z_0)^m \tag{4.69}$$

Because $F(z)$ is an arithmetic combination of $S(z)$ and $T(z)$, we attempt to find the coefficients a_m in terms of σ_k and τ_ℓ.

Addition and subtraction

The coefficients of the series for $F(z)$ can be found straightforwardly from the coefficients of the $S(z)$ and $T(z)$ series when

$$F(z) = S(z) \pm T(z) \tag{4.70a}$$

Renaming the summation indices on the series representations for $S(z)$ and $T(z)$, eq. 4.70a can be written in series form as

$$\sum_{m=-\infty}^{\infty} a_m (z - z_0)^m = \sum_{m=-\infty}^{\infty} \sigma_m (z - z_0)^m \pm \sum_{m=-\infty}^{\infty} \tau_m (z - z_0)^m \tag{4.70b}$$

from which

$$a_m = \sigma_m \pm \tau_m \tag{4.71}$$

Therefore,

$$F(z) = \sum_{m=-\infty}^{\infty} \left(\sigma_m \pm \tau_m \right)(z - z_0)^m \tag{4.72}$$

Example 4.5: Addition/subtraction of two power series

Let

$$S(z) = e^{z^2} \tag{4.73a}$$

and

$$T(z) = \cos z \tag{4.73b}$$

Referring to eq. 4.15, the MacLaurin series for $S(z)$ is

$$e^{z^2} = \sum_{k=0}^{\infty} \frac{z^{2k}}{k!} \tag{4.74a}$$

In addition to using eq. 4.8, the MacLaurin series for $\cos z$ can be obtained by differentiating the MacLaurin series for $\sin z$ given in eq. 4.18. The result is

$$\cos z = \sum_{\ell=0}^{\infty} (-1)^\ell \frac{z^{2\ell}}{(2\ell)!} \tag{4.74b}$$

The sum or difference of $S(z)$ and $T(z)$ is given by

$$e^{z^2} \pm \cos z = \sum_{m=0}^{\infty} \frac{z^{2m}}{m!} \pm \sum_{m=0}^{\infty} (-1)^m \frac{z^{2m}}{(2m)!} = \sum_{m=0}^{\infty} \left(\frac{1}{m!} \pm \frac{(-1)^m}{(2m)!} \right) z^{2m}$$

(4.75)

□

Multiplication

If $S(z)$ and $T(z)$ have Laurent series representations about z_0, then the series form of

$$F(z) = S(z) * T(z)$$

(4.76a)

is

$$\sum_{m=-\infty}^{\infty} a_m (z - z_0)^m = \left[\sum_{k=-\infty}^{\infty} \sigma_k (z - z_0)^k \right] \left[\sum_{\ell=-\infty}^{\infty} \tau_\ell (z - z_0)^\ell \right]$$

$$= \sum_{\substack{k=-\infty \\ \ell=-\infty}}^{\infty} \sigma_k \tau_\ell (z - z_0)^{k+\ell}$$

(4.76b)

where we use the notation of a single summation sign with multiple indices to indicate a multiple sum.

If $S(z)$ has a nonremovable singularity at z, then, in general, all σ_k are nonzero. Thus, its Laurent series is

$$S(z) = \sum_{k=-\infty}^{\infty} \sigma_k (z - z_0)^k$$

(4.77)

with a series of identical form for $T(z)$. If one of these functions has a non-removable singularity, or if both functions have nonremovable singularities that do not cancel, $F(z)$ must have a nonremovable singularity. Then the coefficients of the Laurent series for $F(z)$ are given by

$$a_m = \sum_{k=-\infty}^{\infty} \sigma_k \tau_{m-k} \quad -\infty \leq m \leq \infty$$

(4.78)

If $S(z)$ has a pole of order M at z_0 and $T(z)$ has an Nth order pole at z_0, then $F(z)$ has a pole of order $-(M + N)$, and the coefficients of terms in the double sum with powers of $(z - z_0)$ less than $-(M + N)$ must be zero. With

$$S(z) = \sum_{k=-M}^{\infty} \tau_k (z - z_0)^{-k} = \sigma_{-M} (z - z_0)^{-M} + \sigma_{-M+1} (z - z_0)^{-M+1} + \cdots$$

(4.79a)

and

$$T(z) = \sum_{k=-N}^{\infty} \tau_k (z - z_0)^{-k} = \tau_{-N} (z - z_0)^{-N} + \tau_{-N+1} (z - z_0)^{-N+1} + \cdots$$

(4.79b)

eq. 4.76b yields

$$F(z) = (z - z_0)^{-(M+N)} \left[a_{-(M+N)} + a_{-(M+N)+1} (z - z_0) + \cdots \right]$$
$$(z - z_0)^{-(M+N)} \left[\sigma_{-M} + \sigma_{-M+1} (z - z_0) + \cdots \right] \left[\tau_{-N} + \tau_{-N} (z - z_0) + \cdots \right]$$
$$= (z - z_0)^{-(M+N)} \sum_{k=0}^{\infty} \sum_{\ell=0}^{k} \sigma_{-M+k} \tau_{-N+\ell} (z - z_0)^{-k+\ell}$$

(4.80)

Thus,

$$a_{-(M+N)+k} = \sum_{\ell=0}^{k} \sigma_{-M+\ell} \tau_{-N+k-\ell}$$

(4.81)

If $F(z)$ is analytic at z_0, then $(M + N) = 0$. This means that both $S(z)$ and $T(z)$ are analytic or that one has a pole of some order, and the other has a zero of the same order.

Example 4.6: Multiplication of two power series

The product of the functions used in ex. 4.3 is

$$F(z) = \sum_{m=0}^{\infty} a_m z^m = e^{z^2} \cos z = \sum_{\substack{k=0 \\ \ell=0}}^{\infty} \frac{(-1)^\ell}{[k!][(2\ell)!]} z^{2(k+\ell)}$$

(4.82)

The double sum contains only even powers of z, therefore the coefficients of the series for $F(z)$ satisfy

$$a_m = 0 \quad (m \text{ odd}) \tag{4.83a}$$

and

$$a_m = \sum_{\ell=0}^{m/2} (-1)^\ell \frac{1}{[(\frac{1}{2}m-\ell)!][(2\ell)!]} \quad (m \text{ even}) \tag{4.83b}$$

Therefore,

$$a_0 = 1 \tag{4.84a}$$

$$a_2 = \tfrac{1}{2} \tag{4.84b}$$

$$a_4 = \tfrac{1}{24} \tag{4.84c}$$

and so on. Therefore, the MacLaurin series for $F(z)$ is

$$F(z) = 1 + \tfrac{1}{2}z^2 + \tfrac{1}{24}z^4 + \cdots \tag{4.85}$$

$\square$

Division

If $T(z) \neq 0$, and if

$$F(z) = \frac{S(z)}{T(z)} \tag{4.86a}$$

then

$$F(z) * T(z) = S(z) \tag{4.86b}$$

Following the analysis that leads to eq. 4.74, the coefficients of the series form of eq. 4.82b are related by

$$\sigma_m = \sum_{k=-\infty}^{\infty} a_k \tau_{m-k} \tag{4.87}$$

As examples, we set $m = -1, 0$, and 1 to obtain

$$\sigma_{-1} = \sum_{k=-\infty}^{\infty} a_k \tau_{-1-k} = \cdots + a_{-1}\tau_0 + a_0\tau_{-1} + a_1\tau_{-2} + \cdots \qquad (4.88a)$$

$$\sigma_0 = \sum_{k=-\infty}^{\infty} a_k \tau_{-k} = \cdots + a_{-1}\tau_1 + a_0\tau_0 + a_1\tau_{-1} + \cdots \qquad (4.88b)$$

and

$$\sigma_1 = \sum_{k=-\infty}^{\infty} a_k \tau_{1-k} = \cdots + a_{-1}\tau_2 + a_0\tau_1 + a_1\tau_0 + \cdots \qquad (4.88c)$$

If $T(z)$ has an essential singularity at z_0, then in general, all τ_k are nonzero and the above procedure results in a set of equations in an infinite number of unknowns a_k. It is not possible to solve such a set of equations.

If $F(z)$ is known to have a pole of order N, then $a_k = 0$, $k < -N$. Then eqs. 4.87 and 4.88 become

$$\sigma_m = \sum_{k=-N}^{\infty} a_k \tau_{m-k} \qquad (4.89)$$

from which

$$\sigma_{-1} = \sum_{k=-N}^{\infty} a_k \tau_{-1-k} = a_{-N}\tau_{N-1} + \cdots + a_{-1}\tau_0 + a_0\tau_{-1} + a_1\tau_{-2} + \cdots \quad (4.90a)$$

$$\sigma_0 = \sum_{k=-N}^{\infty} a_k \tau_{-k} = a_{-N}\tau_N + \cdots + a_{-1}\tau_1 + a_0\tau_0 + a_1\tau_{-1} + \cdots \qquad (4.90b)$$

and

$$\sigma_1 = \sum_{k=-N}^{\infty} a_k \tau_{1-k} = a_{-N}\tau_{N+1} \ldots + a_{-1}\tau_2 + a_0\tau_1 + a_1\tau_0 + \ldots \qquad (4.90c)$$

Again, because the τ_k are, in general, nonzero, we still have a set of equations, each containing an infinite number of unknowns a_k, which cannot be solved. Therefore, we conclude that in order to obtain the series representation of $F(z)$ from eq. 4.86b, $T(z)$ must have a pole or be analytic at z_0.

If $T(z)$ has an Mth-order pole at z_0, then

$$\tau_{m-k} = 0. \quad m-k < -M \tag{4.91}$$

and eq. 4.87 becomes

$$\sigma_m = \sum_{k=-\infty}^{M+m} a_k \tau_{m-k} \tag{4.92}$$

This expression still yields a set of equations, each containing an infinite number of unknowns a_k for which there is no solution.

But if, in addition to $T(z)$ having an Mth-order pole at z_0, it is known that $F(z)$ has a pole of order N, then eq. 4.87 becomes

$$\sigma_m = \sum_{k=-N}^{M+m} a_k \tau_{m-k} \tag{4.93}$$

This results in a set of equations containing finite numbers of unknown coefficients a_k which can, in principle, be solved.

From eq. 4.86b, we see that if $F(z)$ has a pole of order N and $T(z)$ has an Mth-order pole at z_0 then $S(z)$ must have a pole at z_0 of order $N + M$. Therefore, in order to find the series representation for $F(z)$ from the series representations of $S(z)$ and $T(z)$, each of these functions must have a pole of some order or be analytic at z_0.

Let $S(z)$ have a pole of order $N + M$ at z_0 and $T(z)$ have a pole of order M at z_0. Then their ratio, $F(z)$, has a pole of order N at z_0. With

$$S(z) = (z-z_0)^{-(N+M)} \left[\sigma_{-(N+M)} + \sigma_{-(N+M)+1}(z-z_0) + \cdots \right] \tag{4.94a}$$

$$T(z) = (z-z_0)^{-M} \left[\tau_{-M} + \tau_{-M+1}(z-z_0) + \cdots \right] \tag{4.94b}$$

and

$$F(z) = (z-z_0)^{-N} \left[a_{-N} + a_{-N+1}(z-z_0) + \cdots \right] \tag{4.94c}$$

Then

$$F(z)*T(z) = S(z) \tag{4.86b}$$

becomes

$$
\begin{aligned}
&\sigma_{-(N+M)} + \sigma_{-(N+M)+1}(z-z_0)+\cdots \\
&= \left[a_{-N} + a_{-N+1}(z-z_0)+\cdots \right]\left[\tau_{-M} + \tau_{-M+1}(z-z_0)+\cdots \right]
\end{aligned}
\tag{4.95}
$$

from which we obtain

$$\sigma_{-(N+M)+m} = \sum_{k=0}^{m} a_{-N+k}\, \tau_{-M+m-k} \tag{4.96a}$$

If all three functions are analytic at z_0, eq. 4.96a becomes

$$\sigma_m = \sum_{k=0}^{m} a_k \tau_{m-k} \tag{4.96b}$$

This yields

$$\sigma_0 = a_0 \tau_0 \tag{4.97a}$$

$$\sigma_1 = a_0 \tau_1 + a_1 \tau_0 \tag{4.97b}$$

$$\sigma_2 = a_0 \tau_2 + a_1 \tau_1 + a_2 \tau_0 \tag{4.97c}$$

and so on. Thus, to find the coefficients of the series for $F(z)$, a_0 is found from eq. 4.97a, then used to find a_1 from eq. 4.97b. From this, eq. 4.97c yields a_2 and so on. With this iterative process, it is usually possible either to determine a sufficient number of coefficients to deduce a general expression for a_k, or to determine a finite sum that is a satisfactory approximation to $F(z)$.

Example 4.7: Division of two power series

From the MacLaurin series for $S(z)$ and $T(z)$ given in eqs. 4.74,

$$e^{z^2} = F(z)\cos z \tag{4.98a}$$

results in

$$\sum_{k=0}^{\infty} \frac{z^{2k}}{k!} = \sum_{\substack{\ell=0 \\ m=0}}^{\infty} (-1)^{\ell} \frac{a_m}{(2\ell)!} z^{2\ell+m} \tag{4.98b}$$

Because $2k$ is an even power of z on the left side of this expression, $2\ell + m$ must be even. Then m must be even. Therefore, a_m must be zero for all odd m. To make this more transparent, we replace m by $2m'$ and ignore the prime to obtain

$$\sum_{k=0}^{\infty} \frac{z^{2k}}{k!} = \sum_{\substack{\ell=0 \\ m=0}}^{\infty} (-1)^{\ell} \frac{a_{2m}}{(2\ell)!} z^{2(\ell+m)} \tag{4.99}$$

from which

$$\sum_{m=0}^{k} a_{2m} \frac{(-1)^{k-m}}{[2(k-m)]!} = \frac{1}{k!} \tag{4.100}$$

This yields

$$a_0 = 1 \tag{4.101a}$$

$$-\tfrac{1}{2}a_0 + a_2 = 1 \tag{4.101b}$$

from which

$$a_2 = \tfrac{3}{2} \tag{4.101c}$$

and so on. Therefore,

$$\frac{e^{z^2}}{\cos z} = 1 - z + \tfrac{3}{2} z^2 - \cdots \tag{4.102}$$

□

4.6 Residues

Contour integral of a function with a pole of order M

Let the only singularity of $F(z)$ be a pole of order M at z_0. Then $F(z)$ can be written as

$$F(z) = \frac{a_{-M}(z_0)}{(z-z_0)^M} + \frac{a_{-(M-1)}(z_0)}{(z-z_0)^{(M-1)}} + \cdots + \frac{a_{-1}(z_0)}{(z-z_0)} + a_0 + a_1(z-z_0) + \cdots$$

$$= \sum_{\ell=1}^{M} \frac{a_{-\ell}(z_0)}{(z-z_0)^\ell} + \sum_{\ell=0}^{\infty} a_\ell(z-z_0)^\ell$$

(4.103)

Because all terms in the second sum have positive exponents,

$$\sum_{\ell=0}^{\infty} a_\ell(z_0)(z-z_0)^\ell \equiv \Phi_A(z)$$

(4.104)

is analytic everywhere in a region R containing z_0.

Consider

$$\oint_C F(z)\,dz = \sum_{\ell=1}^{M} a_{-\ell}(z_0) \oint_C \frac{1}{(z-z_0)^\ell}\,dz + \oint_C \Phi_A(z)\,dz$$

(4.105)

Inasmuch as Φ_A is analytic everywhere in R,

$$\oint_C \Phi_A(z)\,dz = 0$$

(4.106)

Therefore,

$$\oint_C F(z)\,dz = \sum_{\ell=1}^{M} a_{-\ell}(z_0) \oint_C \frac{1}{(z-z_0)^\ell}\,dz$$

(4.107)

Because z_0 is the only singular point inside the contour, we can make the contour a circle of radius ρ centered at z_0. Then, referring to fig. 4.7, any point on the contour can be expressed as

$$z = z_0 + \rho e^{i\phi}$$

(3.83a)

and because ρ is constant,

$$dz = i\rho e^{i\phi}\,d\phi$$

(3.83b)

Therefore,

$$\oint_C \frac{1}{(z-z_0)^\ell}\,dz = \int_{\phi_0}^{\phi_0+2\pi} \frac{i\rho e^{i\phi}}{\rho^\ell e^{i\ell\phi}}\,d\phi = i\rho^{-(\ell-1)} \int_{\phi_0}^{\phi_0+2\pi} e^{-i(\ell-1)\phi}\,d\phi \quad (4.108)$$

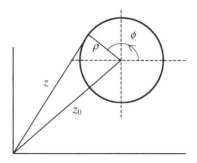

Figure 4.7
Circle around an Mth order pole

If $\ell \neq 1$, then $\ell - 1$ is a nonzero integer so that $e^{-i(\ell-1)2\pi} = 1$. Therefore,

$$\int_{\phi_0}^{\phi_0+2\pi} e^{-i(\ell-1)\phi}d\phi = \left[\frac{e^{-i(\ell-1)\phi_0}e^{-i(\ell-1)2\pi} - e^{-i(\ell-1)\phi_0}}{-i(\ell-1)}\right] = 0 \qquad (4.109a)$$

For $\ell = 1$,

$$i\rho^{-(\ell-1)}\int_{\phi_0}^{\phi_0+2\pi} e^{-i(\ell-1)\phi}d\phi = i\int_{\phi_0}^{\phi_0+2\pi} d\phi = 2\pi i \qquad (4.109b)$$

Therefore, all terms in the sum of eq. 4.107 are zero except the $\ell = 1$ term, and

$$\oint_C F(z)dz = 2\pi i a_{-1}(z_0) \qquad (4.110)$$

If the pole is a simple pole, that is, $M = 1$, the Laurent series for $F(z)$ is

$$F(z) = \frac{a_{-1}(z_0)}{(z - z_0)} + \Phi_A(z) \qquad (4.111)$$

Referring to eq. 3.62b, the residue of a simple pole for this function is given by

$$R(z_0) = \lim_{z \to z_0}(z - z_0)F(z) = a_{-1}(z_0) \qquad (4.112)$$

Because $a_{-1}(z_0)$ satisfies the original definition of the residue of a simple pole, it is called the residue of a pole of any order.

Three methods for determining the residue

If $F(z)$ has an Mth order pole at z_0, its Laurent series is given by

$$F(z) = \frac{a_{-M}(z_0)}{(z-z_0)^M} + \frac{a_{-(M-1)}(z_0)}{(z-z_0)^{(M-1)}} + \cdots + \frac{a_{-1}(z_0)}{(z-z_0)} + \Phi_A(z) \qquad (4.34)$$

$a_{-1}(z_0)$, the residue of this pole, can be found by any one of three methods.

I.   Direct Laurent expansion
It is evident from eq. 4.34, for example, that by expanding the function in its Laurent series, the residue can be identified directly as the coefficient of the $1/(z-z_0)$ term, with no additional computation required.

II.   Derivative method
Multiplying eq. 4.34 by $(z - z_0)^M$, we obtain

$$(z-z_0)^M F(z) = a_{-M}(z_0) + a_{-(M-1)}(z_0)(z-z_0) + \cdots + a_{-1}(z_0)(z-z_0)^{M-1}$$
$$+ (z-z_0)^M * \Phi_A(z)$$
$$\hspace{10cm} (4.113)$$

We see that because $a_{-1}(z_0)$ is multiplied by $(z-z_0)^{M-1}$

$$\frac{d^{(M-1)}}{dz^{(M-1)}}\left[(z-z_0)^M F(z)\right] =$$
$$(M-1)! a_{-1}(z_0) + \frac{M!}{1!} a_0(z_0)(z-z_0) + \frac{(M+1)!}{2!} a_1(z_0)(z-z_0)^2 + \cdots$$
$$\hspace{10cm} (4.114)$$

which yields

$$a_{-1}(z_0) = \frac{1}{(M-1)!} \lim_{z \to z_0} \frac{d^{(M-1)}}{dz^{(M-1)}}\left[(z-z_0)^M F(z)\right] \qquad (4.115)$$

Let $F(z)$ be a function with a Mth-order pole at z_0, defined in terms of a function $G(z)$ that is analytic at z_0, by

$$F(z) = \frac{G(z)}{(z - z_0)^M} \tag{4.116}$$

It was shown in eqs. 3.102 that

$$\oint_C F(z)\,dz = 2\pi i \oint_C \frac{G(z)}{(z - z_0)^M}\,dz = \frac{2\pi i}{(M-1)!} \frac{d^{M-1}G(z)}{dz^{M-1}}\bigg|_{z=z_0}$$

$$= \frac{2\pi i}{(M-1)!} \frac{d^{M-1}}{dz^{M-1}}\Big[(z - z_0)^M F(z)\Big]_{z_0} = 2\pi i a_{-1}(z_0) \tag{4.117}$$

That is, the closed contour integral of a function with a pole of any order is $2\pi i * $Residue at z_0.

If $F(z)$ has several poles of different orders, then, with $G(z)$ analytic everywhere inside and on the contour C,

$$\oint_C \frac{G(z)}{(z - z_0)^M (z - z_1)^N \cdots (z - z_k)^P}\,dz = 2\pi i \sum_{\ell=0}^{k} a_{-1}(z_\ell) \tag{4.118}$$

where the residue of any pole involves the derivative of that part of the integrand that is analytic at that pole. For example, for the integrand of eq. 4.118,

$$a_{-1}(z_0) = \frac{1}{(M-1)!} \frac{d^{M-1}}{dz^{M-1}}\left[\frac{G(z)}{(z - z_1)^N \cdots (z - z_k)^P}\right]_{z=z_0} \tag{4.119}$$

III.   Ratio method

In some cases, a function that has a pole of order M at z_0 can be written as the ratio of two analytic functions,

$$F(z) = \frac{P(z)}{Q(z)} \tag{4.54}$$

We introduce the notation $O[(z - z_0)^q]$ to represent a sum of terms in powers of $(z - z_0)$, the lowest power of which is q. That is,

$$O\left[(z-z_0)^q\right] = \sum_{m=0}^{\infty} c_m (z-z_0)^{q+m} \tag{4.120}$$

If a function $R(z)$ that is analytic at z_0 satisfies

$$R(z_0) = R'(z_0) = \cdots = R^{(M-1)}(z_0) = 0 \tag{4.121a}$$

and

$$R^{(M)}(z_0) \neq 0 \tag{4.121b}$$

the Taylor series for $R(z)$ is given by

$$R(z) = \frac{R^{(M)}(z_0)}{M!}(z-z_0)^M + O\left[(z-z_0)^{M+1}\right] \tag{4.122}$$

As noted earlier, such a function has an Mth-order zero at z_0.

A simple pole of $F(z)$ at z_0 arises from the fact that $P(z)$ has an Mth-order zero and $Q(z)$ has a zero of order $(M+1)$. Then

$$F(z) = \frac{\dfrac{P^{(M)}(z_0)}{(M)!}(z-z_0)^M + O\left[(z-z_0)^{M+1}\right]}{\dfrac{Q^{(M+1)}(z_0)}{(M+1)!}(z-z_0)^{M+1} + O\left[(z-z_0)^{M+2}\right]} \tag{4.123}$$

For such a function then,

$$\begin{aligned}
a_{-1}(z_0) &= \lim_{z \to z_0} (z-z_0) \left[\frac{\dfrac{P^{(M)}(z_0)}{M!}(z-z_0)^M + O\left[(z-z_0)^{M+1}\right]}{\dfrac{Q^{(M+1)}(z_0)}{(M+1)!}(z-z_0)^{M+1} + O\left[(z-z_0)^{M+2}\right]} \right] \\
&= (M+1)\frac{P^{(M)}(z_0)}{Q^{(M+1)}(z_0)}
\end{aligned} \tag{4.124}$$

For example, if $P(z_0) \neq 0$, and $Q(z_0) = 0$ with $Q'(z_0) \neq 0$, $F(z)$ will have a simple pole at z_0 (and thus, $M = 0$). Then

$$a_{-1}(z_0) = \lim_{z \to z_0} (z - z_0) \left[\frac{P(z_0) + O\left[(z - z_0)\right]}{Q'(z_0)(z - z_0) + O\left[(z - z_0)^2\right]} \right] = \frac{P(z_0)}{Q'(z_0)}$$

$$(4.125)$$

The ratio method can be applied to functions that can be written as the ratio of two analytic functions and has a pole of order greater than 1.

However, the process quickly becomes cumbersome. To see this, we let

$$F(z) = \frac{P(z)}{Q(z)} \tag{4.54}$$

have a second-order pole at z_0. Such a pole arises when $P(z)$ has zero of order M and $Q(z)$ has a zero of order $(M + 2)$ at z_0. Then

$$F(z) =$$

$$\frac{\dfrac{P^{(M)}(z_0)}{M!}(z - z_0)^M + \dfrac{P^{(M+1)}(z_0)}{(M+1)!}(z - z_0)^{M+1} + O\left[(z - z_0)^{M+2}\right]}{\dfrac{Q^{(M+2)}(z_0)}{(M+2)!}(z - z_0)^{M+2} + \dfrac{Q^{(M+3)}(z_0)}{(M+3)!}(z - z_0)^{M+3} + O\left[(z - z_0)^{M+4}\right]}$$

$$(4.126)$$

from which

$$a_{-1}(z_0) = \lim_{z \to z_0} \frac{d}{dz}\left[(z - z_0)^2 F(z)\right]$$

$$= \frac{M+2}{(M+3)}\left[\frac{(M+3)P^{(M+1)}(z_0)Q^{(M+2)}(z_0) - (M+1)P^{(M)}(z_0)Q^{(M+3)}(z_0)}{\left(Q^{(M+2)}(z_0)\right)^2} \right]$$

$$(4.127)$$

For example, if the second-order pole of $F(z)$ arises from $P(z_0) \neq 0$ (so $M = 0$) and a second-order zero of $Q(z)$, the residue is given by eq. 4.127 to be

$$a_{-1}(z_0) = \frac{2}{3}\left[\frac{3P'(z_0)Q''(z_0) - P(z_0)Q^{(3)}(z_0)}{(Q''(z_0))^2}\right] \qquad (4.128)$$

The ratio method is also applicable to a function that can be written as the ratio of two functions with poles. Let

$$F(z) = \frac{R(z)}{S(z)} \qquad (4.129)$$

where $R(z)$ has a pole of order $(M + N)$ at z_0 and $S(z)$ has a pole of order M at z_0. This results in $F(z)$ having a pole of order N at z_0.

Defining

$$P(z) \equiv (z - z_0)^{M+N} R(z) \qquad (4.130a)$$

and

$$Q(z) \equiv (z - z_0)^{M+N} S(z) \qquad (4.130b)$$

then $P(z)$ and $Q(z)$ are analytic at z_0, and $Q(z)$ has a zero of order N at z_0.

Let the Laurent series for $R(z)$ and $S(z)$ be

$$R(z) = \sum_{k=-(M+N)}^{\infty} r_k (z - z_0)^k \qquad (4.131a)$$

and

$$S(z) = \sum_{k=-M}^{\infty} s_k (z - z_0)^k \qquad (4.131b)$$

Then, the Taylor series for $P(z)$ and $Q(z)$ are

$$P(z) = r_{-(M+N)}(z_0) + r_{-(M+N-1)}(z_0)(z - z_0) + \cdots \qquad (4.132a)$$

and

$$Q(z) = s_{-M}(z_0)(z - z_0)^N + s_{-(M-1)}(z_0)(z - z_0)^{N+1} + \cdots \qquad (4.132b)$$

If the pole of $F(z)$ is a simple pole, then $N = 1$ and, from eq. 4.125, the residue is

$$a_{-1}(z_0) = \frac{r_{-(M+1)}(z_0)}{s_{-M}(z_0)} \tag{4.133a}$$

As the reader will show in prob. 25, if the pole of $F(z)$ is a second-order pole, then the residue is

$$a_{-1}(z_0) = \frac{r_{-(M+1)}(z_0)s_{-M}(z_0) - r_{-(M+2)}(z_0)s_{-(M-1)}(z_0)}{\left[s_{-M}(z_0)\right]^2} \tag{4.133b}$$

Example 4.8: Three methods for determining residues

It is evident that

$$F(z) = \frac{e^z}{z(z+2)^2} \tag{4.134}$$

has a simple pole at $z = 0$ and a second-order pole at $z = -2$.

(I) *Residues by direct Laurent expansion*
   (a) To find the residue at $z = 0$ by direct Laurent expansion, we must determine the MacLaurin series for $e^z/(z + 2)^2$. It was determined above that

$$e^z = \sum_{n=0}^{\infty} \frac{z^n}{n!} = 1 + z + \frac{z^2}{2!} + \cdots \tag{4.15}$$

The MacLaurin series for $1/(z + 2)^2$ is obtained from

$$\frac{1}{(z+2)^2} = \frac{1}{4}\left[\left(1 + \frac{z}{2}\right)^{-1}\right]^2 = \frac{1}{4}\left(1 - z + \tfrac{3}{4}z^2 - \cdots\right)$$

$$= \frac{1}{4}\sum_{m=0}^{\infty}(-1)^m(m+1)\left(\frac{z}{2}\right)^m \tag{4.135}$$

Therefore, from

$$\frac{e^z}{z(z+2)^2} = \frac{1}{4z}\left(1+z+\frac{z^2}{2!}+\cdots\right)\left(1-z+\frac{3z^2}{4}-\cdots\right) = \frac{1}{4z}+\frac{z}{16}+\cdots$$

(4.136)

we see that

$$a_{-1}(0) = \frac{1}{4}$$

(4.137)

(b) To derive the Laurent series expansion around $z = -2$, we refer to eq. 4.14a to express

$$e^z = e^{-2}\sum_{n=0}^{\infty}\frac{(z+2)^n}{n!} = e^{-2}\left(1+(z+2)+\frac{(z+2)^2}{2!}+\frac{(z+2)^3}{3!}+\cdots\right)$$

(4.138)

The Taylor series for $1/z$ around $z = -2$ is given by

$$\frac{1}{z} = -\frac{1}{2}\left(\frac{1}{1-\frac{1}{2}(z+2)}\right) = -\sum_{m=0}^{\infty}\frac{(z+2)^m}{2^{m+1}}$$

$$= -\frac{1}{2} - \frac{(z+2)}{4} - \frac{(z+2)^2}{8} - \frac{(z+2)^3}{16} - \cdots$$

(4.139)

Thus, the Laurent series for $F(z)$ around $z = -2$ is

$$\frac{e^z}{z(z+2)^2} = -\frac{1}{2}e^{-2}\left[\frac{1}{(z+2)^2} + \frac{3}{2}\frac{1}{(z+2)} + \frac{5}{4} + \cdots\right]$$

(4.140)

from which

$$a_{-1}(-2) = -\frac{3}{4}e^{-2}$$

(4.141)

(II) *Residues by the derivative method*

(a) The residue of the simple pole at $z = 0$ by this method is given by

$$a_{-1}(0) = \lim_{z \to 0}\left(z \frac{e^z}{z(z+2)^2} \right) = \frac{1}{4} \tag{4.142}$$

as obtained above.
    (b) The residue at $z = -2$ is

$$a_{-1}(-2) = \frac{1}{1!} \lim_{z \to -2} \frac{d}{dz}\left((z+2)^2 \frac{e^z}{z(z+2)^2} \right) = -\frac{3}{4}e^{-2} \tag{4.143}$$

(III) *Residues by the ratio method*
    For this example, the numerator and denominator functions can be chosen in different ways.
    (a) For the pole at $z = 0$, we could take

$$P(z) = e^z \tag{4.144a}$$

and

$$Q(z) = z(z+2)^2 \tag{4.144b}$$

or we could choose

$$P(z) = \frac{e^z}{(z+2)^2} \tag{4.145a}$$

and

$$Q(z) = z \tag{4.145b}$$

For either choice, the residue for the simple pole is given by

$$a_{-1}(0) = \frac{P(0)}{Q'(0)} = \frac{1}{4} \tag{4.146}$$

    (b) For the pole at $z = -2$, we can choose either

$$P(z) = e^z \tag{4.147a}$$

and

$$Q(z) = z(z+2)^2 \tag{4.147b}$$

as before, or

$$P(z) = \frac{e^z}{z} \tag{4.147c}$$

and

$$Q(z) = (z+2)^2 \tag{4.147d}$$

Then, independent of how numerator and denominator are chosen, we refer to eq. 4.128 to obtain

$$a_{-1}(z_0) = \frac{2}{3}\left[\frac{3P'(-2)Q''(-2) - P(-2)Q^{(3)}(-2)}{(Q''(-2))^2} \right] = -\frac{3}{4}e^{-2} \tag{4.148}$$
$\square$

Problems

1. For each function below, determine the MacLaurin series by the specified method. Determine the radius of convergence of each series.
   (a) sinh z using eq. 4.8.
   (b) cosh z using eq. 4.8.
   (c) $\ell n(1+z)$ using eq. 4.8.
   (d) (i) cos z from the MacLaurin series for cosh z.
      (ii) sin z from the MacLaurin series for sinh z.
      (iii) sin z by differentiating the MacLaurin series for cos z.
      (iv) cosh z by differentiating the MacLaurin series for sinh z.
      (v) $\ell n(1+z)$ by integrating the geometric series given in eq. A3.4.

2. Find the Taylor series representation for

$$F(z) = \frac{1}{(a-z)}$$

   around $z = b$ for $a \neq b$. What is the radius of convergence of this series?

3. For $a \neq b$, use the MacLaurin series representation for $1/(1-z)$ given in appendix 3 to determine the MacLaurin series representation of

$$F(z) = \frac{1}{(b-az)}$$

What is the radius of convergence of this series?

4. Find the Taylor series expansion of e^z around
    (a) $z = i\pi/2$ (b) $z = 2\pi i$

5. Find the Taylor series expansion of $\ell n (1 + z)$ around
    (a) $z = i\pi/2$ (b) $z = i\pi$ (c) $z = 2\pi i$
6. Using the geometric series, find the Taylor series representation of

$$F(z) = \frac{1}{(az - bz^2)}$$

    expanded around
            (a) $z = a$ (b) $z = b$ (c) $z = 1$
    for a and b real and $b > a > 1$. Determine the radius of convergence of
    each series.

7. Find the first two nonzero terms in the Taylor series for

$$F(z) = e^{z \sin z}$$

    expanded around
            (a) $z = 0$ (b) $z = \pi/2$ (c) $z = \pi$

8. Determine the first three terms in the MacLaurin series (those containing
    the lowest powers of z that have nonzero coefficients) for

    (a) $F(z) = \ell n \left(2 - e^{-z} \right)$ (b) $F(z) = \ell n (1 + \sin z)$ (c) $F(z) = \ell n (\sec z)$

    Determine the radius of convergence of each series.

9. Determine the MacLaurin series for each function below. If a general
    expression for the series cannot be determined, find the first four terms
    in the series (those containing the lowest powers of z that have nonzero
    coefficients).

    (a) $F(z) = z^k e^z$ (integer $k > 0$)   (b) $F(z) = e^z \sin z$

(c) $F(z) = \dfrac{\sinh z}{\cos z}$ (d) $F(z) = \ell n(1+z)\cosh z$

(e) $F(z) = \dfrac{\ell n(1+z)}{\cosh z}$

Determine the radius of convergence of each series.
(Hint: Refer to the results of problem 1.)

10. What is the Taylor series representation of

$$F(z) = z + \ell n(z)$$

expanded around $z = 1$? What is its radius of convergence?
11. Find the three lowest power terms in the MacLaurin series for

$$F(z) = \dfrac{ze^z}{\left(e^z - 1\right)}$$

12. Use the Cauchy ratio to determine the radius of convergence of

(a) $S_a(z) = \displaystyle\sum_{n=0}^{\infty} \dfrac{1}{n!(2n)!}\dfrac{z^{2n}}{9^n}$ (b) $S_b(z) = \displaystyle\sum_{n=0}^{\infty} \dfrac{(n+2)}{\sqrt{(n+1)}}z^n$

(c) $S_c(z) = \displaystyle\sum_{n=0}^{\infty} \dfrac{4^n}{\sqrt{(n+1)}(n+3)^2}z^n$ (d) $S_d(z) = \displaystyle\sum_{n=0}^{\infty} e^{n(z-z_0)}$

13. Estimate the value of each of the following integrals by approximating each integrand as indicated.

(a) $\displaystyle\int_0^1 e^{-z^4}\,dz$

Approximate e^{-w} by the first two terms in its MacLaurin series. Then set $w \to z^4$.

(b) $\displaystyle\int_0^{1/2} \sin\sqrt{z}\,dz$

Approximate $\sin w$ by the first two terms in its MacLaurin series. Then set $w \to \sqrt{z}$.

(c) $\int_0^{1/2} \left[\ell n \left(1 + \sqrt{z} \right) \right]^2 dz$

Approximate $[\ell n (1 + w)]^2$ by the first two terms in its MacLaurin series. Then set $w \to \sqrt{z}$.

(d) $\int_0^1 \sqrt{z} e^{-z^2} dz$

Approximate e^{-w} by the first two terms in its MacLaurin series. Then set $w \to z^2$.

(e) $\int_0^{1/2} e^z \ell n (1 + z) dz$

Approximate $\ell n (1 + z)$ by the first two terms in its MacLaurin series.

(f) $\int_1^2 z e^{1/z} dz$

Approximate e^w by the first two terms in its MacLaurin series. Then set $w \to 1/z$.

14. (a) Determine if

$$F(z) = \frac{\ell n (1 + z)}{z}$$

has a pole at $z = 0$ by determining the Laurent series for $f(z)$ around $z = 0$.
(Hint: See the results of problem 1, part (c).)

(b) Use the results of part (a) of this problem to determine the order of the pole at $z = 0$ of the function

$$F(z) = \frac{z - \ell n (1 + z)}{z^4}$$

15. Determine the indicated Laurent series for each function below.

(a) $F(z) = \dfrac{\cos z}{z^2}$ expanded about $z = 0$.

(b) $F(z) = \dfrac{\sin z}{\left(z - \frac{\pi}{2}\right)^2}$ expanded about $z = \pi/2$.

(c) $F(z) = \dfrac{\sin z}{(z - \pi)^2}$ expanded about $z = \pi$.

(d) $F(z) = \dfrac{\ell n(1 + z)}{z^4}$ expanded about $z = 0$.

16. Determine the three lowest power terms with nonzero coefficients in the Laurent series for

(a) $F(z) = \dfrac{e^{z \sin z}}{\left(z - \frac{\pi}{2}\right)}$ around $z = \pi/2$

(b) $F(z) = \dfrac{e^{z \sin z}}{z(z - \pi)}$ around $z = \pi$

17. Expanding in powers of $1/z$ for $z > 1$, determine the Laurent series representation of

$$F(z) = \dfrac{1}{(1 - z)}$$

18. Determine the Laurent series representation of

$$F(z) = \dfrac{1}{(z^2 - 1)}$$

expanded in powers of

(a) $(z + 1)$ (b) $(z - 1)$ (c) z (d) $(z + 1/2)$ (e) $1/z^2$

19. Determine the Laurent series representation, expanded about $z = 0$, of each function below. In each case, identify the coefficient of $1/z$.

(a) $F(z) = z^3 \cosh\left(\dfrac{1}{z^2}\right)$ (b) $F(z) = z^2 \ell n\left(1 + \dfrac{1}{z}\right)$ (c) $F(z) = z^2 e^{1/z}$

20. Let

$$F(z) = \dfrac{G(z)}{(z - z_0)^M}$$

where $G(z)$ is analytic at z_0. For $M = 3$, what is the residue of the pole of $F(z)$ at z_0 in terms of $G^{(n)}(z)$, the nth derivative of $G(z)$?

21. A function $F(z)$ can be written as the ratio of two functions $P(z)$ and $Q(z)$ each of which is analytic in a region R containing z_0. $F(z)$ has a pole of order M at z_0 and $P(z)$ has a zero of order K at z_0.
   (a) Write the expression for the Taylor series representation of $Q(z)$ in powers of $(z - z_0)$.
   (b) What are the coefficients of $(z - z_0)^{-M}$ and $(z - z_0)^{-M+1}$ in the Laurent series of $F(z)$?

22. Determine each of the following limits using l'Hopital's rule:

(a) $\lim\limits_{z \to 0} \dfrac{\tan^2 z}{z^2}$ (b) $\lim\limits_{z \to 0} \dfrac{\tan(2z)}{\tan(z)}$ (c) $\lim\limits_{z \to z_0} \dfrac{\sin^2 z - \sin^2 z_0}{(z^2 - z_0^2)}$

(d) $\lim\limits_{z \to 0} \dfrac{z}{\ell n(1+z)}$ (e) $\lim\limits_{z \to 0} \dfrac{z^2}{[\ell n(1+z)]^2}$ (f) $\lim\limits_{z \to 1}(z^2 - 1)\tan\left(\dfrac{\pi}{2}z\right)$

23. For each pair of functions $S(z)$ and $T(z)$ given below,
   (a) Find the expression for the general coefficient of the MacLaurin series for

$$F(z) = S(z) + T(z)$$

   (b) Find the coefficients of the first three nonzero terms in the MacLaurin series for

$$F(z) = S(z) * T(z)$$

(c) Find the coefficients of the first three nonzero terms in the MacLaurin series for

$$F(z) = \frac{S(z)}{T(z)}$$

when

(i) $S(z) = e^z$ and $T(z) = \dfrac{1}{(1+z)}$

(ii) $S(z) = \ell n(1+z)$ and $T(z) = e^z$

(iii) $S(z) = \ell n(1+z)$ and $T(z) = \dfrac{1}{(1-z^2)}$

24. Expand each of the following functions in a series around $z = 0$ to evaluate the following limits.

(a) $\displaystyle\lim_{z \to 0} \left[\frac{1}{z} - \frac{1}{\tan z} \right]$ (b) $\displaystyle\lim_{z \to 0} \left[\frac{1}{z} - \frac{1}{\ell n(1+z)} \right]$ (c) $\displaystyle\lim_{z \to 0} \left[\frac{z^2 - \sin^2 z}{z^2 \sin^2 z} \right]$

25. The function $F(z)$ can be written as

$$F(z) = \frac{R(z)}{S(z)}$$

$R(z)$ has a pole of order $(M + 2)$ at z_0 and $S(z)$ has a pole of order M at z_0. Show that the residue of the second-order pole of $F(z)$ is given by

$$a_{-1}(z_0) = \frac{r_{-(M+1)}(z_0)s_{-M}(z_0) - r_{-(M+2)}(z_0)s_{-(M-1)}(z_0)}{\left[s_{-M}(z_0) \right]^2}$$

where r_k and s_k are the coefficients of the Laurent series' expansions about z_0 of $R(z)$ and $S(z)$, respectively.

26. Find the residue of

$$F(z) = \frac{R(z)}{S(z)}$$

when

(a) $R(z) = \dfrac{4}{z^4} - \dfrac{3}{z^3} + \dfrac{2}{z^2} - \dfrac{1}{z} - z + 2z^2 - 3z^3 + \cdots$

and

$$S(z) = \frac{3}{z^3} + \frac{2}{z^2} + \frac{1}{z} + z + 2z^2 + 3z^3 + \cdots$$

(b) $R(z) = -\dfrac{5}{z^5} + \dfrac{4}{z^4} - \dfrac{3}{z^3} + \dfrac{2}{z^2} - \dfrac{1}{z} - z + 2z^2 - 3z^3 + \cdots$

and

$$S(z) = \frac{3}{z^3} + \frac{2}{z^2} + \frac{1}{z} + z + 2z^2 + 3z^3 + \cdots$$

27. For each function below, find the residue of the indicated pole by the method(s) noted.
  (I) Expansion of the function in a Laurent series and identifying the appropriate coefficient
  (II) The derivative method
  (III) The ratio method

(a) $F(z) = \dfrac{\cos^2 z}{z^3}$ at $z = 0$ by methods (I) and (II)

(b) $F(z) = \dfrac{e^{z^2}}{z^5}$ at $z = 0$ by methods (I) and (II)

(c) $F(z) = \dfrac{\sin^2 z}{(z - \pi/2)^2}$ at $z = \pi/2$ by all three methods

(d) $F(z) = \dfrac{1}{\tan z}$ at $z = \pi$ by method (III)

(e) $F(z) = \dfrac{e^z}{z \sin z}$ at $z = 0$ by method (II)

(f) $F(z) = \dfrac{\ell n(1 + z)}{z^3}$ at $z = 0$ by all three methods

(g) $F(z) = \dfrac{e^{z \sin z}}{z(z - \pi)^2}$ at $z = \pi$ by methods (II) and (III)

28. (a) Let $F(z)$ have a pole of order M at z_0. Using the process developed for finding the residue by the derivative method, derive an expression for any coefficient $a_k(z_0)$ with $k \geq -M$ that is equivalent to

$$a_{-1}(z_0) = \frac{1}{(M-1)!} \frac{d^{M-1}}{dz^{M-1}} \left[(z - z_0)^M F(z) \right]_{z=z_0}$$

(Hint: Determine $a_k(z_0)$ for $-M \leq k \leq -1$ separately from $a_k(z_0)$ for $k \geq 0$.)

(b) Apply the result of part (a) to derive an expression for $a_k(\pi)$, the kth coefficient in the Laurent series for

$$F(z) = \frac{\cos^2 z}{(z - \pi)^3}$$

(Hint: Consider $a_k(z_0)$ for $-M \leq k \leq -1$ separately from $a_k(z_0)$ for $k \geq 0$.)

(c) Apply the result of part (a) to derive an expression for the coefficients of the four lowest power terms in the Laurent series expansion around $z = \pi/2$ for

$$F(z) = \frac{\sin^2 z}{\left(z - \frac{\pi}{2} \right)^3}$$

29. Let $M > 0$ be an integer and let $F(z)$ have a pole of order $2M$ at the origin. Prove that if $F(z) = F(-z)$, the residue of the pole is zero for every M.

30. Use the ratio method to determine the residue of the indicated pole of each of the following functions.

(a) $F(z) = \dfrac{z^3}{\sin^2\left(z^2\right)}$ at $z = 0$

(b) $F(z) = \dfrac{\tan^2 z}{z^4}$ at $z = 0$

(c) $F(z) = \tan^2 z$ at $z = \pi/2$

(d) $F(z) = \dfrac{\ell n(1+z)}{\left(e^z - 1\right)^2}$ at $z = 0$ and at $z = 2\pi i$

(e) $F(z) = \dfrac{\ell n(1+z)}{\left(e^z - 1\right)^3}$ at $z = 0$

31. N is an integer. Use Cauchy's residue theorem to prove that for any circular contour centered at the origin of radius $\rho > 0$,

$$\oint z^N dz = \begin{cases} 2\pi i & N = -1 \\ 0 & N \neq -1 \end{cases} = 2\pi i \delta_{N,-1}$$

where $\delta_{N,-1}$ is the Kronecker δ symbol defined by

$$\delta_{m,n} \equiv \begin{cases} 1 & m = n \\ 0 & m \neq n \end{cases}$$

32. (a) For the integer $N > 0$, determine the Laurent series expanded around $z = 0$, for

$$F(z) = \dfrac{e^{z^2}}{z^{2N+3}}$$

(b) Use the results of part (a) to evaluate

$$\oint \frac{e^{z^2}}{z^{2N+3}} dz$$

around a circle centered at the origin.

33. (a) The function

$$F(z) = \frac{e^z}{z(z^2 - 1)(z + 1)}$$

has poles at $z = 0, \pm 1$. Find the residue of each pole by the method stated below.

    $a_{-1}(0)$ by direct Laurent expansion
    $a_{-1}(-1)$ by the derivative method
    $a_{-1}(1)$ by the ratio method

(b) Use the results of part (a) to evaluate

$$\oint \frac{e^z}{z(z^2 - 1)(z + 1)} dz$$

around a circle centered at the origin, of radius (i) 1/2   (ii) 3/2.

Chapter 5

EVALUATION OF INTEGRALS

Certain types of integrals can be evaluated using Cauchy's residue theorem. The integrands of these integrals can have poles in the complex plane, which we assume for now lie either inside or outside a chosen contour, but not on the contour. The evaluation of an integral when the integrand has poles on the contour is discussed in section 5.3.

5.1 Integrals Along the Entire Real Axis

Integrals of the form

$$I_x = \int_{-\infty}^{\infty} F(x)\,dx \tag{5.1a}$$

can be evaluated by Cauchy's theorem by considering the integral

$$I_z = \oint F(z)\,dz \tag{5.1b}$$

around a closed contour that includes the entire real axis. Two common choices for such contours are shown in fig. 5.1.

The segments labeled C_R are large enough that all poles of the integrand in the half-plane are enclosed by the contour. Thus the segments can be shaped into semicircles of large (ultimately infinite) radius R.

We note that when closed in the upper half-plane as in fig. 5.1a, the contour is traversed in the counterclockwise direction. We see from fig. 5.1b that when closed in the lower half-plane, the contour is traversed in the

clockwise direction. Therefore, using Cauchy's residue theorem, the closed
contour integral around either of these contours is

$$\oint_{C_{a\,or\,b}} F(z)\,dz = \pm 2\pi i \sum a_{-1}(z_{\pm}) \tag{5.2}$$

where $\Sigma a_{-1}(z_+)$ is the sum of residues of the poles in the upper (positive) half-
plane and $\Sigma a_{-1}(z_-)$ is the sum of residues of the poles in the lower (negative)
half-plane.

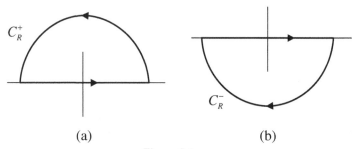

C_R^+

C_R^-

(a)   (b)

Figure 5.1

Contours for evaluating an integral along the entire real axis

Because points on the real axis are defined by $z = x$,

$$\oint_{C_{a\,or\,b}} F(z)\,dz = \int_{-\infty}^{\infty} F(x)\,dx + \lim_{R\to\infty} \int_{C_R^{\pm}} F(z)\,dz = \pm 2\pi i \sum a_{-1}(z_{\pm}) \tag{5.3}$$

We see that to determine the integral of eq. 5.1a by Cauchy's theorem, we
must be able to evaluate the integral along the large semicircular segment in
one of the half-planes.

The behavior of a function at large (essentially infinite) values of its
argument is referred to as the function's *asymptotic behavior*. If the integral
of eq. 5.1a is finite, $F(z)$ must be zero asymptotically. However,

$$\lim_{|z|\to\infty} F(z) = 0 \tag{5.4}$$

is not a sufficient condition to ensure that the integral is finite. When $F(z)$
satisfies

$$\lim_{|z|\to\infty} \int F(z)\,dz = 0 \tag{5.5}$$

we say that $F(z)$ approaches zero asymptotically *fast enough*.

Closing in one of the half-planes, we evaluate the integral along the infinite semicircle by writing

$$z = Re^{i\phi} \tag{5.6a}$$

Because R is constant,

$$dz = iRe^{i\phi}d\phi \tag{5.6b}$$

Therefore, with $\phi \, \varepsilon \, [0,\pi]$ along the semicircle in the upper half-plane and with $\phi \varepsilon \, [2\pi,\pi]$ along the semicircle in the lower half-plane, we have

$$\int_{C_{+\infty}} F(z)\,dz = \lim_{R\to\infty}\int_{C_R^+} F(z)\,dz = \lim_{R\to\infty} i\int_0^\pi RF(Re^{i\phi})e^{i\phi}d\phi \tag{5.7a}$$

and

$$\int_{C_{-\infty}} F(z)\,dz = \lim_{R\to\infty}\int_{C_R^-} F(z)\,dz = \lim_{R\to\infty} i\int_{2\pi}^\pi RF(Re^{i\phi})e^{i\phi}d\phi \tag{5.7b}$$

Because $e^{i\phi}$ is a complex quantity of magnitude 1, it does not affect the asymptotic value of the integrand. This is determined by the behavior of $RF(R)$ in the limit of infinite R. That is,

$$\lim_{R\to\infty}\int_0^\pi RF\left(Re^{i\phi}\right)e^{i\phi}d\phi \sim \lim_{R\to\infty}\int_{2\pi}^\pi RF\left(Re^{i\phi}\right)e^{i\phi}d\phi \sim \lim_{R\to\infty} RF(R) \tag{5.8}$$

The symbol "$\sim$" means that the quantity on the left side of "$\sim$" has the same behavior as that of the quantity on the right side of "$\sim$". Thus, if $F(R)$ asymptotically approaches zero so that,

$$\int_{C_{\pm\infty}} F(z)\,dz \sim \lim_{R\to\infty} RF(R) = 0 \tag{5.9}$$

this is what is meant by the statement that $F(R)$ asymptotically approaches zero "fast enough". Then

$$\lim_{R\to\infty}\oint_{C_\pm} F(z)\,dz = \int_{-\infty}^\infty F(x)\,dx = \pm 2\pi i \sum a_{-1}(z_\pm) \tag{5.10}$$

Functions with inverse power asymptotic behavior

Let α and β be real positive numbers that are large and finite. We write the integral of eq. 5.1a as

$$I_x = \int_{-\infty}^{-\alpha} F(x)\,dx + \int_{-\alpha}^{\beta} F(x)\,dx + \int_{\beta}^{\infty} F(x)\,dx \qquad (5.11)$$

Let the asymptotic behavior of $F(x)$ be

$$F(x) \sim \frac{1}{x^n} \qquad n > 0 \qquad (5.12)$$

Then, eq. 5.11 can be written as

$$I_x \sim \int_{-\infty}^{-\alpha} \frac{1}{x^n}\,dx + \int_{-\alpha}^{\beta} F(x)\,dx + \int_{\beta}^{\infty} \frac{1}{x^n}\,dx \qquad (5.13)$$

Because

$$\int \frac{1}{x^n}\,dx \sim \begin{cases} x^{(1-n)} & n \neq 1 \\ \ell n(|x|) & n = 1 \end{cases} \qquad (5.14)$$

we see that when $n \leq 1$, evaluating the first and third integrals of eq. 5.13 at $\pm\infty$ yields nonfinite results. Thus, if the asymptotic behavior of $F(x)$ is $1/x^n$, the function approaches zero fast enough when $n > 1$. Replacing x by z, we see that if $F(z)$ behaves asymptotically like the inverse of the magnitude of its argument raised to a power greater than 1, then $F(z)$ approaches zero fast enough.

Example 5.1: Evaluating an integral over the entire real axis by Cauchy's residue theorem

The integral

$$I = \int_{-\infty}^{\infty} \frac{1}{(1+x^2)}\,dx \qquad (5.15)$$

can be easily evaluated without using Cauchy's theorem by substituting

$$x = \tan\phi \qquad (5.16)$$

Then

$$\int_{-\infty}^{\infty} \frac{1}{(1+x^2)} dx = \int_{-\pi/2}^{\pi/2} \frac{\sec^2\phi}{(1+\tan^2\phi)} d\phi = \pi \tag{5.17}$$

To evaluate this by Cauchy's theorem, we close the contour in the upper half-plane and note that $1/(1 + z^2)$ has poles at $z = \pm i$ as shown in fig. 5.2a.

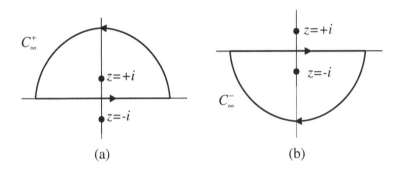

(a)   (b)

Figure 5.2

Contours and poles for evaluating the integral of eq. 5.15

Then

$$\oint \frac{1}{(1+z^2)} dz = \int_{-\infty}^{\infty} \frac{1}{(1+x^2)} dx + \int_{C_{+\infty}} \frac{1}{(1+z^2)} dz \tag{5.18}$$

On $C_{+\infty}$

$$\frac{1}{(1+z^2)} \rightarrow \frac{1}{R^2} e^{-2i\phi} \sim \frac{1}{R^2} \tag{5.19}$$

so that $1/(1 + z^2) \rightarrow 0$ fast enough. Thus,

$$\lim_{R\to\infty} \int_{C_{+\infty}} \frac{1}{(1+z^2)} dz = i \lim_{R\to\infty} \int_{0}^{\pi} \frac{1}{R} e^{-i\phi} d\phi = 0 \tag{5.20}$$

Therefore, because the pole at $z = +i$ is enclosed by the contour,

$$\oint \frac{1}{(1+z^2)} dz = \int_{-\infty}^{\infty} \frac{1}{(1+x^2)} dx = 2\pi i a_{-1}(+i)$$

$$= 2\pi i \lim_{z \to i} \left[(z-i) \frac{1}{(z-i)(z+i)} \right] = \pi$$

(5.21)

This result is also obtained by closing the contour in the lower half-plane as shown in fig. 5.2b. Recognizing that the integral along $C_{-\infty}$ is zero, that the contour is traversed in the clockwise direction, and noting that the pole at $-i$ is enclosed by this contour, we have

$$\oint \frac{1}{(1+z^2)} dz = \int_{-\infty}^{\infty} \frac{1}{(1+x^2)} dx = -2\pi i a_{-1}(-i)$$

$$= -2\pi i \lim_{z \to -i} \left[(z+i) \frac{1}{(z-i)(z+i)} \right] = \pi$$

(5.22)

□

If $F(x)$ is an even function that asymptotically behaves like $1/x^n$ with $n > 1$, then the integral of such a function over half the real axis can be evaluated using Cauchy's residue theorem by writing

$$\int_0^{\infty} F(x)dx = \frac{1}{2}\int_{-\infty}^{\infty} F(x)dx = \frac{1}{2}\oint_{C_{\pm}} F(z)dz = \pm i\pi \sum a_{-1}(z_{\pm})$$

(5.23)

Fourier exponential integrals

The *Fourier exponential integral* is of the form

$$I = \int_{-\infty}^{\infty} f(x)e^{\pm ikx} dx$$

(5.24)

where k is a positive real constant and $f(z)$, in general, has poles in both half-planes.

For functions that approach zero asymptotically such as $1/z^n$ with $n > 1$, the contour can be closed in either half-plane. However, to evaluate the Fourier exponential integral by Cauchy's residue theorem, the contour must be closed in a specific half-plane, and the required half-plane depends on the sign of the exponent.

To understand why this is so, let us consider a Fourier integral with a positive exponential. Closing in one of the half-planes, we have

$$\oint_{C_\pm} f(z)\,e^{ikz}\,dz = \int_{-\infty}^{\infty} f(x)\,e^{ikx}\,dx + \lim_{R\to\infty} \int_{C_{\pm\infty}} f(z)\,e^{ikz}\,dz \qquad (5.25)$$

Representing the points on $C_{\pm\infty}$ as

$$z = Re^{i\phi} = R(\cos\phi + i\,\sin\phi) \qquad (2.30a)$$

the integral over $C_{\pm\infty}$ can be written

$$\lim_{R\to\infty} \int_{C_\pm} f(z)\,e^{ikz}\,dz = \lim_{R\to\infty} \int_{\phi_1}^{\phi_2} f\left(Re^{i\phi}\right) e^{ikR\cos\phi}\, e^{-R\sin\phi}\, iRe^{i\phi}\,d\phi \qquad (5.26)$$

where, we see from fig. 5.1, $[\phi_1, \phi_2] = [0, \pi]$ for points on $C_{+\infty}$ and on $C_{-\infty}$, $[[\phi_1, \phi_2] = [2\pi, \pi]$. Points on $C_{+\infty}$ are characterized by $\sin\phi > 0$. Thus, on this segment of the contour,

$$\sin\phi = |\sin\phi| \qquad (5.27)$$

from which

$$e^{ikz} = e^{ikR\cos\phi} e^{-kR|\sin\phi|} \qquad (5.28)$$

Therefore, the integral around $C_{+\infty}$ becomes

$$\lim_{R\to\infty} \int_{C_{+\infty}} f(z)\,e^{ikz}\,dz = i\lim_{R\to\infty} \int_0^\pi Rf\left(Re^{i\phi}\right) e^{ikR\cos\phi}\, e^{-kR|\sin\phi|}\, e^{i\phi}\,d\phi$$

$$(5.29)$$

$e^{ikR\cos\phi}$ and $e^{i\phi}$ are quantities of magnitude 1 and do not affect the asymptotic behavior of the integrand. This asymptotic behavior is determined by

$$\lim_{R\to\infty} Rf(R)e^{-kR|\sin\phi|}$$

If this limit is zero, the integral around $C_{+\infty}$ will be zero. Because $e^{-kR|\sin\phi|} = 1$ at $\phi = 0$ and π, this also requires

$$\lim_{R\to\infty} Rf(R) = 0 \qquad (5.30)$$

Along $C_{-\infty}$, $\sin\phi < 0$ so that

$$\sin\phi = -|\sin\phi| \tag{5.31}$$

Therefore, closing in the lower half-plane

$$\lim_{R\to\infty} \int_{C_{-\infty}} f(z)e^{ikz}dz = i\lim_{R\to\infty} \int_{2\pi}^{\pi} Rf\left(Re^{i\phi}\right)e^{ikR\cos\phi}\, e^{+kR|\sin\phi|}\, e^{i\phi}d\phi \tag{5.32}$$

Because of the factor $e^{+kR|\sin\phi|}$, this integral around $C_{-\infty}$, will be infinite for functions $f(R)$ that satisfy

$$\lim_{R\to\infty} f(R)e^{kR|\sin\phi|} = \infty \tag{5.33}$$

Therefore, as long as $f(z)$ approaches zero asymptotically fast enough (eq. 5.30), and the Fourier integrand contains e^{ikz} with $k > 0$, closing in the upper half-plane guarantees that the integral around the infinite semicircle will be zero, and Cauchy's theorem can be used to evaluate the Fourier exponential integral with a positive exponential. If the Fourier integrand contains e^{-ikz} with $k > 0$, closing around $C_{+\infty}$, along which $\sin\phi > 0$, would yield.

$$\lim_{R\to\infty} \int_{C_+} f(z)e^{-ikz}dz = i\lim_{R\to\infty} \int_{0}^{\pi} Rf\left(Re^{i\phi}\right)e^{ikR\cos\phi}\, e^{kR|\sin\phi|}\, e^{i\phi}d\phi = \infty$$

$$\tag{5.34a}$$

and when closed around $C_{-\infty}$, along which $\sin\phi < 0$, we would obtain

$$\lim_{R\to\infty} \int_{C_-} f(z)e^{-ikz}dz = i\lim_{R\to\infty} \int_{2\pi}^{\pi} Rf\left(Re^{i\phi}\right)e^{ikR\cos\phi}e^{-kR|\sin\phi|}\, e^{i\phi}d\phi = 0$$

$$\tag{5.34b}$$

Therefore, with $k > 0$, the Fourier exponential integral can be evaluated by Cauchy's theorem as

$$\oint_{C_\pm} f(z)e^{\pm ikz}dz = \int_{-\infty}^{\infty} f(x)e^{\pm ikx}dx = \pm 2\pi i \sum a_{-1}(z_\pm) \tag{5.35}$$

Example 5.2: Evaluating a Fourier exponential integral by Cauchy's residue theorem

To evaluate the integral

$$I = \int_{-\infty}^{\infty} \frac{e^{2ix}}{(1+x^2)} dx \qquad (5.36)$$

we note that by extending the integration variable into the complex plane, the integrand has poles at $z = \pm i$. Because the exponent of the exponential is positive, the contour must be closed in the upper half-plane, enclosing the pole at $z = +i$. Therefore, the description of the poles and the contour for evaluating this integral is shown in fig. 5.2a. Then,

$$\int_{-\infty}^{\infty} \frac{e^{2ix}}{(1+x^2)} dx = 2\pi i a_{-1}(+i) = \pi e^{-2} \qquad (5.37)$$

We note that this Fourier exponential integral is real. Therefore, writing

$$\int_{-\infty}^{\infty} \frac{e^{2ix}}{(1+x^2)} dx = \int_{-\infty}^{\infty} \frac{\cos(2x)}{(1+x^2)} dx + i\int_{-\infty}^{\infty} \frac{\sin(2x)}{(1+x^2)} dx = \pi e^{-2} \qquad (5.38)$$

and equating real and imaginary parts we obtain

$$\int_{-\infty}^{\infty} \frac{\cos(2x)}{(1+x^2)} dx = \pi e^{-2} \qquad (5.39a)$$

and

$$\int_{-\infty}^{\infty} \frac{\sin(2x)}{(1+x^2)} dx = 0 \qquad (5.39b)$$
□

Fourier sine and cosine integrals

The integrals of eqs. 5.39 are examples of the *Fourier cosine* and *Fourier sine* integrals. The general forms of these are

$$I_c = \int_{-\infty}^{\infty} f(x)\cos(kx) dx \qquad (5.40a)$$

and

$$I_s = \int_{-\infty}^{\infty} f(x)\sin(kx) dx \qquad (5.40b)$$

Referring to eqs. 2.55, we see that the Fourier sine and Fourier cosine integrals can be written as combinations of Fourier exponential integrals. Writing $\cos(kx)$ and $\sin(kx)$ as

$$\left.\begin{array}{c}\cos(kx)\\\sin(kx)\end{array}\right\} = \frac{e^{ikx} + \sigma^2 e^{-ikx}}{2\sigma} \qquad (5.41)$$

we see that when $\sigma = 1$, the combination of exponentials is $\cos(kx)$. When $\sigma = i$, we obtain the exponential representation of $\sin(kx)$. Therefore, both the Fourier sine and Fourier cosine integrals can be expressed as

$$I = \frac{1}{2\sigma}\left[\int_{-\infty}^{\infty} f(x)e^{ikx}dx + \sigma^2 \int_{-\infty}^{\infty} f(x)e^{-ikx}dx\right] \qquad (5.42)$$

Each of these exponential integrals can then be evaluated by closing the contour in the appropriate half-plane and using Cauchy's theorem.

If $f(x)$ is real, the Fourier sine and cosine integrals can be expressed as a single exponential integral;

$$\int_{-\infty}^{\infty} f(x)\cos(kx)\,dx = \int_{-\infty}^{\infty} f(x)\,\mathrm{Re}\left(e^{ikx}\right)dx = \mathrm{Re}\left(\int_{-\infty}^{\infty} f(x)e^{ikx}dx\right)$$

$$(5.43a)$$

and

$$\int_{-\infty}^{\infty} f(x)\sin(kx)\,dx = \int_{-\infty}^{\infty} f(x)\,\mathrm{Im}\left(e^{ikx}\right)dx = \mathrm{Im}\left(\int_{-\infty}^{\infty} f(x)e^{ikx}dx\right) \qquad (5.43b)$$

If $f(x)$ is imaginary, it can be written as $ig(x)$ with $g(x)$ real. Then, this analysis is applied to $g(x)$ to obtain

$$\int_{-\infty}^{\infty} f(x)\cos(kx)dx = i\,\mathrm{Re}\left(\int_{-\infty}^{\infty} g(x)e^{ikx}dx\right) \qquad (5.44a)$$

and

$$\int_{-\infty}^{\infty} f(x)\sin(kx)\,dx = i\,\mathrm{Im}\left(\int_{-\infty}^{\infty} g(x)e^{ikx}dx\right) \qquad (5.44b)$$

Thus, if $f(x)$ is entirely real or entirely imaginary, the Fourier sine and cosine integrals can be evaluated by a single Fourier exponential integral. When $f(x)$ has both real and imaginary parts, one must evaluate two Fourier exponential integrals.

5.2 Integrals of Functions of Sine and Cosine

Integrals of the form

$$I = \int_0^{2\pi} f(\cos\theta, \sin\theta)\, d\theta \tag{5.45}$$

can be evaluated using Cauchy's residue theorem. We write

$$\cos\theta = \frac{e^{i\theta} + e^{-i\theta}}{2} \tag{2.55a}$$

and

$$\sin\theta = \frac{e^{i\theta} - e^{-i\theta}}{2i} \tag{2.55b}$$

and make the substitution

$$z = e^{i\theta} \tag{5.46}$$

so that

$$e^{-i\theta} = \frac{1}{z} \tag{5.47}$$

Then eqs. 2.55 become

$$\left.\begin{array}{c} \cos\theta \\ \sin\theta \end{array}\right\} = \frac{z + \sigma^2 z^{-1}}{2\sigma} \qquad \sigma = \left\{\begin{array}{l} 1 \\ i \end{array}\right. \tag{5.48}$$

Because $|z| = |e^{i\theta}| = 1$ and $\theta \; \varepsilon \; [0,2\pi]$, the points over which the integral is taken lie on the unit circle. Therefore, with

$$dz = ie^{i\theta}d\theta = iz\,d\theta \tag{5.49}$$

eq. 5.45 can be written

$$I = \oint_{\substack{unit \\ circle}} f\left(\frac{z+z^{-1}}{2}, \frac{z-z^{-1}}{2i}\right)\frac{1}{iz}dz \tag{5.50a}$$

As long as none of the poles of the integrand lie on the unit circle, Cauchy's theorem yields

$$\int_0^{2\pi} f(\cos\theta, \sin\theta)\,d\theta = \oint_{\substack{unit \\ circle}} f\left(\frac{z+z^{-1}}{2}, \frac{z-z^{-1}}{2i}\right)\frac{1}{iz}dz$$

$$= 2\pi i \sum a_{-1}(poles\ inside\ unit\ circle) \tag{5.50b}$$

Example 5.3: Evaluating an integral of a function of $\sin\theta$ and $\cos\theta$ by Cauchy's residue theorem

Applying the approach described above, it is straightforward to show that

$$\int_0^{2\pi} \frac{1}{(10+6\sin\theta)}d\theta = \frac{1}{3}\oint_{\substack{unit \\ circle}} \frac{1}{(z+\frac{i}{3})(z+3i)}dz \tag{5.51}$$

It is evident that this integrand has poles at $z = -i/3$ which is inside the unit circle and $z = -3i$ which is outside the unit circle. Therefore,

$$\int_0^{2\pi} \frac{1}{(10+6\sin\theta)}d\theta = 2\pi i a_{-1}(-i/3) = \frac{\pi}{4} \tag{5.52}$$

$\square$

5.3 Cauchy's Principal Value Integral and the Dirac δ Symbol

If a function has a pole at a point z_0, the integral of that function over a contour is not defined if z_0 is on the contour. A nonrigorous way to understand this is by viewing an integral as a sum of an infinite number of infinitesimal terms. Because the term obtained from evaluating the function at z_0 is undefined, the sum containing that term is undefined. Therefore, the integral will be defined only if z_0 is not on the contour.

Let $F(z)$ have a simple pole at z_0. We consider an integral of $F(z)$ around a contour that comes infinitesimally close to, but does not access z_0, as shown in fig. 5.3.

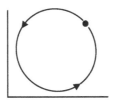

Figure 5.3

A pole that is infinitesimally "on" a contour

In order for the integral around this contour to have a defined value, the contour must be traversed in such a way as to avoid the pole. Two ways to achieve this are shown in fig. 5.4.

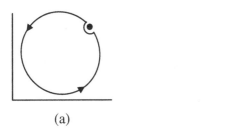
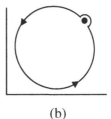

(a)   (b)

Figure 5.4

Contours that avoid a pole

As an example, let us illustrate the method of avoiding a pole on the contour by considering an integral of the form

$$I = \int_{-\infty}^{\infty} \frac{f(x)}{(x - x_0)} dx \qquad (5.53)$$

where x_0 is real and the path from $-\infty$ to ∞ is taken along the real axis except at x_0. To obtain a defined result, the contour must be traversed in such a way as to avoid the pole at x_0.

We take $f(z)$ to possibly have poles at points off the real axis, but none on the real axis. Thus, the only pole on the real axis arises from $1/(x - x_0)$. We also assume that $f(z)$ approaches zero asymptotically fast enough so that the contour can be closed in at least one of the half-planes. We close in the upper half-plane, but the analysis is unchanged if the contour were closed in the lower half-plane.

To avoid the pole at x_0, the contour can be traversed around x_0 in either of two ways as shown in fig. 5.5.

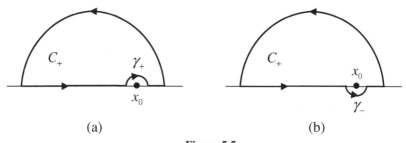

(a)   (b)

Figure 5.5
Contours that avoid a pole on the real axis

There are no singularities of $f(z)$ between x_0 and either of the small segments γ_+ and γ_-. Therefore, γ_+ and γ_- can be shaped into semicircles of infinitesimal radius ρ centered at x_0, and the values of the integrals around these small semicircles will not be affected when we take $\rho \to 0$.

Referring to fig. 5.5a,

$$\oint_{C_+} \frac{f(z)}{(z - x_0)} dz = \int_{-\infty}^{x_0 - \rho} \frac{f(x)}{(x - x_0)} dx + \int_{\gamma_+} \frac{f(z)}{(z - x_0)} dz + \int_{x_0 + \rho}^{\infty} \frac{f(x)}{(x - x_0)} dx$$

(5.54a)

where the integral around the infinite semicircle is assumed to be zero and has been omitted. Applying Cauchy's residue theorem to the integral around C_+, we note that x_0 is excluded from the contour and so the residue of the integrand at x_0 is not included in the sum of residues. Therefore, eq. 5.54a becomes

$$\oint_{C_+} \frac{f(z)}{(z-x_0)} dz = \int_{-\infty}^{x_0-\rho} \frac{f(x)}{(x-x_0)} dx + \int_{\gamma_+} \frac{f(z)}{(z-x_0)} dz + \int_{x_0+\rho}^{\infty} \frac{f(x)}{(x-x_0)} dx$$

$$= 2\pi i \sum a_{-1}(z_+)$$

(5.54b)

where $\Sigma a_{-1} (z_+)$ is the sum of residues of $f(z)/(z-x_0)$ arising from the poles of $f(z)$ in the upper half-plane.

Referring to fig. 5.6a, points on γ_+ are described by

$$z = x_0 + \rho e^{i\phi}$$

(5.55)

where ρ is infinitesimal (ultimately zero) and the direction of traversal of the contour defines the range of ϕ to be from π to 0.

$$\gamma_+$$

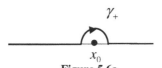

$$x_0$$

Figure 5.6a

Infinitesimal semicircle around x_0

Therefore,

$$\lim_{\rho \to 0} \int_{\gamma_+} \frac{f(z)}{(z-x_0)} dz = \lim_{\rho \to 0} \int_{\pi}^{0} \frac{f(x_0+\rho e^{i\phi})}{\rho e^{i\phi}} i\rho e^{i\phi} d\phi$$

(5.56)

$$= \int_{\pi}^{0} f(x_0) i d\phi = -i\pi f(x_0)$$

and because x_0 is outside the closed contour, eq. 5.54b becomes

$$\lim_{\rho \to 0} \left[\int_{-\infty}^{x_0-\rho} \frac{f(x)}{(x-x_0)} dx + \int_{x_0+\rho}^{\infty} \frac{f(x)}{(x-x_0)} dx \right] = 2\pi i \sum a_{-1}(z_+) + i\pi f(x_0)$$

(5.57)

The quantity on the left-hand side of eq. 5.57 is called the *Cauchy principal value integral* and it is denoted by

$$\lim_{\rho \to 0} \left[\int_{-\infty}^{x_0-\rho} \frac{f(x)}{(x-x_0)} dx + \int_{x_0+\rho}^{\infty} \frac{f(x)}{(x-x_0)} dx \right] \equiv P \int_{-\infty}^{\infty} \frac{f(x)}{(x-x_0)} dx \quad (5.58a)$$

We introduce another notation for the principal value that is used inter-
changeably with that of eq. 5.58a:

$$\lim_{\rho \to 0} \left[\int_{-\infty}^{x_0-\rho} \frac{f(x)}{(x-x_0)} dx + \int_{x_0+\rho}^{\infty} \frac{f(x)}{(x-x_0)} dx \right] \equiv \int_{-\infty}^{\infty} \frac{f(x)}{(x-x_0)_P} dx \quad (5.58b)$$

Then, the closed contour integral becomes

$$\lim_{\rho \to 0} \oint_{C_+} \frac{f(z)}{(z-x_0)} dz = P \int_{-\infty}^{\infty} \frac{f(x)}{(x-x_0)} dx - i\pi f(x_0) \quad (5.59)$$

and the principal value integral is given by

$$P \int_{-\infty}^{\infty} \frac{f(x)}{(x-x_0)} dx = 2\pi i \sum a_{-1}(z_+) + i\pi f(x_0) \quad (5.60)$$

From the definition of the principal value integral, we see that the integral
along the real axis accesses all points from $-\infty$ to the point immediately
before x_0 and all points immediately beyond x_0 to ∞. Therefore, the principal
value integral is independent of the choice of segment γ_+ or γ_- and so
contains no indication about how one avoids the pole.

To see this, let us integrate around the contour of fig. 5.5b, avoiding the
pole by traversing along γ_-. Then,

$$\lim_{\rho \to 0} \oint_{C_+} \frac{f(z)}{(z-x_0)} dz = P \int_{-\infty}^{\infty} \frac{f(x)}{(x-x_0)} dx + \lim_{\rho \to 0} \int_{\gamma_-} \frac{f(z)}{(z-x_0)} dz \quad (5.61)$$

Again, the integral over the infinite semicircle is zero and has been omitted.
Referring to fig. 5.6b, points on γ_- are defined by

$$z = x_0 + \rho e^{i\phi} \quad (5.55)$$

with ϕ ranging from π to 2π.

Figure 5.6b
Infinitesimal semicircle around x_0

Therefore,

$$\lim_{\rho\to 0}\int_{\gamma_-}\frac{f(z)}{(z-x_0)}dz=\lim_{\rho\to 0}\int_{\pi}^{2\pi}\frac{f(x_0+\rho e^{i\phi})}{\rho e^{i\phi}}i\rho e^{i\phi}d\phi$$

$$=i\lim_{\rho\to 0}\int_{\pi}^{2\pi}f(x_0+\rho e^{i\phi})d\phi=i\pi f(x_0)$$

(5.62)

By closing in the upper half-plane, the pole at x_0 is enclosed by the contour. Thus, the sum of residues of poles enclosed by the contour must include the residue of the integrand at x_0, which is

$$a_{-1}(x_0)=\lim_{z\to x_0}\left[(z-x_0)\frac{f(z)}{(z-x_0)}\right]=f(x_0)$$

(5.63)

Therefore, applying Cauchy's theorem to eq. 5.61, we obtain

$$P\int_{-\infty}^{\infty}\frac{f(x)}{(x-x_0)}dx+i\pi f(x_0)=2\pi i f(x_0)+2\pi i\sum a_{-1}(z_+)$$

(5.64)

from which we obtain the identical result given in eq. 5.59,

$$P\int_{-\infty}^{\infty}\frac{f(x)}{(x-x_0)}dx=2\pi i\sum a_{-1}(z_+)+i\pi f(x_0)$$

(5.60)

The fact that the principal value integral found by avoiding the pole around γ_+ is the same as that found by avoiding the pole around γ_- demonstrates that the principal value integral is independent of how one avoids this pole.

However, when avoiding the pole around γ_-, the closed contour integral is

$$\lim_{\rho\to 0}\oint_{C_+}\frac{f(z)}{(z-x_0)}dz=P\int_{-\infty}^{\infty}\frac{f(x)}{(x-x_0)}dx+i\pi f(x_0)$$

(5.65)

Comparing this to eq. 5.59, we see that the closed contour integral does depend on how one avoids the pole at x_0.

Displacement of the pole and the Dirac δ symbol

Referring to fig. 5.6a, when the contour avoids the pole at x_0 by traversing around γ_+, the contour is taken slightly above the pole, and thus the pole is slightly below the contour. Therefore, avoiding the pole by integrating around γ_+ is equivalent to taking a contour that does not deviate from the real axis, and displacing the pole infinitesimally below the real axis. This is accomplished by giving x_0 an infinitesimal negative imaginary part as shown in fig. 5.7a.

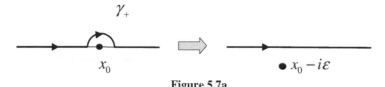

Figure 5.7a

Equivalence of the contour displaced above a pole on the real axis and
the pole displaced below the real axis with the contour along the real axis

Therefore,

$$\lim_{\rho \to 0} \oint_{C_+} \frac{f(z)}{(z-x_0)} dz = \lim_{\varepsilon \to 0} \int_{-\infty}^{\infty} \frac{f(x)}{(x-x_0+i\varepsilon)} dx \qquad (5.66)$$

where again the integral around $C_{+\infty}$ is zero and has been omitted. Using the notation of eq. 5.58b, we can express eq. 5.59 as

$$\oint_{C_+} \frac{f(z)}{(z-x_0)} dz = \lim_{\varepsilon \to 0} \int_{-\infty}^{\infty} \frac{f(x)}{(x-x_0+i\varepsilon)} dx = \int_{-\infty}^{\infty} \frac{f(x)}{(x-x_0)_P} dx - i\pi f(x_0)$$

$$(5.67a)$$

Avoiding the pole at x_0 by taking the contour along γ_- which is below the pole is equivalent to taking the contour along the entire real axis and displacing the pole infinitesimally above the real axis by assigning it an infinitesimal positive imaginary part as shown below in fig. 5.7b.

Figure 5.7b

Equivalence of the contour displaced below a pole on the real axis and
the pole displaced above the real axis with the contour along the real axis

An analysis identical to that above yields

$$\oint_{C_+} \frac{f(z)}{(z-x_0)} dz = \lim_{\varepsilon \to 0} \int_{-\infty}^{\infty} \frac{f(x)}{(x-x_0-i\varepsilon)} dx = \int_{-\infty}^{\infty} \frac{f(x)}{(x-x_0)_P} dx + i\pi f(x_0)$$

(5.67b)

The Dirac δ symbol, denoted by $\delta(x-y)$, is a mathematical entity called a *distribution*. (For a more extensive treatment of distributions, the reader is referred to texts such as Bremermann, 1965; Constantinescu, 1980; Zemanian, 1987.) A function has specified values at given values of the variable(s) on which it depends. A distribution has meaning only when multiplied by a function and integrated.

Let $F(x)$ and $G(y)$ be functions. A distribution $D(x, y)$ is defined such that

$$F(x) = \int_a^b D(x, y)G(y)dy$$

(5.68)

The Dirac δ symbol is the *unit distribution*, which means that if $D(x, y) = \delta(x-y)$, then $G(x) = F(x)$, and $\delta(x-y)$ is defined by

$$\int_a^b \delta(x-y)F(y)dy = \begin{cases} F(x) & a < x < b \\ 0 & x < a, \ x > b \end{cases}$$

(5.69)

Using this property, eqs. 5.67 can be written

$$\lim_{\varepsilon \to 0} \int_{-\infty}^{\infty} \frac{f(x)}{(x-x_0 \pm i\varepsilon)} dx = \int_{-\infty}^{\infty} \frac{f(x)}{(x-x_0)_P} dx \mp i\pi \int_{-\infty}^{\infty} f(x)\delta(x-x_0)dx$$

(5.70)

Because this is valid for any function $f(x)$ that has the required asymptotic behavior, the integrands must be equal. Therefore,

$$\lim_{\varepsilon \to 0} \frac{1}{(x-x_0 \pm i\varepsilon)} = \frac{1}{(x-x_0)_P} \mp i\pi \delta(x-x_0)$$

(5.71)

We see from this that the entities

$$\lim_{\varepsilon \to 0} \frac{1}{(x-x_0 \pm i\varepsilon)} \quad \text{and} \quad \frac{1}{(x-x_0)_P}$$

are also distributions. We also note that

$$\frac{1}{(x-x_0)_P} = \lim_{\varepsilon \to 0} \text{Re}\left[\frac{1}{(x-x_0 \pm i\varepsilon)}\right] \tag{5.72a}$$

and

$$\delta(x-x_0) = \mp\frac{1}{\pi}\lim_{\varepsilon \to 0}\text{Im}\left[\frac{1}{(x-x_0 \pm i\varepsilon)}\right] \tag{5.72b}$$

If $f(x)$ is real, using eq. 5.72a we can express a principal value integral as

$$P\int_{-\infty}^{\infty}\frac{f(x)}{(x-x_0)}dx = \lim_{\varepsilon \to 0}\int_{-\infty}^{\infty}f(x)\text{Re}\left(\frac{1}{(x-x_0 \pm i\varepsilon)}\right)dx$$
$$= \lim_{\varepsilon \to 0}\text{Re}\int_{-\infty}^{\infty}\frac{f(x)}{(x-x_0 \pm i\varepsilon)}dx \tag{5.73}$$

Closing the contour in one of the half-planes, the pole at x_0 can be displaced so it is excluded from the contour by taking the appropriate sign of $i\varepsilon$. Then the residue at x_0 is not included in the sum of residues.

If $f(x)$ is complex, expressing it in terms of its real and imaginary parts $f_1(x)$ and $f_2(x)$ yields

$$P\int_{-\infty}^{\infty}\frac{f(x)}{(x-x_0)}dx$$
$$= \lim_{\varepsilon \to 0}\text{Re}\int_{-\infty}^{\infty}\frac{f_1(x)}{(x-x_0 \pm i\varepsilon)}dx + i\lim_{\varepsilon \to 0}\text{Re}\int_{-\infty}^{\infty}\frac{f_2(x)}{(x-x_0 \pm i\varepsilon)}dx \tag{5.74}$$

Each of these integrals is like that involving a real function as shown in eq. 5.73.

A second approach for complex $f(x)$ is obtained from eq. 5.72a, by writing

$$\frac{1}{(x-x_0)_P} = \frac{1}{2}\lim_{\varepsilon \to 0}\left[\frac{1}{(x-x_0 + i\varepsilon)} + \frac{1}{(x-x_0 - i\varepsilon)}\right] \tag{5.75a}$$

Then

$$P\int_{-\infty}^{\infty} \frac{f(x)}{(x-x_0)}\,dx$$

$$= \frac{1}{2}\lim_{\varepsilon \to 0}\left[\int_{-\infty}^{\infty}\frac{f(x)}{(x-x_0+i\varepsilon)}\,dx + \int_{-\infty}^{\infty}\frac{f(x)}{(x-x_0-i\varepsilon)}\,dx\right]$$

(5.75b)

We note that the pole at $x_0 - I\varepsilon$ in the integrand of the first integral is below the real axis. If the contour is closed in the upper half-plane, this pole is excluded from the contour. Likewise, the pole at $x_0 + i\varepsilon$ in the integrand of the second integral is outside the contour if the contour can be closed in the lower half-plane.

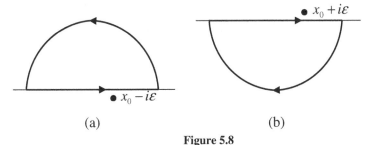

(a)   (b)

Figure 5.8

Closing the contour to exclude the poles at $x_0 \pm i\varepsilon$

Example 5.4: Evaluating a principal value integral by Cauchy's residue theorem

With x_0 and x_1 real, we consider

$$I = P\int_{-\infty}^{\infty}\frac{1}{\left(x^2+x_1^2\right)(x-x_0)}\,dx$$

(5.76)

(a) To evaluate this integral by the prescription given in eq. 5.60, we identify

$$f(x) = \frac{1}{\left(x^2+x_1^2\right)}$$

(5.77)

which has poles at $\pm i x_1$. Then, closing the contour in the upper half-plane, and noting that $f(x)$ behaves asymptotically as $1/x^2$ which approaches zero at infinity fast enough, eq. 5.60 yields

$$P\int_{-\infty}^{\infty} \frac{1}{\left(x^2+x_1^2\right)\left(x-x_0\right)} dx = 2\pi i a_{-1}(ix_1)+i\pi f(x_0) = \frac{-\pi x_0}{x_1\left(x_0^2+x_1^2\right)}$$

(5.78)

We see that this result is real as expected because the integrand is real.

(b) To evaluate the principal value integral by displacing the pole, we write

$$\frac{1}{(x-x_0)_P} = \mathrm{Re}\left[\frac{1}{(x-x_0\pm i\varepsilon)}\right]$$

(5.72a)

Because $1/(x^2+x_1^2)$ is real, we can write the principal value integral as

$$P\int_{-\infty}^{\infty} \frac{1}{\left(x^2+x_1^2\right)\left(x-x_0\right)} dx = \lim_{\varepsilon\to 0}\mathrm{Re}\left[\int_{-\infty}^{\infty} \frac{1}{\left(x^2+x_1^2\right)\left(x-x_0\pm i\varepsilon\right)} dx\right]$$

(5.79)

If the contour is closed in the upper half-plane, it is advantageous to choose $+i\varepsilon$. We see that this choice places the pole at $x_0 - i\varepsilon$ which is in the lower half-plane, and is therefore excluded from the contour as in fig. 5.8a. In this way, it is not necessary to determine the residue at x_0.

With the integral on the infinite semicircle omitted,

$$\lim_{\varepsilon\to 0}\oint_{C_+} \frac{1}{\left(z^2+x_1^2\right)\left(z-x_0+i\varepsilon\right)} dz = \lim_{\varepsilon\to 0}\int_{-\infty}^{\infty} \frac{1}{\left(x^2+x_1^2\right)\left(x-x_0+i\varepsilon\right)} dx$$

$$= 2\pi i \lim_{\varepsilon\to 0}a_{-1}(ix_1) = \lim_{\varepsilon\to 0}\frac{\pi}{x_1}\frac{1}{\left(i(x_1+\varepsilon)-x_0\right)}$$

(5.80)

Because ε is infinitesimally small, we can take $\varepsilon \to 0$ without affecting the imaginary part of the denominator. Therefore, eq. 5.80 yields

$$P\int_{-\infty}^{\infty} \frac{1}{\left(x^2 + x_1^2\right)\left(x - x_0\right)} dx = \lim_{\varepsilon \to 0} \mathrm{Re} \int_{-\infty}^{\infty} \frac{1}{\left(x^2 + x_1^2\right)\left(x - x_0 + i\varepsilon\right)} dx$$

$$= \frac{\pi}{x_1} \mathrm{Re} \frac{1}{\left(ix_1 - x_0\right)} = \frac{-\pi x_0}{x_1\left(x_1^2 + x_0^2\right)}$$

(5.81)

□

The Dirac δ symbol as a function

Although the δ symbol is not a function, it is possible to treat it as such. That is, we can assign values to $\delta(x - x_0)$ at all x. For this reason, $\delta(x - x_0)$ is often referred to as the *Dirac δ function*.

Let ε be an infinitesimal, real, and positive quantity and take $G(x)$ to be an arbitrary continuous function that asymptotically approaches zero fast enough. We consider the integrals

$$\lim_{\varepsilon \to 0} \int_{-\infty}^{x_0 - \varepsilon} G(x)\,\delta(x - x_0)\,dx \quad \text{and} \quad \lim_{\varepsilon \to 0} \int_{x_0 + \varepsilon}^{\infty} G(x)\,\delta(x - x_0)\,dx$$

Because x_0 is outside the range of integration of both integrals, x never accesses the point x_0. Therefore, from the definition of the δ symbol given in eq. 5.69,

$$\lim_{\varepsilon \to 0} \int_{-\infty}^{x_0 - \varepsilon} G(x)\,\delta(x - x_0)\,dx = \lim_{\varepsilon \to 0} \int_{x_0 + \varepsilon}^{\infty} G(x)\,\delta(x - x_0)\,dx = 0 \quad (5.82)$$

Inasmuch as $G(x)$ is arbitrary, these integrals are zero only if we interpret

$$\delta(x - x_0) = 0 \quad x \neq x_0 \tag{5.83}$$

We now set $G(x) = 1$. Then, because $x_0 - \varepsilon < x_0 < x_0 + \varepsilon$,

$$\lim_{\varepsilon \to 0} \int_{x_0 - \varepsilon}^{x_0 + \varepsilon} \delta(x - x_0)\,dx = 1 \tag{5.84}$$

Interpreting the integral as the area under the functional curve between the endpoints, we see that the area under the δ function curve is 1. When $\varepsilon \to 0$, the interval $x_0 - \varepsilon$ to $x_0 + \varepsilon \to 0$, with the area remaining 1. Therefore, the "width" of the δ function is zero, and so its "height" at x_0 must be infinite so the area can be 1. Thus, the Dirac δ-function is defined as

$$\delta(x - x_0) = \begin{cases} 0 & x \neq x_0 \\ \infty & x = x_0 \end{cases} \tag{5.85}$$

A graphical representation of $\delta(x-x_0)$ is shown in fig. 5.9.

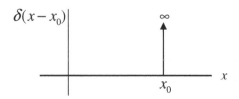

Figure 5.9

Graphical representation of $\delta(x - x_0)$ when viewed as a function

It is possible to obtain an expression that describes this functionlike form of $\delta(x - x_0)$. We have shown that the δ function can be written as

$$\delta(x - x_0) = \mp \frac{1}{\pi} \lim_{\varepsilon \to 0} \text{Im} \left[\frac{1}{(x - x_0 \pm i\varepsilon)} \right] \tag{5.72b}$$

After rationalizing the denominator, this becomes

$$\delta(x - x_0) = \frac{1}{\pi} \lim_{\varepsilon \to 0} \text{Im} \left[\frac{\varepsilon}{(x - x_0)^2 + \varepsilon^2} \right] \tag{5.86}$$

We see that if $x \neq x_0$, this limit is zero. And when $x = x_0$, the limit is infinite. Thus, eq. 5.86 satisfies the properties of $\delta(x - x_0)$ described in fig. 5.9.

5.4 Miscellaneous Integrals

The integrals discussed so far are taken along the entire real axis, and are evaluated using Cauchy's theorem by closing the contour by an infinite semicircle in one of the half-planes. The choice of the half-plane to close in becomes evident from the integrand. If the integrand behaves asymptotically as x^{-n} with $n > 1$, the infinite semicircle can be chosen in either half-plane. For a Fourier exponential integral, the half-plane in which the contour must be closed depends on the sign of the exponent in the exponential.

There are other integrals that can be evaluated using Cauchy's theorem which do not traverse the entire real axis and/or do not require closing with an infinite semicircle in one of the half-planes. The contours for these integrals are unique to the integral being evaluated. As such, their evaluation using Cauchy's theorem is illustrated by examples.

Example 5.5: Evaluating an integral by Cauchy's theorem using a rectangular contour

To evaluate

$$I_x \equiv \int_{-\infty}^{\infty} \frac{xe^x}{\left(1+e^{2x}\right)^2} dx \tag{5.87a}$$

we consider

$$I_z \equiv \oint \frac{ze^z}{\left(1+e^{2z}\right)^2} dz \tag{5.87b}$$

around the rectangular contour shown in fig. 5.10. The integrand has second order poles at the values of z for which

$$e^{2z} = -1 \tag{5.88a}$$

These values are

$$z = \pm i\pi/2, \pm i3\pi/2, \ldots = \pm i(N+\tfrac{1}{2})\pi \quad N = 0,1,2\ldots \tag{5.88b}$$

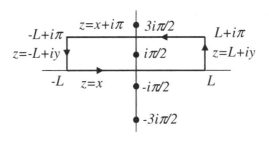

Figure 5.10

Contour appropriate for the integral of eq. 5.87a

Beginning at the lower left-hand corner of the rectangle,

$$\oint \frac{ze^z}{\left(1+e^{2z}\right)^2}dz = \int_{-L}^{L} \frac{xe^x}{\left(1+e^{2x}\right)^2}dx + \int_{0}^{\pi} \frac{(L+iy)e^{(L+iy)}}{\left(1+e^{2(L+iy)}\right)^2}idy$$

$$+\int_{L}^{-L} \frac{(x+i\pi)e^{(x+i\pi)}}{\left(1+e^{2(x+i\pi)}\right)^2}dx + \int_{\pi}^{0} \frac{(-L+iy)e^{(-L+iy)}}{\left(1+e^{2(-L+iy)}\right)^2}idy$$

$$(5.89)$$

Because the limits of the integral of eq. 5.87a are $\pm\infty$, we eventually take $L \to \infty$. Therefore, we can view these integrands for very large L. In this limit, the magnitude of $e^{2(L+iy)}$ is e^{2L} which is large compared to 1. Thus, ignoring 1 in the denominator,

$$\lim_{L\to\infty}\left[\frac{(L+iy)e^{(L+iy)}}{\left(1+e^{2(L+iy)}\right)^2}\right] \sim e^{-3iy}\lim_{L\to\infty}(L+iy)e^{-3L}=0 \qquad (5.90a)$$

Similarly, for large L, $e^{2(-L+iy)}$ has a very small magnitude relative to 1. Therefore, ignoring the exponential in the denominator

$$\lim_{L\to\infty}\left[\frac{(-L+iy)e^{(-L+iy)}}{\left(1+e^{2(-L+iy)}\right)^2}\right] \sim e^{iy}\lim_{L\to\infty}(-L+iy)e^{-L}=0 \qquad (5.90b)$$

Thus, the integrals along the sides of the rectangle are zero in the limit of infinite L, and eq. 5.89 becomes

$$\lim_{L\to\infty}\oint \frac{ze^z}{\left(1+e^{2z}\right)^2}dz = \int_{-\infty}^{\infty} \frac{xe^x}{\left(1+e^{2x}\right)^2}dx + \int_{\infty}^{-\infty} \frac{(x+i\pi)e^{(x+i\pi)}}{\left(1+e^{2(x+i\pi)}\right)^2}dx \quad (5.91)$$

Using

$$e^{(x+i\pi)} = -e^x \qquad (5.92a)$$

and

$$e^{2(x+i\pi)} = e^{2x} \qquad (5.92b)$$

and referring to fig. 5.10, we note that only the pole at $z = i\pi/2$ is enclosed in the rectangle. Therefore, eq. 5.91 becomes

$$\lim_{L\to\infty} \oint \frac{ze^z}{\left(1+e^{2z}\right)^2} dz = \int_{-\infty}^{\infty} \frac{xe^x}{\left(1+e^{2x}\right)^2} dx + \int_{-\infty}^{\infty} \frac{(x+i\pi)e^x}{\left(1+e^{2x}\right)^2} dx$$

$$= 2\int_{-\infty}^{\infty} \frac{xe^x}{\left(1+e^{2x}\right)^2} dx + i\pi \int_{-\infty}^{\infty} \frac{e^x}{\left(1+e^{2x}\right)^2} dx = 2\pi i a_{-1}\left(\frac{i\pi}{2}\right)$$

$$(5.93)$$

With the substitution $w = e^z$, we can evaluate

$$\int_{-\infty}^{\infty} \frac{e^x}{\left(1+e^{2x}\right)^2} dx = \int_0^{\infty} \frac{1}{\left(1+w^2\right)^2} dw = \frac{1}{2}\int_{-\infty}^{\infty} \frac{1}{\left(1+w^2\right)^2} dw \qquad (5.94)$$

Closing the contour in the upper half of the w-plane, this integral can be evaluated straightforwardly using Cauchy's theorem. We obtain

$$\int_{-\infty}^{\infty} \frac{e^x}{\left(1+e^{2x}\right)^2} dx = \frac{\pi}{4} \qquad (5.95)$$

The residue of the pole at $i\pi/2$ can be evaluated by standard methods. Recognizing that this pole is second-order, it is straightforward to show that

$$a_{-1}\left(\frac{i\pi}{2}\right) = \frac{(\pi+2i)}{8} \qquad (5.96)$$

from which

$$\int_{-\infty}^{\infty} \frac{xe^x}{\left(1+e^{2x}\right)^2} dx = -\frac{\pi}{4} \qquad (5.97)$$

$\square$

The method described earlier to evaluate Fourier integrals cannot be applied to integrals over the real axis from $-\infty$ to ∞ if the integrand contains an imaginary exponential that is not linear in the integration variable. However, Cauchy's theorem can be used to evaluate certain integrals that contain exponentials that are not linear.

Example 5.6: Evaluating selected integrals by Cauchy's residue theorem using contours that include portions of the arc of a circle

(a) With k a real positive constant, the integral

$$\int_{-\infty}^{\infty} e^{ikx^2}\, dx = 2\int_0^{\infty} e^{ikx^2}\, dx \tag{5.98a}$$

can be evaluated by considering

$$I = \oint e^{ikz^2}\, dz \tag{5.98b}$$

around the contour shown in fig. 5.11.

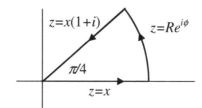

Figure 5.11

Contour for evaluating the integral of eq. 5.98a

On the segment along the real axis,

$$z = x \tag{5.99a}$$

Along the large (ultimately infinite) arc

$$z = Re^{i\phi} \qquad 0 \le \phi \le \pi/4 \tag{5.99b}$$

and along the slanted segment, defined by $y = x$,

$$z = x(1+i) = x\sqrt{2}\, e^{i\pi/4} \tag{5.99c}$$

Therefore, with $(1 + i)^2 = 2i$,

$$\lim_{R \to \infty} \oint e^{ikz^2} dz = \int_0^\infty e^{ikx^2} dx + \lim_{R \to \infty} \int_0^{\pi/4} e^{ikR^2 e^{2i\phi}} iRe^{i\phi} \, d\phi$$
$$+ (1+i) \int_\infty^0 e^{-2kx^2} \, dx \qquad (5.100)$$

On the large arc,

$$e^{ikR^2 e^{2i\phi}} = e^{ikR^2 \cos(2\phi)} e^{-kR^2 \sin(2\phi)} \qquad (5.101)$$

In the range $0 \le \phi \le \pi/4$,

$$\sin(2\phi) = |\sin(2\phi)| \qquad (5.102)$$

Therefore,

$$\lim_{R \to \infty} e^{-kR^2 \sin(2\phi)} = 0 \qquad (5.103)$$

and the integral along the infinite arc is zero. As shown in appendix 5, the integral along the slanted segment is given by

$$-(1+i) \int_0^\infty e^{-2kx^2} dx = -(1+i) \frac{1}{2} \sqrt{\frac{\pi}{2k}} \qquad (5.104)$$

Therefore, because there are no poles inside the contour of fig. 5.13, the integral around the closed contour is zero. Thus, eq. 5.100 becomes

$$\int_0^\infty e^{ikx^2} dx = \frac{(1+i)}{2} \sqrt{\frac{\pi}{2k}} \qquad (5.105)$$

Writing the exponential in trigonometric form, this becomes,

$$\int_0^\infty \left[\cos(kx^2) + i \sin(kx^2) \right] dx = \frac{(1+i)}{2} \sqrt{\frac{\pi}{2k}} \qquad (5.106)$$

Equating real parts and imaginary parts, we obtain

$$\int_0^\infty \cos(kx^2)dx = \int_0^\infty \sin(kx^2)dx = \frac{1}{2}\sqrt{\frac{\pi}{2k}} \qquad (5.107)$$

These integrals are called the *Fresnel cosine* and *sine integrals*. They arise in the study of a phenomenonexhibited by all types of waves (light, sound, etc.) called *diffraction*.

(b) To evaluate

$$I_x = \int_0^\infty \frac{1}{(1+x^3)}dx \qquad (5.108a)$$

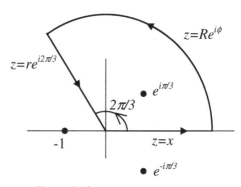

Figure 5.12

Contour for the integral of eq. 5.108a

using the residue theorem, we consider

$$I_z = \oint \frac{1}{(1+z^3)}dz \qquad (5.108b)$$

around the contour shown in fig. 5.12.

The poles of the integrand of eq. 5.108b are at the cube roots of -1, which were shown in ex. 2.5b to be

$$z_1 = e^{i\pi/3} = \frac{1}{2} + i\frac{\sqrt{3}}{2} \qquad (5.109a)$$

$$z_2 = e^{3i\pi/3} = -1 \tag{5.109b}$$

and

$$z_3 = e^{5i\pi/3} = \frac{1}{2} - i\frac{\sqrt{3}}{2} \tag{5.109c}$$

Referring to fig. 5.12, the integral around the closed contour becomes

$$\oint \frac{1}{(1+z^3)} dz =$$

$$\int_0^\infty \frac{1}{(1+x^3)} dx + \lim_{R\to\infty} \int_{C_\infty} \frac{1}{(1+z^3)} dz + e^{2i\pi/3} \int_\infty^0 \frac{1}{(1+r^3 e^{2\pi i})} dr \tag{5.110}$$

On C_∞ , points are described by

$$z = Re^{i\phi} \tag{5.111}$$

with $0 \le \phi \le 2\pi/3$. Then

$$\lim_{R\to\infty} \int_{C_\infty} \frac{1}{(1+z^3)} dz = \lim_{R\to\infty} \int_0^{2\pi/3} \frac{iRe^{i\phi}}{(1+R^3 e^{3i\phi})} d\phi = 0 \tag{5.112}$$

In the integral along the slanted segment, with $e^{2\pi i} = 1$, we replace r with x, and write

$$\int_\infty^0 \frac{1}{(1+r^3 e^{2\pi i})} dr = -\int_0^\infty \frac{1}{(1+x^3)} dx \tag{5.113}$$

Referring to fig. 5.12, we see that only the pole at $z = e^{i\pi/3}$ is enclosed by the contour. Therefore,

$$\oint \frac{1}{(1+z^3)} dz = \left(1 - e^{2\pi i/3}\right) \int_0^\infty \frac{1}{(1+x^3)} dx = 2\pi i a_{-1}\left(e^{i\pi/3}\right) \tag{5.114}$$

With

$$a_{-1}\left(e^{i\pi/3}\right)=\tfrac{1}{3}e^{-2\pi i/3} \tag{5.115a}$$

and

$$\left(1-e^{2\pi i/3}\right)=e^{i\pi/3}\left(e^{-i\pi/3}-e^{i\pi/3}\right)=-2i\,e^{i\pi/3}\sin\left(\frac{\pi}{3}\right)=-i\sqrt{3}\,e^{i\pi/3} \tag{5.115b}$$

we obtain

$$\int_0^\infty \frac{1}{(1+x^3)}dx=\frac{2\pi}{3\sqrt{3}} \tag{5.116}$$

□

Problems

Using Cauchy's residue theorem and the methods developed in this chapter (unless otherwise specified), evaluate the integrals given below.

1. $\displaystyle\int_{-\infty}^{\infty}\frac{1}{(1+x^4)}dx$

2. $\displaystyle\int_{-\infty}^{\infty}\frac{1}{(1+x^2)(1+x^4)}dx$

3. $\displaystyle\int_{-\infty}^{\infty}\frac{x^2}{(1+x^4)}dx$

4. $\displaystyle\int_{-\infty}^{\infty}\frac{x^2}{\left(1+x^2\right)^2}dx$

5. (a) $\displaystyle\int_{-\infty}^{\infty}\frac{e^{3ix}}{(1+x^2)}dx$

(b) Use the results of part (a) to evaluate

(i) $\int_{-\infty}^{\infty} \frac{\cos(3x)}{(1+x^2)} dx$ (ii) $\int_{-\infty}^{\infty} \frac{\sin(3x)}{(1+x^2)} dx$

6. (a) $\int_{-\infty}^{\infty} \frac{e^{-2ix}}{(1+x^2)} dx$

(b) Use the results of part (a) to evaluate

$\int_{-\infty}^{\infty} \frac{\cos^2 x}{(1+x^2)} dx$

7. (a) $\int_{-\infty}^{\infty} \frac{e^{2ix}}{(x-i)(x+2i)^2} dx$

(b) $\int_{-\infty}^{\infty} \frac{e^{-2ix}}{(x-i)(x+2i)^2} dx$

(c) Use the results of parts (a) and (b) to evaluate

(i) $\int_{-\infty}^{\infty} \frac{\cos(2x)}{(x-i)(x+2i)^2} dx$ (ii) $\int_{-\infty}^{\infty} \frac{\sin(2x)}{(x-i)(x+2i)^2} dx$

8. $\int_{-\infty}^{\infty} \frac{\sin(3x)}{(x-2i)^2(1+x^2)} dx$

9. (a) $\int_{-\infty}^{\infty} \frac{e^{ix}}{(x-i)(x+2i)} dx$

(b) Use the results of part (a) to evaluate

$\int_{-\infty}^{\infty} \frac{(x^2+2)\cos(x)+x\sin(x)}{(x^2+1)(x^2+4)} dx$

10. With α and β real and $\beta > \alpha$, use Cauchy's residue theorem to evaluate

$$\int_{-\infty}^{\infty} \frac{\sinh(\alpha x)}{\sinh(\beta x)} dx$$

11. Use Cauchy's residue theorem to evaluate

(a) $\displaystyle\int_0^{2\pi} \frac{1}{\left(\cos\theta + i\sqrt{3}\right)} d\theta$ (b) $\displaystyle\int_0^{2\pi} \frac{1}{\left(\sin\theta + i\sqrt{3}\right)} d\theta$

(c) $\displaystyle\int_0^{2\pi} \frac{1}{\left(\sin\theta + \frac{1}{2}(1-i)\cos\theta\right)} d\theta$ (d) $\displaystyle\int_0^{2\pi} \frac{1}{\left(1 + \frac{1}{2}(1-i)\tan\theta\right)} d\theta$

(e) $\displaystyle\int_0^{2\pi} \frac{\sin\theta}{(2 - \cos\theta)} d\theta$ (f) $\displaystyle\int_0^{2\pi} \frac{(2 - \sin\theta)}{(2 - \cos\theta)} d\theta$

(g) $\displaystyle\int_0^{2\pi} \frac{1}{(5 + 3\sin\theta)^2} d\theta$ (h) $\displaystyle\int_0^{\pi} \frac{\cos^2\theta}{(5 + 3\cos\theta)} d\theta$

Note: You must be certain there are no poles on the contour.

12. w is a real variable satisfying $1 < w < \infty$. Use Cauchy's residue theorem to find the function

(a) $\displaystyle F(w) \equiv \int_0^{2\pi} \frac{1}{(w - \cos\theta)^2} d\theta$ (b) $\displaystyle F(w) \equiv \int_0^{2\pi} \frac{1}{(w + \sin\theta)^2} d\theta$

Demonstrate that w cannot satisfy $0 < w < 1$.

13. It shown in prob. 31 of chapter 4, that for integer N

$$\oint z^N dz = \begin{cases} 2\pi i & N = -1 \\ 0 & N \neq -1 \end{cases}$$

With this result, use Cauchy's residue theorem to evaluate

(a) $\int_0^{2\pi} \cos^3\theta\, d\theta$ (b) $\int_0^{2\pi} \sin^4\theta\, d\theta$ (c) $\int_0^{2\pi} \sin^{2M}\theta\, d\theta$

(d) $\int_0^{2\pi} \sin^{2M+1}\theta\, d\theta$ (e) $\int_0^{2\pi} \cos^{2M}\theta\, d\theta$ (f) $\int_0^{2\pi} \cos^{2M+1}\theta\, d\theta$

where M is a positive integer

14. Use Cauchy's residue theorem to evaluate

$$\int_0^{2\pi} \frac{\cos^2(3\theta)-2}{5+4\cos(2\theta)}\, d\theta$$

15. By closing the contour in the lower half-plane, derive an expression for

$$P\int_{-\infty}^{\infty} \frac{f(x)}{(x-x_0)}\, dx$$

analogous to eq. 5.60, in terms of residues of poles in the lower half-plane and $f(x_0)$.

16. Evaluate

$$P\int_{-\infty}^{\infty} \frac{1}{(x^2+1)^2(x-2)}\, dx$$

(a) By the prescription given in eq. 5.60
(b) By displacing the pole from the real axis

17. Evaluate

$$P\int_{-\infty}^{\infty} \frac{1}{(x^3-1)}\, dx$$

(a) By the prescription given in eq. 5.60
(b) By displacing the pole from the real axis

18. With $k > 0$ a real constant, evaluate

$$P\int_{-\infty}^{\infty} \frac{e^{ikx}}{x}\,dx$$

   (a) By the prescription given in eq. 5.60.
   (b) By displacing the pole from the real axis.
   (c) Use the results of part (a) or (b) to determine

   (i) $P\int_{-\infty}^{\infty} \frac{\cos(x)}{x}\,dx$ (ii) $\int_{-\infty}^{\infty} \frac{\sin(x)}{x}\,dx$

   Explain why the integral in (ii) is not a principal value integral.

19. Generalize the results of ex. 5.7 by using Cauchy's residue theorem to evaluate

$$\int_{0}^{\infty} \frac{1}{(1+x^m)}\,dx$$

   with integer $m \geq 4$, using the contour shown in fig. P5.1.

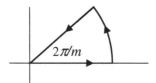

Figure P5.1
Contour for the integral of problem 19

20. With α real and $0 < \alpha < 1$, evaluate

$$\int_{-\infty}^{\infty} \frac{e^{\alpha x}}{\left(1+e^x\right)}\,dx$$

   by Cauchy's residue theorem, integrating around the rectangular contour shown in fig. P5.2.

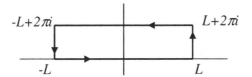

Figure P5.2

Contour for the integral of problem 20

21. Let t be real and positive. It is clear that on the semicircle of radius R in the left half of the z-plane shown in fig. P5.3a

$$\lim_{R \to \infty} e^{zt} = 0$$

and on the semicircle of radius R in the right half of the z-plane shown in fig. P5.3b

$$\lim_{R \to \infty} e^{zt} = \infty$$

Use this and Cauchy's residue theorem to evaluate

$$\int_{-i\infty}^{i\infty} \frac{e^{zt}}{(z+\alpha)} dz \qquad \mathrm{Re}(\alpha) > 0$$

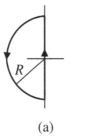
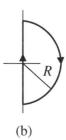

(a)   (b)

Figure P5.3

Large semicircular contours in the (a) left and (b) right half-planes

22. Let $w > 0$ be real. By integrating

$$\oint e^{-z^2} dz$$

around the contour of fig. P5.4, in the limit of infinite L, evaluate

$$\int_{-\infty}^{\infty} e^{-x^2}\cos(2wx)dx \quad\text{and}\quad \int_{-\infty}^{\infty} e^{-x^2}\sin(2wx)dx$$

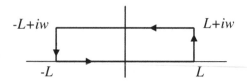

Figure P5.4
Contour for problem 22

Chapter 6

MULTIVALUED FUNCTIONS, BRANCH POINTS, AND CUTS

Because $e^{i2\pi} = 1$, it is straightforward to see that if the argument of a complex variable z is increased by 2π, one obtains the same value of the complex variable. That is, for a given r and θ, we write

$$z(r,\theta) = re^{i\theta} \tag{6.1a}$$

Then

$$z(r,\theta+2\pi) = re^{i\theta}e^{2i\pi} = re^{i\theta} = z(r,\theta) \tag{6.1b}$$

A function $F(z)$ that satisfies

$$F\left[z(r,\theta+2\pi)\right] = F\left[z(r,\theta)\right] \tag{6.2}$$

is called a *single-valued function*. We see from eqs. 6.1 that $F(z) = z$ is a single-valued function.

In chapter 2, it was shown that fractional roots of unity have more than one value. In ex. 3.2, it was indicated that the function $z^{3/2}$ is not analytic because it has multiple values at every point z. A function $F(z)$ that has more than one value at a given z is called a *multivalued function*. (Some authors do not refer to $F(z)$ as a function if it is multivalued. We do not adhere to that more rigorous terminology in this treatment. $F(z)$ is called a function even if it is multivalued.) As indicated in chapter 2, a function that has multiple values at a point is singular at that point.

6.1 NonInteger Power, Logarithm Functions

n^{th} power function

Consider

$$\left[z(r,\theta)\right]^n = r^n e^{in\theta} \tag{6.3a}$$

Increasing θ by 2π, this becomes

$$\left[z(r,\theta+2\pi)\right]^n = r^n e^{in\theta} e^{2\pi in} = e^{2\pi in}\left[z(r,\theta)\right]^n \tag{6.3b}$$

When n is an integer, $e^{2\pi in} = 1$ and z^n is single-valued. If n is not an integer, $e^{2\pi in} \neq 1$ and z^n has multiple values at z. If n is an irreducible rational fraction, z^n is called a *fractional root function*. If n is an irrational number, z^n is referred to as an *irrational power function*.

Example 6.1: Multiple values of the square root and cube root functions

(a) For $n = 1/2$,

$$\left[z(r,\theta)\right]^{1/2} = r^{1/2} e^{i\theta/2} \tag{6.4a}$$

Increasing θ by 2π, we obtain

$$\left[z(r,\theta+2\pi)\right]^{1/2} = r^{1/2} e^{i\theta/2} e^{2\pi i/2} = -\left[z(r,\theta)\right]^{1/2} \tag{6.4b}$$

A second increase by 2π, 4π in total, results in

$$\left[z(r,\theta+4\pi)\right]^{1/2} = r^{1/2} e^{i\theta/2} e^{4\pi i/2} = \left[z(r,\theta)\right]^{1/2} \tag{6.4c}$$

Thus, $F(z) = z^{1/2}$ has two values at any z and so is a *double-valued function*.
(b) With $n = 1/3$, the values of the cube root function are

$$\left[z(r,\theta)\right]^{1/3} = r^{1/3} e^{i\theta/3} \tag{6.5a}$$

$$\left[z(r,\theta+2\pi)\right]^{1/3} = r^{1/3} e^{i\theta/3} e^{2\pi i/3} = \frac{\left(-1+i\sqrt{3}\right)}{2}\left[z(r,\theta)\right]^{1/3} \tag{6.5b}$$

and

$$[z(r,\theta+4\pi)]^{1/3} = r^{1/3}e^{i\theta/3}e^{4\pi i/3} = \frac{(-1-i\sqrt{3})}{2}[z(r,\theta)]^{1/3} \qquad (6.5c)$$

Increasing θ by an additional 2π, a total of 6π, results in

$$[z(r,\theta+6\pi)]^{1/3} = r^{1/3}e^{i\theta/3}e^{6\pi i/3} = [z(r,\theta)]^{1/3} \qquad (6.5d)$$

Therefore, $z^{1/3}$ has three different values at any z, and is therefore a *triple-valued function*. □

General fractional root function

Using ex. 6.1 as a guide, we can deduce that when the power of z is an irreducible rational fraction, the fractional root function $F(z)$ has a finite number of values at a given z. The different values of $F(z)$ can be obtained by considering the function in the form $F[z(r,\theta+2\pi k)]$.

Let M and N be integers with $N \geq 2$ such that M/N is an irreducible rational fraction. At a given z, the multiple values of

$$[z(r,\theta)]^{M/N} = r^{M/N}e^{iM\theta/N} \qquad (6.6)$$

are found by varying the integer k in the expression

$$[z(r,\theta+2\pi k)]^{M/N} = r^{M/N}e^{i\theta M/N}e^{i2\pi k M/N} = e^{i2\pi k M/N}[z(r,\theta)]^{M/N} \qquad (6.7)$$

We note that because $e^{i2\pi NM/N} = e^{i2\pi M} = 1$,

$$[z(r,\theta+2\pi N)]^{M/N} = r^{M/N}e^{i\theta M/N} = [z(r,\theta)]^{M/N} \qquad (6.8)$$

Therefore, all values of $z^{M/N}$ at a given z are found from eq. 6.7 with a value of k in the range $0 \leq k \leq (N-1)$. Because there are N different values of k in this range, $z^{M/N}$ is an N-valued function.

We see from eq. 6.7 that the elements in the set of factors

$$\{e^{i2\pi k M/N}\} = \{1, e^{i2\pi M/N}, e^{i4\pi M/N}, \dots, e^{i2\pi M(N-1)/N}\} \qquad (6.9a)$$

define the different values of $z^{M/N}$ at a given value of z.

Except for the ordering of the elements, these factors are independent of M. To see this, we compare the elements of the set of eq. 6.9a to the set of factors

$$\left\{e^{i2\pi k/N}\right\} = \left\{1,\ e^{i2\pi/N},\ e^{i4\pi/N},\ ...,e^{i2\pi(N-1)/N}\right\} \tag{6.9b}$$

which define the different values of $z^{1/N}$. Because kM is an integer, we can write this product as

$$kM = \mu N + \lambda \tag{6.10}$$

with μ and λ integers and with $0 \le \lambda \le (N-1)$. For example, for $N = 3$ which restricts $0 \le k \le 2$, it is straightforward to see that $kM = 2 = 0 * N + 2$ and $kM = -8 = -3 * N + 1$. Therefore, in general

$$e^{i2\pi kM/N} = e^{i2\pi\mu}e^{i2\pi\lambda/N} = e^{i2\pi\lambda/N} \tag{6.11}$$

Because λ spans the same range of values as k,

$$\left\{e^{i2\pi kM/N}\right\} = \left\{e^{i2\pi\lambda/N}\right\} = \left\{e^{i2\pi k/N}\right\} \tag{6.12}$$

Thus, the elements in the set of values defining $z^{M/N}$ with $M \ne 1$ are identical to the elements in the set of values defining $z^{1/N}$ at a given z.

Example 6.2: Multiple values of several cube root functions

Setting $N = 3$, we consider the cube root functions for $M = 1, 2, 4$, and -4. From the discussion above, we see that such functions are triple-valued. Therefore, the multiple values of each function are found by considering $\arg(z) = \theta + 2\pi k$ with $k = 0, 1$, and 2. These values can be represented in terms of the following sets of factors.

$$\left\{\left[z(r,\theta+2\pi k)\right]^{1/3}\right\} = r^{1/3}e^{i\theta/3}\left\{1,\ e^{i2\pi/3},\ e^{i4\pi/3}\right\}$$

$$= r^{1/3}e^{i\theta/3}\left\{1,\ \tfrac{1}{2}\left(-1+i\sqrt{3}\right),\ \tfrac{1}{2}\left(-1-i\sqrt{3}\right)\right\} \tag{6.13a}$$

$$\left\{\left[z(r,\theta+2\pi k)\right]^{2/3}\right\} = r^{2/3}e^{i2\theta/3}\left\{1,\ e^{i4\pi/3},\ e^{i8\pi/3}\right\}$$

$$= r^{2/3}e^{i2\theta/3}\left\{1,\ \tfrac{1}{2}\left(-1-i\sqrt{3}\right),\ \tfrac{1}{2}\left(-1+i\sqrt{3}\right)\right\} \tag{6.13b}$$

$$\left\{ \left[z(r,\theta+2\pi k) \right]^{4/3} \right\} = r^{4/3} e^{i4\theta/3} \left\{ 1, \ e^{i8\pi/3}, \ e^{i16\pi/3} \right\}$$

$$= r^{4/3} e^{i4\theta/3} \left\{ 1, \ \tfrac{1}{2}\left(-1+i\sqrt{3} \right), \ \tfrac{1}{2}\left(-1-i\sqrt{3} \right) \right\} \qquad (6.13c)$$

$$\left\{ \left[z(r,\theta+2\pi k) \right]^{-4/3} \right\} = r^{-4/3} e^{i-4\theta/3} \left\{ 1, \ e^{-i8\pi/3}, \ e^{-i16\pi/3} \right\}$$

$$= r^{-4/3} e^{-i4\theta/3} \left\{ 1, \ \tfrac{1}{2}\left(-1-i\sqrt{3} \right), \ \tfrac{1}{2}\left(-1+i\sqrt{3} \right) \right\} \qquad (6.13d)$$

Thus, the elements in the sets of factors defining the multiple values of $z^{M/3}$ are the same for $M = 1, 2, 4,$ and -4. □

Irrational power function

Let α be an irrational number. The irrational power function z^{α} has analytic structure that is similar to the fractional root function. With

$$\left[z(r,\theta+2\pi k) \right]^{\alpha} = r^{\alpha} e^{i\alpha\theta} e^{i2\pi\alpha k} \qquad (6.14)$$

the set of factors describing the different values of z^{α} is

$$\left\{ e^{i2\pi\alpha k} \right\} = \left\{ 1, \ e^{i2\pi\alpha}, \ e^{i4\pi\alpha}, \ ... \right\} \qquad (6.15)$$

Because αk is an irrational number, $e^{i2\pi\alpha k} \neq 1$ for all k. Therefore, the set of factors defining the different values of z^{α} contains an infinite number of elements. Thus, the irrational power function is an infinite-valued function.

Logarithm function

Expressing z in polar form, we have

$$\ell n\left[z(r,\theta) \right] = \ell n(r) + i\theta \qquad (6.16)$$

Changing θ by $2\pi k$ results in

$$\ell n\left[z(r,\theta+2\pi k) \right] = \ell n(r) + i\theta + i2\pi k = \ell n\left[z(r,\theta) \right] + i2\pi k \qquad (6.17)$$

It is easy to see that there is no value of $k \neq 0$ for which $\ln[z(r, \theta + 2\pi k)]$ and $\ln[z(r, \theta)]$ are equal. Therefore, the set of factors for the logarithm function, defined by

$$\left\{\ell n\left[z(r, \theta + 2\pi k)\right] - \ell n\left[z(r, \theta)\right]\right\} = \left\{i2\pi k\right\}$$
$$= \left\{..., -i4\pi k, -i2\pi k, 0, i2\pi k, i4\pi k, ...\right\} \tag{6.18}$$

is an infinite set, and the logarithm function is an infinite-valued function.

6.2 Riemann Sheets, Branch Points, and Cuts

In 1851, Georg Riemann proposed the idea of viewing a complex plane comprised of a stack of complex sheets in order to distinguish between the different values of a multivalued function. These sheets are arranged such that the origins and the coordinate axes are aligned one on top of the other when viewed along a direction perpendicular to the complex plane. Such sheets are referred to as *Riemann sheets*.

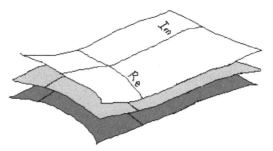

Figure 6.1
Multisheeted complex plane

Let $F(z)$ be a multivalued function and let k_m and k_n be two different integers. Then, for a given r and θ, $z(r, \theta + 2\pi k_m)$ and $z(r, \theta + 2\pi k_n)$ have the same value on all Riemann sheets. If

$$F\left[z(r, \theta + 2\pi k_m)\right] \neq F\left[z(r, \theta + 2\pi k_n)\right] \tag{6.19}$$

these two values of $F(z)$ are represented on two different sheets of a multi-sheeted complex plane.

For example, it was shown in eqs. 6.4 that the square root function has two values at a given value of z;

$$z^{1/2} = \pm r e^{i\theta} \tag{6.20}$$

Therefore, the square root function is defined uniquely in a complex plane comprised of two sheets. At points on one sheet, the function has values given by $+re^{i\theta}$ and on the other sheet, the values of the function are given by $-re^{i\theta}$. Similarly, the cube root function described in eqs. 6.5 has three distinct values at a given z and is therefore described in a three-sheeted complex plane, as is illustrated for $z^{1/3}$ in fig. 6.2.

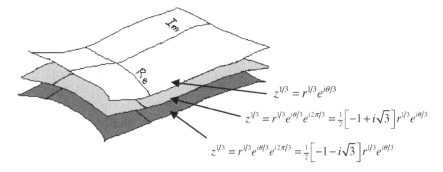

$$z^{1/3} = r^{1/3}e^{i\theta/3}$$

$$z^{1/3} = r^{1/3}e^{i\theta/3}e^{i2\pi/3} = \tfrac{1}{2}\left[-1+i\sqrt{3}\right]r^{1/3}e^{i\theta/3}$$

$$z^{1/3} = r^{1/3}e^{i\theta/3}e^{i2\pi/3} = \tfrac{1}{2}\left[-1-i\sqrt{3}\right]r^{1/3}e^{i\theta/3}$$

Figure 6.2

Values of $z^{1/3}$ in a three-sheeted complex plane

From these examples we see that a multivalued function can be viewed as being single-valued in a multisheeted complex plane.

Branch point

The *branch point* or *point of accumulation* is defined as the point with the smallest magnitude for which a function is multivalued. For example, as shown in eq. 6.7, the general N^{th} root function can be written as

$$z^{M/N} = \left[z(r,\theta+2\pi k)\right]^{M/N} = r^{M/N}e^{iM\theta/N}e^{i2\pi k M/N} \tag{6.21}$$

We see that this function has multiple values for all $0 < r < \infty$. That is, the multivaluedness starts at $r = 0$, and therefore at $z = 0$. As such, the general N^{th} root function is said to have a branch point at $z = 0$. The same argument holds for the irrational power function.

Writing

$$\ell n(z) = \ell n(r) + i\theta + i2\pi k \tag{6.22}$$

we see that the logarithm function is multivalued for $0 < r < \infty$, and so also has a branch point at $z = 0$.

The examples of multivalued functions that have been discussed so far have branch points at the origin. By replacing z by $z - z_0$, the branch point can be translated to the point z_0. Therefore, for example,

$$F_{M/N}(z) = (z - z_0)^{M/N} \tag{6.23}$$

has a fractional root branch point at z_0 and

$$F_{\ell n}(z) = \ell n(z - z_0) \tag{6.24}$$

has a logarithm branch point at z_0.

Branch cut

Let $F(z)$ be a multivalued function with a branch point at z_0. We let θ increase so that z varies from $z[r, \theta + 2\pi k]$ to $z[r, \theta + 2\pi(k + 1)]$. In doing so, values of $F(z)$ migrate from the kth sheet, defined by $F\{z[r, \theta + 2\pi k]\}$, to the $(k + 1)^{th}$ sheet, defined by $F\{z[r, \theta + 2\pi(k + 1)]\}$. In order for these values of $F(z)$ to vary continuously, we envision the kth sheet to be cut along some line called the *branch line* or *branch cut*, which extends from the branch point to ∞. This branch cut allows access to the $(k + 1)^{th}$ sheet from the kth sheet.

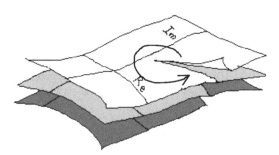

Figure 6.3
Cut in a Riemann sheet permitting continuous access to an adjacent sheet

We note from the discussion above that the increase of θ by 2π can begin at any value of θ. Therefore, the cut can be oriented at any angle θ_0 to the positive real axis. All sheets are cut in this way to allow access from points on any one sheet to points on any adjacent sheet. The sheet defined by $k = 0$, for which $\theta_0 < \theta < \theta_0 + 2\pi$, is called the *principal sheet* or the *principal branch* of $F(z)$. The second sheet is defined by $\theta_0 + 2\pi < \theta < \theta_0 + 4\pi$, and so on.

Construction of a physical model of a
multisheeted complex plane

The reader may be able to visualize this idea more clearly by constructing a physical model of a multisheeted complex plane. To do this, take a stack of papers with coordinate axes drawn on each sheet. Three or four sheets should be enough. Sheets of different colors may help in this visualization.

- Arrange the papers so that the axes on the different sheets are aligned with each other and the origins of the different sheets lie one on top of the other. This can be done easily by drawing each real axis parallel to one edge and each imaginary axis parallel to an adjacent edge on each piece of paper. Place the origin at the center of each sheet. Then align the sheets in the stack.
- Starting from one edge, cut the stack of sheets along any arbitrary line, extending the cut to some point in from the edge. Make sure that the alignment of the axes and origins is maintained.
- Lift the section of the top sheet that is above the cut, then tape the section below the cut in the first sheet to the section of the second sheet that is above the cut.
- Tape the part of the second sheet that is below the cut in the second sheet to the part of the third sheet that is above the cut in the third sheet.
- Continue this process until all the sheets have been taped together, the section below the cut in the kth sheet being taped to the section above the cut in the $(k+1)^{\text{th}}$ sheet.

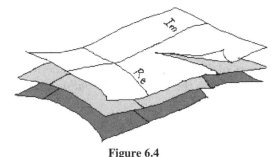

Figure 6.4

Physical model of a multisheeted complex plane

To represent a multisheeted complex plane with a finite number of sheets such as is required for the fractional root function, one must imagine that the section below the cut in the last sheet is taped to the section above the cut in the first sheet. This is not possible physically but is a valid mathematical construction. To represent an infinitely sheeted complex plane such as that required to describe the logarithm or irrational power function, one must imagine a model containing an infinite number of sheets of paper, all connected as described above.

Discontinuity across the cut

On each Riemann sheet, we define *the top of the cut* or *the region above the cut* to be the side of the cut at which we would have to start so that by increasing the argument of z by 2π in the counterclockwise direction, we would arrive at a point on the opposite side of the cut across from the starting point. This second side of the cut is referred to as *the bottom of the cut* or *the region below the cut*.

Consider a multisheeted complex plane defined by a multivalued function $F(z)$. We take the cut to extend from a branch point at z_0 to ∞ along a line oriented at an angle θ_0 to the positive real axis. The values of $F(z)$ on the two sides of the cut are shown in fig. 6.5.

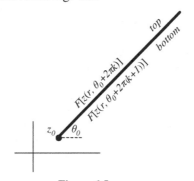

Figure 6.5

Top and bottom of a cut in the kth sheet for a multivalued function $F(z)$

Because the values of $F(z)$ are not the same on the two sides of the cut, $F(z)$ is not continuous across the cut. The *discontinuity across the cut* in the kth sheet is defined by

$$\Delta_k(z) \equiv F_{top}(z) - F_{bottom}(z)$$
$$= F\left[z\left(r, \, \theta_0 + 2\pi k\right)\right] - F\left[z\left(r, \, \theta_0 + 2\pi(k+1)\right)\right] \neq 0 \tag{6.25}$$

Clearly the discontinuity across the cut in the principal sheet is

$$\Delta_0(z) = F\left[z\left(r, \theta_0\right)\right] - F\left[z\left(r, \theta_0 + 2\pi\right)\right] \tag{6.26}$$

Because $F(z)$ is discontinuous everywhere across the cut, the cut is also referred to as a *line of discontinuity*.

6.3 Branch Structure

Let $F(z)$ be a multivalued function which is single-valued in a multisheeted complex plane. The *branch structure* or *cut structure* of $F(z)$ is determined by

- Specifying the position of the branch point(s)
- Specifying the orientation chosen for the cut(s) associated with the branch point(s)
- Identifying the top and bottom of each cut
- Determining the values of the function at points infinitesimally above and below the cut(s), and from these values, the discontinuity of the function across the cut(s)

Unless otherwise specified, all analysis of branch structure refers to the structure of the function on the principal sheet. Much of the analysis is accomplished using diagrams such as the one shown in fig. 6.6.

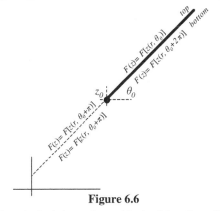

Figure 6.6

Cut structure on the principal sheet of a multivalued function with one branch point

Referring to fig. 6.6, we see that the argument of each point z along the top of the cut is $\theta = \theta_0$. Points along the bottom of the cut have argument $\theta = \theta_0 + 2\pi$. Points along the extension of the cut into the region defined by $|z| < |z_0|$ are along the dashed line. The argument of these points is $\theta = \theta_0 + \pi$.

In many applications, the branch point is on the real axis at a point $z_0 = x_0$, and the cut associated with that branch point is chosen to extend along the real axis to either $+\infty$ or to $-\infty$. Then, with

$$|x - x_0| = (x_0 - x) \qquad x < x_0 \tag{6.27a}$$

$$|x - x_0| = (x - x_0) \qquad x > x_0 \tag{6.27b}$$

the cut structure is that shown in fig. 6.7a.

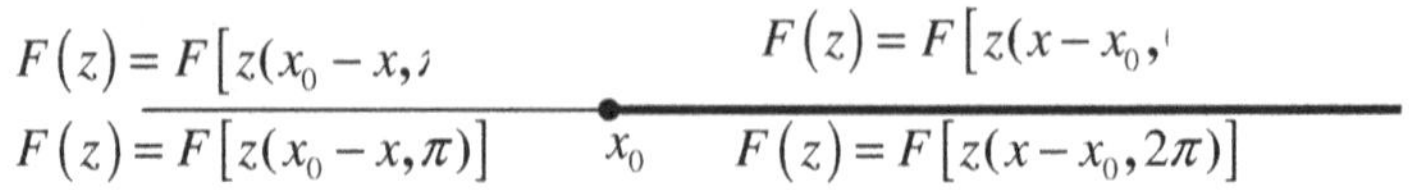

$$F(z) = F\left[z(x_0 - x, i\right. \qquad\qquad F(z) = F\left[z(x - x_0, i\right.$$

───────────────────────────────────────

$$F(z) = F\left[z(x_0 - x, \pi)\right] \qquad x_0 \qquad F(z) = F\left[z(x - x_0, 2\pi)\right]$$

Figure 6.7a

Cut structure for a cut along the real axis to $+\infty$

If the cut is taken to extend from x_0 to $+\infty$, as shown in fig. 6.7a, we see that to get from one side of the cut to the other by a counterclockwise rotation of 2π, one must start at a point z infinitesimally above the real axis with $\mathrm{Re}(z) = x > x_0$. A counterclockwise rotation of 2π accesses a point in the region defined by $x > x_0$ infinitesimally below the real axis. That is, the top and bottom of the cut are infinitesimally above and below the real axis, respectively. The values of the multivalued function just above and just below the real axis in the regions $x < x_0$ and $x > x_0$ are those shown in fig. 6.7a. The discontinuity across the cut in the region $x > x_0$ is given by

$$\Delta(z) = F\left[z(x - x_0, 0)\right] - F\left[z(x - x_0, 2\pi)\right] \tag{6.28a}$$

Because $F(z)$ is continuous across the real axis in the region $x < x_0$, the function at a point x on the real axis in this region has the value

$$F(z) = F\left[z(x_0 - x, \pi)\right] \tag{6.28b}$$

independent of whether the point is accessed from above or below the real axis.

When the cut associated with the branch point at x_0 is taken to extend to $-\infty$ as shown in fig. 6.7b, $F(z)$ is continuous across the real axis for $x > x_0$.

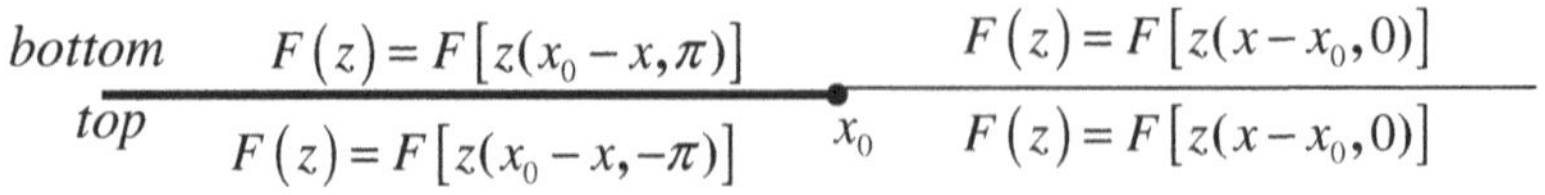

bottom $\qquad F(z) = F\left[z(x_0 - x, \pi)\right] \qquad\qquad F(z) = F\left[z(x - x_0, 0)\right]$

───────────────────────────────────────

top $\qquad F(z) = F\left[z(x_0 - x, -\pi)\right] \qquad x_0 \qquad F(z) = F\left[z(x - x_0, 0)\right]$

Figure 6.7b

Cut structure for a cut along the real axis to $-\infty$

To get from one side of the cut to the other by a counterclockwise rotation of 2π, we must start at a point in the region $x < x_0$ infinitesimally below the real axis so that a counterclockwise rotation of 2π accesses a point that is infinitesimally above the real axis in the region $x < x_0$. Thus, the top and bottom of the cut are infinitesimally below and above the real axis, respectively. The values of the multivalued function just above and just below the cut are those shown in fig. 6.7b. The discontinuity across the cut is given by

$$\Delta(z) = F\left[z(x_0 - x, -\pi)\right] - F\left[z(x_0 - x, \pi)\right] \qquad (6.29a)$$

The value of the function at a point on the real axis in the region $x > x_0$ is

$$F(z) = F\left[z(x - x_0, 0)\right] \qquad (6.29b)$$

independent of whether the point is accessed from above or below the real axis in this region.

Points on different sheets

Let $F(z)$ be a multivalued function with a branch point at z_0. In order to simplify the discussion, we take z_0 to be zero (or transform to another complex plane by replacing $z - z_0$ by z, which translates the branch point to the origin). We take the associated cut to extend from the origin to ∞ along a line that is oriented at an angle θ_0 with the positive real axis. The principal sheet is defined by $\theta_0 \leq \theta \leq \theta_0 + 2\pi$ and, the second sheet is defined by $\theta_0 + 2\pi \leq \theta \leq \theta_0 + 4\pi$.
Let

$$z_1 = r_1 e^{i\theta_1} \qquad (6.30a)$$

and

$$z_2 = r_2 e^{i\theta_2} \qquad (6.30b)$$

be points on the principal sheet. Then θ_1 and θ_2 are each in the range $[\theta_0, \theta_0 + 2\pi]$. If $\theta_1 + \theta_2 > \theta_0 + 2\pi$, the point defined by

$$z = z_1 z_2 = r_1 r_2 e^{i(\theta_1 + \theta_2)} \qquad (6.31)$$

is not on the principal sheet.

Example 6.3: Argand diagram with multiple sheets

Let the points

$$z_1 = i = e^{i\pi/2} \tag{6.32a}$$

and

$$z_2 = -1 + i = \sqrt{2}e^{i3\pi/4} \tag{6.32b}$$

be two points on the principal sheet of a multisheeted Argand diagram that describes a multivalued function $F(z)$ with a branch point at the origin. We consider the two cases in which the associated cut extends from the origin to $+\infty$ along the positive real axis, and from the origin to $-\infty$ along the negative real axis.

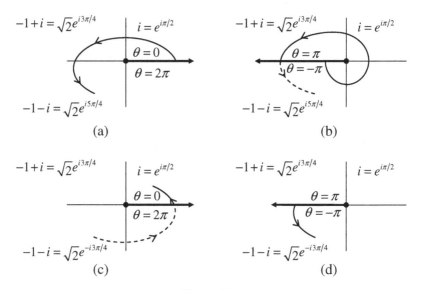

Figure 6.8
Complex numbers i and $-1 + i$ on the principal sheet
of two possible multisheeted complex planes

We see that both $\pi/2$ and $3\pi/4$ are within $[0, 2\pi]$, the range defining the principal sheet for the cut of fig. 6.8a, and $[-\pi, \pi]$ which defines the principal sheet for the cut shown in fig. 6.8b. Therefore, whichever cut we choose, both points are on the principal sheet.

The product of the points, as represented in eqs. 6.32, is given by

$$z_1 z_2 = i(-1+i) = -1-i = \sqrt{2}\, e^{i5\pi/4} \tag{6.33}$$

When $-1 - i$ is the complex number with an argument of $5\pi/4$, it is in the range $[0,2\pi]$ and is therefore a point on the principal sheet when the cut extends to $+\infty$ as shown in fig. 6.8a. Because $5\pi/4 > \pi$, the point $-1 - i$ of eq. 6.33 is on the second sheet when the cut is chosen to extend to $-\infty$ as in fig. 6.8b.

Another representation of $-1 - i$ is

$$-1-i = \sqrt{2}\, e^{-i3\pi/4} \tag{6.34}$$

which is not the product of z_1 and z_2. Because the argument $-3\pi/4 < 0$, this representation of $-1 - i$ is on the sheet just below the principal sheet (the -1th sheet, defined by $-2\pi < \theta < 0$) when the cut is taken to extend to $+\infty$. Because $-\pi \le -3\pi/4 \le \pi$, this representation of $-1 - i$ is on the principal sheet when the cut is taken to extend to $-\infty$. $\square$

Cut structure for the fractional root function

The general fractional root function

$$(z - z_0)^{M/N} = |z - z_0|^{M/N}\, e^{i\theta_0 M/N} \equiv r^{M/N} e^{i\theta_0 M/N} \tag{6.35}$$

has a branch point at z_0. A possible cut structure for this function is shown in fig. 6.9. As can be seen, along the top of the cut,

$$\arg(z - z_0) = \theta_0 \tag{6.36a}$$

and along the bottom of the cut,

$$\arg(z - z_0) = \theta_0 + 2\pi \tag{6.36b}$$

Therefore,

$$\left[(z - z_0)^{M/N} \right]_{top} = r^{M/N} e^{iM\theta_0/N} \tag{6.37a}$$

and

$$\left[(z - z_0)^{M/N} \right]_{bottom} = r^{M/N} e^{iM\,\theta_0/N} e^{i2\pi M/N} \tag{6.37b}$$

Thus, the discontinuity across the cut is given by

$$\Delta(z) = \left[(z - z_0)^{M/N} \right]_{top} - \left[(z - z_0)^{M/N} \right]_{bottom}$$

$$= \left(1 - e^{i2\pi M/N} \right) r^{M/N} e^{i\theta_0 M/N} = -2i e^{i\pi M/N} r^{M/N} e^{i\theta_0 M/N} \sin\left(\frac{M\pi}{N} \right) \tag{6.38}$$

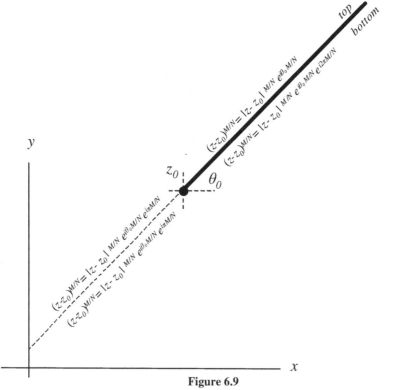

Figure 6.9
General cut structure for the fractional root function

Consider the case when the branch point is on the real axis at $z_0 = x_0$. Setting $\theta_0 = 0$, the cut extends along the axis from x_0 to $+\infty$ so the principal sheet is defined by $0 \le \theta \le 2\pi$. The branch point at x_0 divides the real axis into two regions. Region I is defined by $x < x_0$ and region II is described by $x > x_0$.

- **Region I**
  In this region, the fractional root function is continuous across the real axis. Referring to fig. 6.9, with $\theta_0 = 0$, points along the real axis are defined by $\arg(z - x_0) = \pi$. Therefore, at all $x < x_0$,

$$(z - x_0)^{M/N} = (x_0 - x)^{M/N} e^{iM\pi/N} \tag{6.39}$$

  Thus, the discontinuity across the real axis is

$$\Delta_I(x) = 0 \tag{6.40}$$

- **Region II**
  In this region, the values of the fractional root function along the top of the cut are different from those on the bottom of the cut. Along the top of the cut, $\arg(z - x_0) = 0$, so

$$\left[(z - x_0)^{M/N} \right]_{top} = (x - x_0)^{M/N} \tag{6.41a}$$

  Along the bottom of the cut, $\arg(z - x_0) = 2\pi$. Thus,

$$\left[(z - x_0)^{M/N} \right]_{bottom} = (x - x_0)^{M/N} e^{i2\pi M/N} \tag{6.41b}$$

  Then, from eq. 6.38 with $\theta_0 = 0$, the discontinuity across this cut is given straightforwardly to be

$$\begin{aligned}
\Delta_{II}(x) &= \left[(z - x_0)^{M/N} \right]_{top} - \left[(z - x_0)^{M/N} \right]_{bottom} \\
&= (x - x_0)^{M/N} e^{i\pi M/N} \left(e^{-i\pi M/N} - e^{i\pi M/N} \right) \\
&= -2i e^{i\pi M/N} (x - x_0)^{M/N} \sin\left(\frac{M\pi}{N} \right)
\end{aligned} \tag{6.42}$$

Therefore, the cut structure for the general fractional root function when the cut is taken to extend to $+\infty$ along the real axis is that shown in fig. 6.10a.

$$(z - x_0)^{M/N} = (x_0 - x)^{M/N} e^{i\pi M/N} \qquad\qquad (z - x_0)^{M/N} = (x - x_0)^{M/N}$$

Figure 6.10a

Cut structure for $(z - x_0)^{M/N}$ when the principal branch is defined by $0 \le \theta \le 2\pi$

When the cut extends along the real axis to $-\infty$, the principal branch is defined by $-\pi \le \theta \le \pi$. If the cut of fig. 6.9 is extended to $-\infty$ along the real axis by taking $\theta_0 = \pi$, the argument of points along the top of the cut is π which requires the argument of points along the bottom of the cut to be 3π. With this choice, the principal sheet is defined by $\pi \le \theta \le 3\pi$. In order to define the principal sheet by $-\pi \le \theta \le \pi$, we must define $\theta_0 = -\pi$.

Again the real axis is divided into two regions by the branch point at x_0, with region I defined by $x < x_0$ and region II defined by $x > x_0$.

- Region I
  The cut extends into this region. The top of the cut is infinitesimally below the real axis, and points along the top of the cut are defined by $\arg(z - x_0) = -\pi$. Points along the bottom of the cut, just above the real axis, are described by $\arg(z - x_0) = \pi$. Therefore,

$$\left[(z - x_0)^{M/N} \right]_{top} = (x_0 - x)^{M/N} e^{-i\pi M/N} \tag{6.43a}$$

and

$$\left[(z - x_0)^{M/N} \right]_{bottom} = (x_0 - x)^{M/N} e^{i\pi M/N} \tag{6.43b}$$

Because $\theta_0 = -\pi$, the discontinuity across the cut is found from eq. 6.38 to be

$$\Delta_I(x) = -2i(x_0 - x)^{M/N} \sin\left(\frac{M\pi}{N} \right) \tag{6.44}$$

- Region II
  Points in this region are defined by $\arg(z - x_0) = 0$. Because there is no cut in this region, the fractional root function is continuous across the real axis, with values given by

$$(z - x_0)^{M/N} = (x - x_0)^{M/N} \tag{6.45}$$

and

$$\Delta_{II}(x) = 0 \tag{6.46}$$

Combining these results, the cut structure is that shown in fig. 6.10b.

$$\left(z-x_0\right)^{M/N} = \left(x_0-x\right)^{M/N} e^{i\pi M/N} \qquad\qquad \left(z-x_0\right)^{M/N} = \left(x-x_0\right)^{M/N}$$

$$x_0$$

Figure 6.10b

Cut structure for $(z-x_0)^{M/N}$ when the principal branch is defined by $-\pi \le \theta \le \pi$

Cut structure for the logarithm function

The cut structure for the general logarithm function

$$F_{\ell n}(z) = \ell n(z-z_0) \qquad\qquad (6.47)$$

is straightforward to deduce using the analysis presented for the fractional root function. The cut structure of the logarithm function is shown in fig. 6.11 on the following page.

With $z_0 = x_0$ and $\theta_0 = 0$, the cut associated with the branch point at x_0 extends to $+\infty$ along the real axis and the real axis is divided into two regions by the branch point with region I described by $x < x_0$, and region II defined by $x > x_0$.

- Region I
  In this region, points on the real axis are described by $\arg(z - x_0) = \pi$. Because the cut does not extend into region I, the function is continuous across the real axis. Therefore, on the real axis,

$$\ell n(z-x_0) = \ell n\left(x_0-x\right)+i\pi \qquad\qquad (6.48)$$

and

$$\Delta_1(x) = 0 \qquad\qquad (6.49)$$

- Region II
  Because the top of the cut is just above the real axis and the bottom of the cut is just below the real axis, points along the top of the cut are

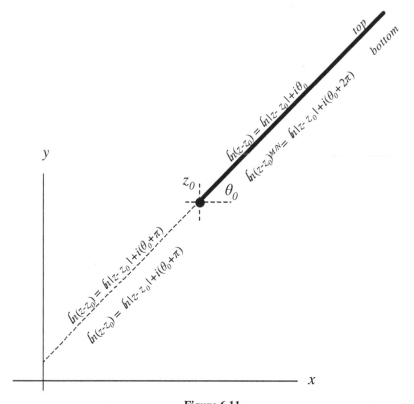

Figure 6.11
General cut structure for the logarithm function

defined by arg$(z - x_0) = 0$, and those along the bottom of the cut are described by arg$(z - x_0) = 2\pi$. Therefore,

$$\left[\ell n(z - x_0)\right]_{top} = \ell n(x - x_0) \tag{6.50a}$$

and

$$\left[\ell n(z - x_0)\right]_{bottom} = \ell n(x - x_0) + 2\pi i \tag{6.50b}$$

Then the discontinuity across the cut is given by

$$\Delta_{II}(x) = \left[\ell n(z - x_0)\right]_{top} - \left[\ell n(z - x_0)\right]_{bottom}$$
$$= \left(\ell n(x - x_0)\right) - \left(\ell n(x - x_0) + 2\pi i\right) = -2\pi i \tag{6.51}$$

This structure is shown in fig. 6.12a.

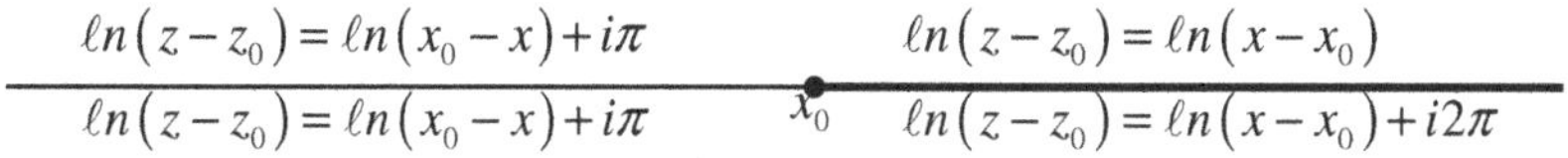

$$\ell n\left(z-z_0\right)=\ell n\left(x_0-x\right)+i\pi \qquad\qquad \ell n\left(z-z_0\right)=\ell n\left(x-x_0\right)$$
$$\overline{\ell n\left(z-z_0\right)=\ell n\left(x_0-x\right)+i\pi} \quad \overset{\bullet}{x_0} \quad \ell n\left(z-z_0\right)=\ell n\left(x-x_0\right)+i2\pi$$

Figure 6.12a

Cut structure for ℓn $(z-x_0)$ with the cut extending to $+\infty$ along the real axis

When the cut is chosen to extend to $-\infty$ along the real axis, the top and bottom of the cut are just below and just above the real axis, respectively.

- Region I
  Points along the top of the cut are defined by $\arg(z-x_0)=-\pi$. Along the bottom of the cut, points are described by $\arg(z-x_0)=\pi$. Therefore,

$$\left[\ell n(z-x_0)\right]_{top} = \ell n\left(x_0-x\right)-i\pi \tag{6.52a}$$

and

$$\left[\ell n(z-x_0)\right]_{bottom} = \ell n\left(x_0-x\right)+i\pi \tag{6.52b}$$

from which the discontinuity across the cut is given by

$$\Delta_1(x) = \left[\ell n(z-x_0)\right]_{top} - \left[\ell n(z-x_0)\right]_{bottom}$$
$$= \left(\ell n(x_0-x)-i\pi\right)-\left(\ell n(x_0-x)+\pi i\right) = -2\pi i \tag{6.53}$$

- Region II
  In this region, the logarithm function is continuous across the real axis, with values at points on the real axis

$$\ell n(z-x_0) = \ell n(x-x_0) \tag{6.54}$$

Thus, the discontinuity across the real axis is

$$\Delta_{II}(x) = 0 \tag{6.55}$$

The structure for this choice of cut is shown in fig. 6.12b.

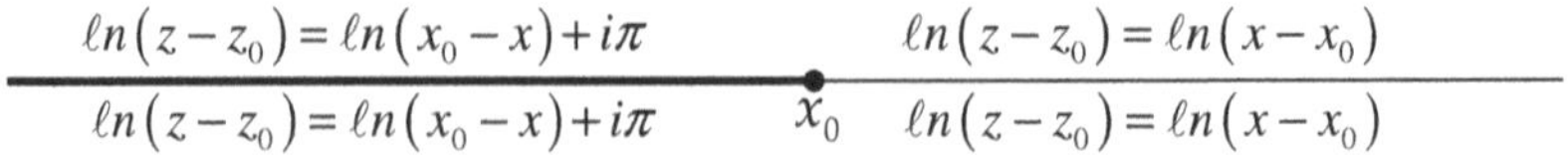

$$\begin{array}{c|c}
\ell n\left(z-z_{0}\right)=\ell n\left(x_{0}-x\right)+i\pi & \ell n\left(z-z_{0}\right)=\ell n\left(x-x_{0}\right) \\
\hline
\ell n\left(z-z_{0}\right)=\ell n\left(x_{0}-x\right)+i\pi \quad x_{0} & \ell n\left(z-z_{0}\right)=\ell n\left(x-x_{0}\right)
\end{array}$$

Figure 6.12b

Cut structure for ℓn $(z–x_0)$ with the cut to $-\infty$ along the real axis

6.4 Multiple Branch Points

When a function has more than one branch point, the orientation of the cut associated with any one branch point is independent of the orientation of any other cut.

Functions with two branch points

Let K, L, M, and N be integers so that K/L and M/N are irreducible fractions with L and N greater than 1. Then the function

$$F(z) = (z - z_1)^{K/L} (z - z_2)^{M/N} \tag{6.56}$$

has fractional root branch points at z_1 and z_2. To make the discussion less cumbersome without compromising the important details of the analysis, we take $K/L = M/N$ and take the branch points to be on the real axis at $z_1 = x_1$ and $z_2 = x_2$ with $x_1 < x_2$. Then

$$F(z) = \left[(z - x_1)(z - x_2)\right]^{M/N} \tag{6.57}$$

We follow the pattern of analysis presented for a function with one branch point. When a function has two branch points on the real axis, those branch points divide the axis into three regions. Region I is defined by points $x < x_1$, region II describes points in the range $x_1 < x < x_2$ and region III is defined by points $x > x_2$.

Thus, in region I

$$\left|x - x_1\right| = (x_1 - x) \tag{6.58a}$$

and

$$|x - x_2| = (x_2 - x) \tag{6.58b}$$

In region II

$$|x - x_1| = (x - x_1) \tag{6.58c}$$

and

$$|x - x_2| = (x_2 - x) \tag{6.58d}$$

and in region III

$$|x - x_1| = (x - x_1) \tag{6.58e}$$

and

$$|x - x_2| = (x - x_2) \tag{6.58f}$$

One common choice for the orientation of the two cuts is along the real axis in the configuration shown in fig. 6.13. By choosing the cut associated with x_1 to extend to $-\infty$, the principal branch for the factor $(z - x_1)^{MIN}$ is defined by $-\pi \le \theta \le \pi$. Similarly, because the cut associated with x_2 extends to $+\infty$, the principal branch for $(z - x_2)^{MIN}$ is defined by $0 \le \theta \le 2\pi$.

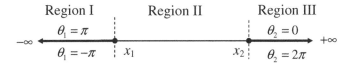

Figure 6.13
Cuts associated with two fractional root branch points

Referring to figs. 6.10, we deduce the values of $F(z)$ just above and just below the real axis in each region as follows:

- Region I
  Because the cut in this region is associated with x_1, $(z - x_2)^{MIN}$ is continuous across the real axis, but $(z - x_1)^{MIN}$ is not. Referring to fig. 6.10b and

eqs. 6.43, we see that the top of the cut is just below the real axis and at points along the top of this cut

$$(z - x_1)^{M/N} = (x_1 - x)^{M/N} e^{-i\pi M/N} \tag{6.59a}$$

Along the bottom of this cut, just above the real axis,

$$(z - x_1)^{M/N} = (x_1 - x)^{M/N} e^{i\pi M/N} \tag{6.59b}$$

Because $(z - x_2)^{M/N}$ is analytic in this region, we see from fig. 6.10a that

$$(z - x_2)^{M/N} = (x_2 - x)^{M/N} e^{i\pi M/N} \tag{6.59c}$$

Therefore, from eqs. 6.59a and 6.59c, we find

$$\left[\left[(z - x_1)(z - x_2)\right]^{M/N}\right]_{top} = \left[(x_1 - x)(x_2 - x)\right]^{M/N} \tag{6.60a}$$

and from eqs. 6.59b and 6.59c,

$$\left[\left[(z - x_1)(z - x_2)\right]^{M/N}\right]_{bottom} = \left[(x_1 - x)(x_2 - x)\right]^{M/N} e^{i2\pi M/N} \tag{6.60b}$$

Thus, the discontinuity across this cut is

$$\begin{aligned}
\Delta_1(x) &= \left[(x_1 - x)(x_2 - x)\right]^{M/N} \left(1 - e^{i2\pi M/N}\right) \\
&= -2i e^{i\pi M/N} \left[(x_1 - x)(x_2 - x)\right]^{M/N} \sin\left(\frac{\pi M}{N}\right)
\end{aligned} \tag{6.61}$$

- Region II
  Because neither cut extends into this region, both $(z - x_1)^{M/N}$ and $(z - x_2)^{M/N}$ are continuous across the real axis. Referring to fig. 6.10a, we see that at points on the real axis, the two factors can be expressed as

$$(z - x_1)^{M/N} = (x - x_1)^{M/N} \tag{6.62a}$$

and from fig. 6.10b we have

$$(z - x_2)^{M/N} = (x_2 - x)^{M/N} e^{i\pi M/N} \qquad (6.62b)$$

Therefore, at all points on the real axis in region II,

$$\left[(z - x_1)(z - x_2)\right]^{M/N} = \left[(x - x_1)(x_2 - x)\right]^{M/N} e^{i\pi M/N} \qquad (6.63)$$

and the discontinuity across the real axis is

$$\Delta_{II}(x) = 0 \qquad (6.64)$$

- Region III
  Because the cut in this region is associated with x_2, $(z - x_1)^{M/N}$ is continuous across the real axis, but $(z - x_2)^{M/N}$ is not. Therefore, referring to fig. 6.10b,

$$(z - x_1)^{M/N} = (x - x_1)^{M/N} \qquad (6.65a)$$

at all x in this region. From fig. 6.10a, we obtain

$$(z - x_2)^{M/N} = (x - x_2)^{M/N} \qquad (6.65b)$$

along the top of the cut, and along the bottom of the cut,

$$(z - x_2)^{M/N} = (x - x_2)^{M/N} e^{i2\pi M/N} \qquad (6.65c)$$

Therefore, in region III,

$$\left[\left[(z - x_1)(z - x_2)\right]^{M/N} \right]_{top} = \left[(x - x_1)(x - x_2)\right]^{M/N} \qquad (6.66a)$$

and

$$\left[\left[(z - x_1)(z - x_2)\right]^{M/N} \right]_{bottom} = \left[(x - x_1)(x - x_2)\right]^{M/N} e^{i2\pi M/N} \qquad (6.66b)$$

and the discontinuity across the real axis is given by

$$\Delta_{III}(x) = \left[(x - x_1)(x - x_2)\right]^{M/N} \left(1 - e^{i2\pi M/N}\right)$$
$$= -2i e^{i\pi M/N} \left[(x - x_1)(x - x_2)\right]^{M/N} \sin\left(\frac{\pi M}{N}\right) \qquad (6.67)$$

From this analysis, we deduce the cut structure of $F(z)$ to be that shown in fig. 6.14.

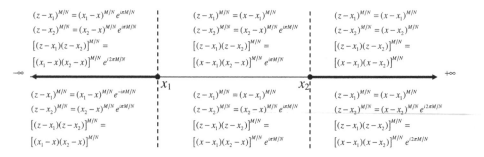

Figure 6.14

Analytic structure for a choice of cuts for a function with two fractional root branch points

Another common choice is to extend both cuts along the real axis to $+\infty$ as shown in fig. 6.15.

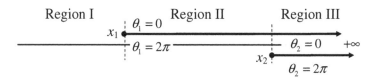

Figure 6.15

Choice of cuts associated with two fractional root branch points

For this orientation of the cuts, the values of the two factors of $F(z)$ in each of the regions are found from fig. 6.10a.

- Region I

 Because neither cut extends into region I, both factors are continuous across the real axis, and each factor has an argument of π. Referring to fig. 6.10a, we see that at points along the real axis

$$(z - x_1)^{M/N} = (x_1 - x)^{M/N} e^{i\pi M/N} \tag{6.68a}$$

 and

$$(z - x_2)^{M/N} = (x_2 - x)^{M/N} e^{i\pi M/N} \tag{6.68b}$$

 so that at all points on the real axis in region I

$$\left[(z - x_1)(z - x_2)\right]^{M/N} = \left[(x_1 - x)(x_2 - x)\right]^{M/N} e^{i2\pi M/N} \tag{6.68c}$$

and the discontinuity across the real axis is

$$\Delta_1(x) = 0 \tag{6.69}$$

- Region II
  The cut that extends into this region is associated with x_1. Therefore, the values of $(z - x_1)^{M/N}$ just above and just below the cut are given by

$$\left[(z-x_1)^{M/N}\right]_{top} = (x-x_1)^{M/N} \tag{6.70a}$$

and

$$\left[(z-x_1)^{M/N}\right]_{bottom} = (x-x_1)^{M/N} e^{i2\pi M/N} \tag{6.70b}$$

respectively. Because $(z - x_2)^{M/N}$ is analytic in region II, its value at all points on the real axis is given by

$$(z-x_2)^{M/N} = (x_2-x)^{M/N} e^{i\pi M/N} \tag{6.70c}$$

Therefore, in region II,

$$\left[\left[(z-x_1)(z-x_2)\right]^{M/N}\right]_{top} = \left[(x-x_1)(x_2-x)\right]^{M/N} e^{i\pi M/N} \tag{6.71a}$$

and

$$\left[\left[(z-x_1)(z-x_2)\right]^{M/N}\right]_{bottom} = \left[(x-x_1)(x_2-x)\right]^{M/N} e^{i3\pi M/N} \tag{6.71b}$$

from which the discontinuity across the cut is given by

$$\Delta_{II}(x) = \left[(x-x_1)(x_2-x)\right]^{M/N} \left(e^{i\pi M/N} - e^{i3\pi M/N}\right)$$
$$= -2ie^{i2\pi M/N} \left[(x-x_1)(x_2-x)\right]^{M/N} \sin\left(\frac{\pi M}{N}\right) \tag{6.72}$$

- **Region III**

  Both cuts extend into region III, so both factors are discontinuous across the real axis in this region. Again referring to fig. 6.10a, we have

$$\left[(z-x_1)^{M/N}\right]_{top} = (x-x_1)^{M/N} \tag{6.73a}$$

$$\left[(z-x_1)^{M/N}\right]_{bottom} = (x-x_1)^{M/N} e^{i2\pi M/N} \tag{6.73b}$$

$$\left[(z-x_2)^{M/N}\right]_{top} = (x-x_2)^{M/N} \tag{6.73c}$$

and

$$\left[(z-x_2)^{M/N}\right]_{bottom} = (x-x_2)^{M/N} e^{i2\pi M/N} \tag{6.73d}$$

From these, we obtain

$$\left[\left[(z-x_1)(z-x_2)\right]^{M/N}\right]_{top} = \left[(x-x_1)(x_2-x)\right]^{M/N} \tag{6.74a}$$

and

$$\left[\left[(z-x_1)(z-x_2)\right]^{M/N}\right]_{bottom} = \left[(x-x_1)(x_2-x)\right]^{M/N} e^{i4\pi M/N} \tag{6.74b}$$

Thus, the discontinuity across the real axis is

$$\begin{aligned}
\Delta_{\mathrm{III}}(x) &= \left[(x-x_1)(x_2-x)\right]^{M/N} \left(1 - e^{i4\pi M/N}\right) \\
&= -2i e^{i2\pi M/N} \left[(x-x_1)(x_2-x)\right]^{M/N} \sin\left(\frac{2\pi M}{N}\right)
\end{aligned} \tag{6.75}$$

The cut structure for this choice of orientation of the cuts is shown below in fig. 6.16.

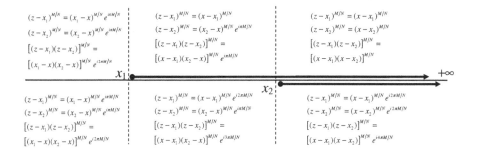

Figure 6.16

Analytic structure for a choice of cuts for a function with
two fractional root branch points

Example 6.4: Cut structure of a function with two square root branch points

Referring to the analysis of the general fractional root function given above, we consider the cut structure for a function with two square root branch points on the real axis. We take both cuts to extend to $+\infty$. This function is given by

$$F(z) = \left[(z - x_1)(z - x_2) \right]^{M/2} \tag{6.76}$$

We note that in order for $M/2$ to be an irreducible fraction, M must be an odd integer.

- Region I
  Because M is odd, we see from eq. 6.68c that

$$\begin{aligned}
\left[(z - x_1)(z - x_2) \right]^{M/2} &= \left[(x_1 - x)(x_2 - x) \right]^{M/2} e^{i2\pi M/2} \\
&= -\left[(x_1 - x)(x_2 - x) \right]^{M/2}
\end{aligned} \tag{6.77}$$

at all points on the real axis and the discontinuity across the real axis in this region is

$$\Delta_1(x) = 0 \tag{6.78}$$

- Region II
  From eqs. 6.76 and 6.77 we see that in region II,

$$\left[\left[(z-x_1)(z-x_2)\right]^{M/2}\right]_{top} = \left[(x-x_1)(x_2-x)\right]^{M/2} e^{i\pi M/2}$$

$$= i(-1)^{(M-1)/2}\left[(x-x_1)(x_2-x)\right]^{M/2} \tag{6.79a}$$

and

$$\left[\left[(z-x_1)(z-x_2)\right]^{M/2}\right]_{bottom} = \left[(x-x_1)(x_2-x)\right]^{M/2} e^{i3\pi M/2}$$

$$= -i(-1)^{(M-1)/2}\left[(x-x_1)(x_2-x)\right]^{M/2} \tag{6.79b}$$

Therefore, the discontinuity across the cut is given by

$$\Delta_{II}(x) = 2i(-1)^{(M-1)/2}\left[(x-x_1)(x_2-x)\right]^{M/2} \tag{6.80}$$

- Region III
  Referring to eqs. 6.74, the functional values along the top and bottom of the two cuts are

$$\left[\left[(z-x_1)(z-x_2)\right]^{M/2}\right]_{top} = \left[(x-x_1)(x-x_2)\right]^{M/2} \tag{6.81a}$$

and

$$\left[\left[(z-x_1)(z-x_2)\right]^{M/2}\right]_{bottom} = \left[(x-x_1)(x-x_2)\right]^{M/2} e^{i2\pi M}$$

$$= \left[(x-x_1)(x-x_2)\right]^{M/2} \tag{6.81b}$$

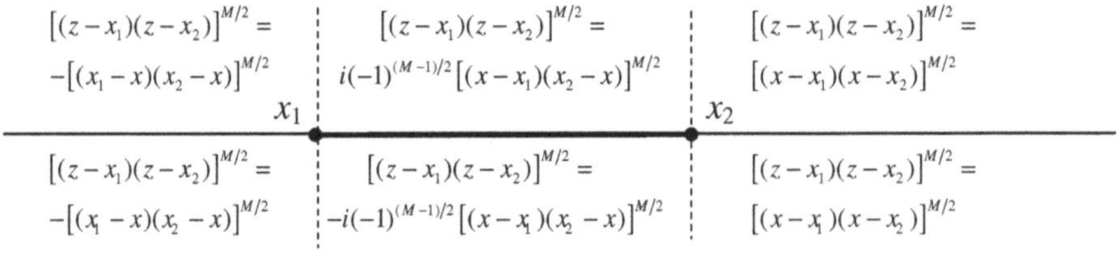

$[(z-x_1)(z-x_2)]^{M/2} =$ $[(z-x_1)(z-x_2)]^{M/2} =$ $[(z-x_1)(z-x_2)]^{M/2} =$

$-[(x_1-x)(x_2-x)]^{M/2}$ $i(-1)^{(M-1)/2}[(x-x_1)(x_2-x)]^{M/2}$ $[(x-x_1)(x-x_2)]^{M/2}$

x_1 x_2

$[(z-x_1)(z-x_2)]^{M/2} =$ $[(z-x_1)(z-x_2)]^{M/2} =$ $[(z-x_1)(z-x_2)]^{M/2} =$

$-[(x_1-x)(x_2-x)]^{M/2}$ $-i(-1)^{(M-1)/2}[(x-x_1)(x_2-x)]^{M/2}$ $[(x-x_1)(x-x_2)]^{M/2}$

Figure 6.17

A cut structure for the square root function with two branch
points, taking both cuts extending to $+\infty$

Thus, the discontinuity across the real axis in region III is

$$\Delta_{\mathrm{III}}(x) = 0 \tag{6.82}$$

It was shown in ex. 6.1a that when we start at a point on the principal sheet of $z^{1/2}$ and increase the argument of z by two factors of 2π (4π in all), two cuts are encountered. In doing so, we return to the value of $z^{1/2}$ on the principal sheet, which indicates that encountering two square root cuts is equivalent to encountering no cut. That is why the discontinuity across the real axis in region III is zero for the general square root function.

In prob. 6a of this chapter, the reader is asked to show if both cuts of the square root function extend to $-\infty$ along the real axis, having two cuts in region I is equivalent to having no cuts in that region and this choice of cuts also results in the cut structure shown in fig. 6.17. □

Example 6.5: Cut structure of a function with two logarithm branch points

The function

$$F(z) = \ell n \; \frac{z - x_1}{z - x_2} = \ell n(z - x_1) - \ell n(z - x_2) \tag{6.83}$$

has logarithm branch points on the real axis at $z = x_1$ and $z = x_2$. We take $x_1 < x_2$ and take the cuts to extend to $-\infty$ and $+\infty$ as in fig. 6.13. For this orientation of cuts, the analysis of the cut structure of $F(z)$ is as follows:

• Region I
  Because the cut in this region is associated with x_1, $\ell n\,(z{-}x_2)$ is continuous across the real axis and $\ell n\,(z{-}x_1)$ is not. Referring to figs. 6.12, the values of $\ln\,(z{-}x_1)$ along the top and bottom of this cut are

$$\left[\ell n(z - x_1)\right]_{top} = \ell n(x_1 - x) - i\pi \tag{6.84a}$$

and

$$\left[\ell n(z - x_1)\right]_{bottom} = \ell n(x_1 - x) + i\pi \tag{6.84b}$$

Because $\ell n\,(x - x_2)$ is continuous across the real axis, its value at all points on the real axis in this region is given by

$$\ell n(z - x_2) = \ell n(x_2 - x) + i\pi \tag{6.84c}$$

Therefore,

$$\ell n \left. \frac{z - x_1}{z - x_2} \right|_{top} = \ell n \frac{x_1 - x}{x_2 - x} - 2\pi i \qquad (6.85a)$$

and

$$\ell n \left. \frac{z - x_1}{z - x_2} \right|_{bottom} = \ell n \frac{x_1 - x}{x_2 - x} \qquad (6.85b)$$

from which the discontinuity across the cut is

$$\Delta_I(x) = -2\pi i \qquad (6.86)$$

- Region II
  Because neither cut extends into this region, both $\ell n\,(x - x_1)$ and $\ell n\,(x - x_2)$ are continuous across the real axis. Referring to figs. 6.12, we see that the values of the two logarithm functions at points on the real axis are given by

$$\ell n(z - x_1) = \ell n(x - x_1) \qquad (6.87a)$$

and

$$\ell n(z - x_2) = \ell n(x_2 - x) + i\pi \qquad (6.87b)$$

Therefore, at points on the real axis in this region

$$\ell n \frac{z - x_1}{z - x_2} = \ell n \frac{x - x_1}{x_2 - x} - i\pi \qquad (6.88)$$

and the discontinuity across the real axis is

$$\Delta_{II}(x) = 0 \qquad (6.89)$$

- Region III
  Because the cut in this region is associated with x_2, $\ell n\,(x - x_1)$ is continuous across the real axis, but $\ell n\,(x - x_2)$ is not. We see from figs. 6.12 that in this region

$$\ell n(z - x_1) = \ell n(x - x_1) \tag{6.90a}$$

$$\left[\ell n(z - x_2)\right]_{top} = \ell n(x - x_2) \tag{6.90b}$$

and

$$\left[\ell n(z - x_2)\right]_{bottom} = \ell n(x - x_2) + 2\pi i \tag{6.90c}$$

Therefore,

$$\ell n\,\frac{z - x_1}{z - x_2}\bigg|_{top} = \ell n\,\frac{x - x_1}{x - x_2} \tag{6.91a}$$

$$\ell n\,\frac{z - x_1}{z - x_2}\bigg|_{bottom} = \ell n\,\frac{x - x_1}{x - x_2} - 2\pi i \tag{6.91b}$$

and

$$\Delta_{III}(x) = 2\pi i \tag{6.92}$$

These results are shown in fig. 6.18 below.

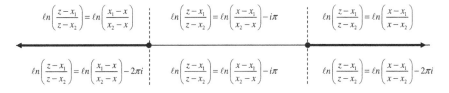

Figure 6.18

Cut structure for the logarithm function of eq. 6.83

in the three regions on the real axis

Another common choice for the orientation of the logarithm cuts is to extend both cuts to $+\infty$ along the real axis as shown in fig. 6.15. Again, the values of $\ell n\,(x - x_1)$ and $\ell n\,(x - x_2)$ in each region are deduced from the cut structure shown in figs. 6.12.

- Region I

  Because there is no cut in this region, both $\ell n\,(x - x_1)$ and $\ell n\,(x - x_2)$ are continuous across the real axis. The values of both logarithms at points on the real axis are given by

$$\ell n(z - x_1) = \ell n(x_1 - x) + i\pi \tag{6.93a}$$

and

$$\ell n(z - x_2) = \ell n(x_2 - x) + i\pi \tag{6.93b}$$

  Therefore, at points along the real axis in this region

$$\ell n\,\frac{z - x_1}{z - x_2} = \ell n\,\frac{x_1 - x}{x_2 - x} \tag{6.94}$$

  and the discontinuity across the real axis is

$$\Delta_1(x) = 0 \tag{6.95}$$

- Region II

  The cut in this region is associated with x_1. Therefore, we see from figs. 6.12 that

$$\left[\ell n(z - x_1)\right]_{top} = \ell n(x - x_1) \tag{6.96a}$$

$$\left[\ell n(z - x_1)\right]_{bottom} = \ell n(x - x_1) + 2\pi i \tag{6.96b}$$

and

$$\ell n(z - x_2) = \ell n(x_2 - x) + i\pi \tag{6.96c}$$

Therefore, and at all x in region II

$$\ell n \left. \frac{z-x_1}{z-x_2} \right|_{top} = \ell n \frac{x-x_1}{x_2-x} - i\pi \tag{6.97a}$$

and

$$\ell n \left. \frac{z-x_1}{z-x_2} \right|_{bottom} = \ell n \frac{x-x_1}{x_2-x} + i\pi \tag{6.97b}$$

from which

$$\Delta_{II}(x) = -2\pi i \tag{6.98}$$

• Region III
  Because both cuts extend into this region, we see from figs. 6.12 that

$$\left[\ell n(z-x_1) \right]_{top} = \ell n(x-x_1) \tag{6.99a}$$

$$\left[\ell n(z-x_2) \right]_{top} = \ell n(x-x_2) \tag{6.99b}$$

$$\left[\ell n(z-x_1) \right]_{bottom} = \ell n(x-x_1) + 2\pi i \tag{6.99c}$$

and

$$\left[\ell n(z-x_2) \right]_{bottom} = \ell n(x-x_2) + 2\pi i \tag{6.99d}$$

Therefore, at points on the real axis in region III

$$\ell n \left. \frac{z-x_1}{z-x_2} \right|_{top} = \ell n \frac{x-x_1}{x-x_2} \tag{6.100a}$$

$$\ell n \left. \frac{z-x_1}{z-x_2} \right|_{bottom} = \ell n \frac{x-x_1}{x-x_2} \tag{6.100b}$$

Thus, the discontinuity of the function across the real axis is

$$\Delta_{\text{III}}(x) = 0 \tag{6.101}$$

Figure 6.19 shows the cut structure of the logarithm function along the real axis in the three regions.

$\ell n\left(\frac{z-x_1}{z-x_2}\right) = \ell n\left(\frac{x_1-x}{x_2-x}\right)$
$\ell n\left(\frac{z-x_1}{z-x_2}\right) = \ell n\left(\frac{x-x_1}{x_2-x}\right) - i\pi$
$\ell n\left(\frac{z-x_1}{z-x_2}\right) = \ell n\left(\frac{x-x_1}{x-x_2}\right)$
$+\infty$
$\ell n\left(\frac{z-x_1}{z-x_2}\right) = \ell n\left(\frac{x_1-x}{x_2-x}\right)$
$\ell n\left(\frac{z-x_1}{z-x_2}\right) = \ell n\left(\frac{x-x_1}{x_2-x}\right) + i\pi$
$\ell n\left(\frac{z-x_1}{z-x_2}\right) = \ell n\left(\frac{x-x_1}{x-x_2}\right)$

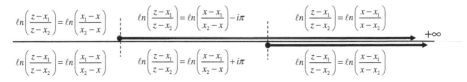

Figure 6.19
Cut structure of the logarithm function of eq. 6.83
in the three regions along the real axis

Because the discontinuity across the real axis in region III is zero, this cut structure for the logarithm function, shown in fig. 6.19, is equivalent to a finite cut extending from x_1 to x_2 as shown in fig. 6.20. □

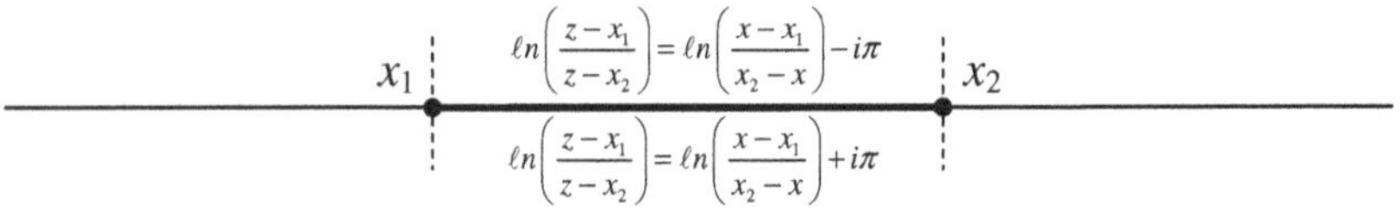

x_1
$\ell n\left(\frac{z-x_1}{z-x_2}\right) = \ell n\left(\frac{x-x_1}{x_2-x}\right) - i\pi$
x_2
$\ell n\left(\frac{z-x_1}{z-x_2}\right) = \ell n\left(\frac{x-x_1}{x_2-x}\right) + i\pi$

Figure 6.20
Cut structure equivalent to that of fig. 6.19

In prob. 7a of this chapter, the reader will show that if both cuts of the logarithm function extend to $-\infty$, having two cuts in region I is equivalent to having no cuts in that region and one again has the cut structure shown in fig. 6.20.

6.5 Evaluation of Integrals

Let $F(z)$ be a multivalued function that has a single branch point at z_0 and an associated cut that extends to ∞ along a line is oriented at an angle θ_0 to the real axis. Let $F(z)$ have poles at $z_1, \ldots, z_N$ none of which is on the cut. At all points along the cut with $|z| > |z_0|$ the discontinuity is given by

$$\Delta(z) = F\left[z(r,\theta_0)\right] - F\left[z(r,\theta_0 + 2\pi)\right] \tag{6.102}$$

where

$$r = |z - z_0|$$ (6.103)

We consider the integral in the counterclockwise direction around the closed contour shown in fig. 6.21.

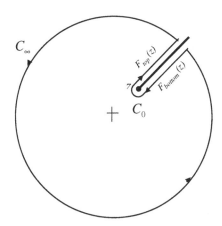

Figure 6.21

Contour for the integral involving a multivalued function

Points on C_∞ have a magnitude R that will be taken to infinity, and points on C_0 have magnitude ρ that approaches zero. Then, writing the closed contour integral as a sum of integrals over various segments, we have

$$\oint F(z)\,dz = \int_{z_0}^\infty F_{top}(z)\,dz + \lim_{R\to\infty} \int_{C_\infty} F(z)\,dz$$

$$+ \int_\infty^{z_0} F_{bottom}(z)\,dz + \lim_{\rho\to 0} \int_{C_0} F(z)\,dz$$ (6.104)

Taking the integrals around C_∞ and C_0 to be zero and inverting the limits on the integral along the bottom of the cut, we apply Cauchy's residue theorem to obtain

$$\oint F(z)\,dz = 2\pi i \sum_{k=1}^N a_{-1}(z_k)$$

$$= \int_{z_0}^\infty F_{top}(z)\,dz - \int_{z_0}^\infty F_{bottom}(z)\,dz = \int_{z_0}^\infty \Delta(z)\,dz$$ (6.105)

Specific examples

Many integrals that can be evaluated using the branch structure of a multivalued function are along a part of the real axis. With x_0 real, such integrals are of the form

$$I_+ = \int_{x_0}^{\infty} F(x)\,dx \tag{6.106a}$$

and

$$I_- = \int_{-\infty}^{x_0} F(x)\,dx \tag{6.106b}$$

To apply this approach to evaluating semi-infinite integrals, a multivalued function $F(z)$ must be found that has

- A branch point at x_0, the finite limit of the integral
- The cut associated with x_0 that extends to $+\infty$ (or $-\infty$)
- The discontinuity across the cut that is a constant multiple of $F(x)$

The approach is illustrated by several examples.

Example 6.6: Evaluating an integral using the cut structure of a logarithm function

It was shown in ex. 5.7 that

$$\int_0^{\infty} \frac{1}{(1+x^3)}\,dx = \frac{2\pi}{3\sqrt{3}} \tag{5.116}$$

To evaluate this integral using the properties of a multivalued function, we note that the lower limit of the integral is 0 and the path of integration is along the real axis to $+\infty$. Thus, we seek a multivalued function $F(z)$ that has

- A branch point at the origin
- An associated cut that can be extended to $+\infty$ along the positive real axis
- A discontinuity across that cut that is a constant multiple of $1/(1+x^3)$

It was shown earlier that the function $\ell n(z)$ has

- A branch point at the origin
- The associated cut that can be extended to $+\infty$ along the real axis
- The discontinuity across the cut that is $-2\pi i$

Therefore,

$$F(z) = \frac{\ell n(z)}{(1+z^3)} \tag{6.107}$$

has the required properties:
- A branch point at the origin
- The associated cut that can be extended to $+\infty$ along the real axis
- The discontinuity across the cut that is $-2\pi i/(1 + x^3)$

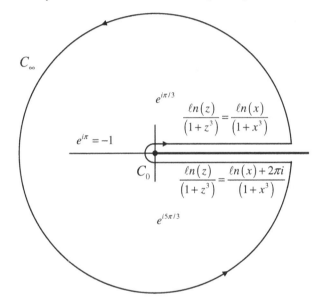

Figure 6.22

Analytic structure and contour for evaluating the integral of eq. 5.116

Therefore, Cauchy's theorem for the integral of the function of eq. 6.107 around the contour of fig. 6.22 is

$$\oint \frac{\ell n(z)}{(1+z^3)} dz = 2\pi i \sum_{k=1}^{3} a_{-1}(z_k)$$

$$= \int_0^\infty \frac{\ell n(x)}{(1+x^3)} dx + \int_C \frac{\ell n(z)}{(1+z^3)} dz + \int_\infty^0 \frac{\ell n(x)+2\pi i}{(1+x^3)} dx + \int_{C_0} \frac{\ell n(z)}{(1+z^3)} dz$$

$$\tag{6.108}$$

where $a_{-1}(z_k)$ is the residue of the kth pole of $\ell n\,(z)/(1 + z^3)$. These poles, shown in fig. 6.22, are on the principal sheet defined by $0 \le \theta \le 2\pi$.

We take the segments C_∞ and C_0 to be circles of very large (ultimately infinite) and very small (ultimately zero) radii R and ρ, respectively. Then, on C_∞

$$\frac{\ell n(z)}{\left(1+z^3\right)} = \frac{\ell n(R)+i\phi}{\left(1+R^3 e^{3i\phi}\right)} \simeq \frac{\ell n(R)}{R^3 e^{3i\phi}} \tag{6.109a}$$

and on C_0

$$\frac{\ell n(z)}{\left(1+z^3\right)} = \frac{\ell n(\rho)+i\phi}{\left(1+\rho^3 e^{3i\phi}\right)} \simeq \ell n(\rho) \tag{6.109b}$$

Therefore,

$$\lim_{R\to\infty} \int_{C_\infty} \frac{\ell n(z)}{(1+z^3)}\,dz = \lim_{R\to\infty} \frac{\ell n(R)}{R^2} \int_0^{2\pi} e^{-2i\phi}\,d\phi = 0 \tag{6.110a}$$

and

$$\lim_{\rho\to 0} \int_{C_0} \frac{\ell n(z)}{(1+z^3)}\,dz = \lim_{\rho\to 0} i\rho\,\ell n(\rho) \int_0^{2\pi} e^{i\phi}\,d\phi = 0 \tag{6.110b}$$

After inverting the limits on the integral along the bottom of the cut, eq. 6.108 becomes

$$-2\pi i \int_0^\infty \frac{1}{(1+x^3)}\,dx = 2\pi i \sum_{k=1}^3 a_{-1}(z_k) \tag{6.111a}$$

from which

$$\int_0^\infty \frac{1}{(1+x^3)}\,dx = -\left[a_{-1}\left(e^{i\pi/3}\right) + a_{-1}\left(e^{i\pi}\right) + a_{-1}\left(e^{i5\pi/3}\right) \right] \tag{6.111b}$$

Defining

$$P(z) = \ell n(z) \tag{6.112a}$$

and

$$Q(z) = (1 + z^3) \tag{6.112b}$$

we recognize that $P(z)$ is analytic at each z_k and $Q(z)$ is an entire function with a first-order zero at each z_k. Therefore, it is straightforward to find the residues using the ratio method given in eq. 4.121. The result is

$$a_{-1}(z_k) = \frac{P(z_k)}{Q'(z_k)} = \frac{\ell n(z_k)}{3z_k^2} \tag{6.113}$$

Then

$$a_{-1}\left(e^{i\pi/3}\right) = \frac{\pi}{18}\left(\sqrt{3} - i\right) \tag{6.114a}$$

$$a_{-1}\left(e^{i\pi}\right) = \frac{i\pi}{3} \tag{6.114b}$$

and

$$a_{-1}\left(e^{i5\pi/3}\right) = -\frac{5\pi}{18}\left(\sqrt{3} + i\right) \tag{6.114c}$$

Thus eq. 6.111 yields the expected result

$$\int_0^\infty \frac{1}{(1+x^3)} dx = \frac{2\pi}{3\sqrt{3}} \tag{5.121}$$
□

To evaluate integrals by the method under discussion, it is necessary to demonstrate that the integrals over the infinite and infinitesimal segments of the contour are zero. It is assumed that the reader will henceforth be able to determine whether these integrals are zero using the techniques employed in the above example. As such, the following examples involve functions for which the integrals over C_∞ and C_0 are zero, and it is assumed that the reader will verify this.

Example 6.7: Evaluating an integral using the cut structure of a logarithm function

By substituting $x = \tan \phi$, it is straightforward to show that

$$\int_1^\infty \frac{1}{(1+x^2)}\, dx = \frac{\pi}{4} \tag{6.115}$$

- We could substitute $x' = x-1$ and evaluate an integral over the interval $[0,\infty]$ using multivalued functions, as in ex. 6.6.

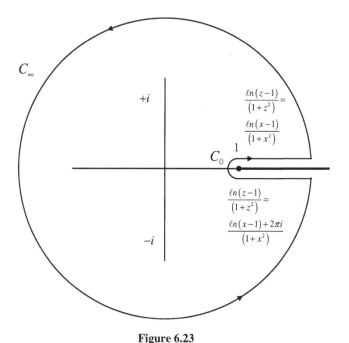

Figure 6.23
Contour and cut structure for the integral of eq. 6.115

To evaluate this integral using the properties of a multivalued function without such a substitution, we must find a function
- With a branch point at $z = 1$
- An associated cut that can be extended from 1 to $+\infty$ along the real axis
- With a discontinuity across the cut that is a constant multiple of $1/(1 + x^2)$

We again use the fact that the logarithm function has a constant discontinuity across the cut of $-2\pi i$ and consider

$$F(z) = \frac{\ell n(z-1)}{(1+z^2)} \tag{6.116}$$

which has

- A branch point at $z = 1$
- An associated cut that extends from 1 to $+\infty$ along the real axis
- A discontinuity across the cut of $-2\pi i/(1 + x^2)$

Then, with the contour of integration and the poles of $F(z)$ shown in fig. 6.23, we have

$$\oint \frac{\ell n(z-1)}{(1+z^2)}dz = 2\pi i\left[a_{-1}(i)+a_{-1}(-i)\right]$$

$$= \int_1^\infty \frac{\ell n(x-1)}{(1+x^2)}dx + \int_\infty^1 \frac{\ell n(x-1)+2\pi i}{(1+x^2)}dx = -2\pi i\int_1^\infty \frac{1}{(1+x^2)}dx \tag{6.117}$$

where the integrals along C_∞ and C_0 are zero and have been omitted.

In order to correctly express the polar form of $z - 1$ in the logarithm at $z = \pm i$, we refer to fig. 6.24, and write

$$z_1 = +i = 1 + \sqrt{2}\, e^{i3\pi/4} \tag{6.118a}$$

and

$$z_2 = -i = 1 + \sqrt{2}\, e^{i5\pi/4} \tag{6.118b}$$

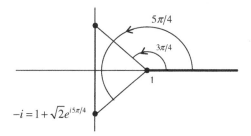

Figure 6.24

Poles of $\ell n \,(z-1)/(1+z^2)$

As before, we use the ratio method to determine the residues. We obtain

$$a_{-1}(z_k) = \frac{\ell n(z_k - 1)}{2z_k} \tag{6.119}$$

so that

$$a_{-1}(z_1) = \frac{1}{2i} \, \ell n\left(\sqrt{2}\right) + \frac{i3\pi}{4} \tag{6.120a}$$

and

$$a_{-1}(z_2) = \frac{1}{-2i} \, \ell n\left(\sqrt{2}\right) + \frac{i5\pi}{4} \tag{6.120b}$$

Substituting these into eq. 6.117, we obtain the correct result

$$\int_1^\infty \frac{1}{(1+x^2)} \, dx = \frac{\pi}{4} \tag{6.115}$$
□

Example 6.8: Evaluating an integral using the cut structure of a fractional root function

Let M and N be integers such that $N \geq 2$ and M/N is an irreducible fraction. To use Cauchy's residue theorem to evaluate

$$I = \int_0^\infty \frac{x^{M/N}}{(1+x^2)} \, dx \tag{6.121}$$

we determine a function with

- An Nth root branch point at the origin
- An associated cut that can be extended from 0 to $+\infty$
- A discontinuity across the cut that is a constant multiple of $x^{M/N}/(1 + x^2)$

It was discussed earlier that $(z - x_0)^{M/N}$ has these branch properties, with the discontinuity across the cut along the real axis from x_0 to $+\infty$ given by

$$\Delta(x) = -2ie^{i\pi M/N}(x - x_0)^{M/N} \sin\left(\frac{M\pi}{N}\right) \tag{6.42}$$

With this as a guide, we consider the function

$$F(z) = \frac{z^{M/N}}{(1+z^2)} \tag{6.122}$$

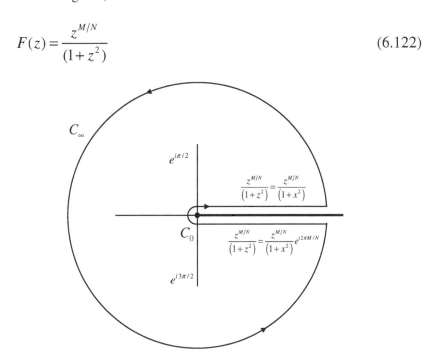

Figure 6.25

Contour and analytic structure for evaluating the integral of eq. 6.121

The analytic structure, along with the contour for evaluating the integral of eq. 6.121 is shown in fig. 6.25. Then

$$\oint \frac{z^{M/N}}{(1+z^2)} dz = 2\pi i \left[a_{-1}(+i) + a_{-1}(-i) \right]$$

$$= \int_0^\infty \frac{x^{M/N}}{(1+x^2)} dx + \int_{C_\infty} \frac{z^{M/N}}{(1+z^2)} dz + e^{i2\pi M/N} \int_\infty^0 \frac{x^{M/N}}{(1+x^2)} dx + \int_{C_0} \frac{z^{M/N}}{(1+z^2)} dz \tag{6.123}$$

The integral of eq. 6.121 is finite only for a specific range of M/N. To determine that range, consider the integral over the segments C_∞ and C_0. Using the analysis that led to eq. 6.109a, we see that for points on C_∞

$$\lim_{R \to \infty} \int_{C_\infty} \frac{z^{M/N}}{(1+z^2)} dz = \left[i \int_0^{2\pi} e^{i\left(\frac{M}{N}-1\right)\phi} d\phi \right] \lim_{R \to \infty} R^{\left(\frac{M}{N}-1\right)} \tag{6.124a}$$

To ensure that this limit is zero, we must require $M/N < 1$.

For points on C_0, analysis of the type that led to eq. 6.109b yields

$$\lim_{\rho \to 0} \int_{C_0} \frac{z^{M/N}}{(1+z^2)} dz = \left[i \int_0^{2\pi} e^{i\left(\frac{M}{N}+1\right)\phi} d\phi \right] \lim_{\rho \to 0} \rho^{\left(\frac{M}{N}+1\right)} \qquad (6.124b)$$

This limit is zero if $M/N > -1$. Therefore, the integral of eq. 6.121 is finite for

$$-1 < M/N < 1 \qquad (6.125)$$

Then the integrals on C_∞ and C_0 are zero and eq. 6.123 becomes

$$\oint \frac{z^{M/N}}{(1+z^2)} dz = 2\pi i \left[a_{-1}(i) + a_{-1}(-i) \right]$$

$$\left(1 - e^{i2\pi M/N} \right) \int_0^\infty \frac{x^{M/N}}{(1+x^2)} dx = -2i e^{i\pi M/N} \sin\left(\frac{M\pi}{N} \right) \int_0^\infty \frac{x^{M/N}}{(1+x^2)} dx$$

$$(6.126)$$

For the cut shown in fig. 6.25, the principal sheet is defined by $0 \le \theta \le 2\pi$. Requiring the poles to be on the principal sheet, $\pm i$ must be expressed as

$$i = e^{i\pi/2} \qquad (6.127a)$$

and

$$-i = e^{i3\pi/2} \qquad (6.127b)$$

To determine the residues of these poles, we again use the ratio method. We define

$$P(z) = z^{M/N} \qquad (6.128a)$$

and

$$Q(z) = (1 + z^2) \qquad (6.128b)$$

both of which are analytic at the poles. Therefore,

$$a_{-1}(z_k) = \frac{P(z_k)}{Q'(z_k)} = \frac{1}{2} z_k^{(\frac{M}{N}-1)}$$

(6.129)

from which

$$a_{-1}(i) = \frac{1}{2} e^{i(\frac{M}{N}-1)\pi/2} = -\frac{1}{2} i e^{i\pi M/2N}$$

(6.130a)

and

$$a_{-1}(-i) = \frac{1}{2} e^{i3(\frac{M}{N}-1)\pi/2} = \frac{1}{2} i e^{i3\pi M/2N}$$

(6.130b)

Thus,

$$a_{-1}(i) + a_{-1}(-i) = -e^{i\pi M/N} \sin\left(\frac{\pi M}{2N}\right)$$

(6.131)

and eq. 6.126 results in

$$\int_0^\infty \frac{x^{M/N}}{(1+x^2)} dx = \frac{\pi}{2} \sec\left(\frac{\pi M}{2N}\right)$$

(6.132)

□

When the integrand of an integral has multiple branch points, the prescription for choosing the multivalued function must be modified somewhat. We conclude this chapter with examples to illustrate the approach for functions with multiple branch points.

Example 6.9: Evaluating an integral of a function with two square root branch points

By substituting

$$w^2 = x^2 - 1$$

(6.133a)

then

$$\sqrt{2} \tan\phi = w$$

(6.133b)

it is straightforward to show that

$$\int_1^\infty \frac{x}{(x^2+1)\sqrt{x^2-1}}\,dx = \frac{\pi}{2\sqrt{2}} \qquad (6.134)$$

To obtain this result using the properties of multivalued functions, we note that the integrand has two square root branch points at $z = x = \pm 1$. Thus, we must determine a function with

- Square root branch points at ± 1
- Associated cuts that extend from these branch points to $\pm\infty$ along the positive and negative real axes, respectively
- Discontinuities across both cuts that are constant multiples of the integrand

Using the analysis presented earlier, we can see that the contour integral appropriate for this problem is

$$\oint \frac{z}{(z^2+1)\sqrt{z^2-1}}\,dz$$

The analytic structure of this integrand and the appropriate closed contour are shown in fig. 6.26.

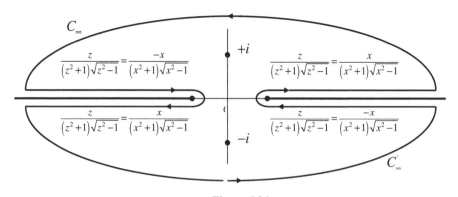

Figure 6.26

Analytic structure and contour for evaluating the integral of eq. 6.134

Then

$$\oint \frac{z}{(z^2+1)\sqrt{z^2-1}}\,dz = 2\pi i\left[a_{-1}(i)+a_{-1}(-i)\right]$$

$$= \int_1^\infty \frac{x}{(x^2+1)\sqrt{x^2-1}}\,dx + \int_{-\infty}^{-1} \frac{x}{(x^2+1)\left(-\sqrt{x^2-1}\right)}\,dx$$

$$+ \int_{-1}^{-\infty} \frac{x}{(x^2+1)\sqrt{x^2-1}}\,dx + \int_\infty^1 \frac{x}{(x^2+1)\left(-\sqrt{x^2-1}\right)}\,dx$$

$$(6.135a)$$

where the integrals around the infinitely large and infinitesimally small segments are zero and have been omitted. Inverting the limits and replacing x by $-x$ in the integrals along the negative real axis, we obtain

$$\int_1^\infty \frac{x}{(x^2+1)\sqrt{x^2-1}}\,dx = \frac{i\pi}{2}\left[a_{-1}(i)+a_{-1}(-i)\right] \qquad (6.135b)$$

With the cuts oriented as shown in fig. 6.26, the values of z at $\pm i$ are obtained as follows.

• The branch point at $+1$ arises from the factor $\sqrt{z-1}$. The principal sheet of this square root function is defined by the range $[0,2\pi]$, so the argument of z must be in this range. Therefore, referring to fig. 6.27a,

$$z_1 = +i = 1+\sqrt{2}e^{i3\pi/4} \qquad (6.136a)$$

and

$$z_2 = -i = 1+\sqrt{2}e^{i5\pi/4} \qquad (6.136b)$$

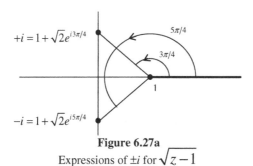

Figure 6.27a
Expressions of $\pm i$ for $\sqrt{z-1}$

- The branch point at -1 arises from the factor $\sqrt{z+1}$. The principal sheet of this square root factor is defined by $[-\pi, \pi]$. Therefore, the argument of z must be in this range. Thus,

$$z_1 = +i = -1 + \sqrt{2}e^{i\pi/4} \tag{6.136c}$$

and

$$z_2 = -i = -1 + \sqrt{2}e^{-i\pi/4} \tag{6.136d}$$

as shown in fig. 6.27b.

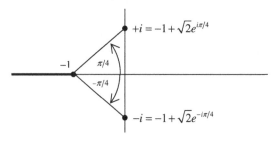

Figure 6.27b
Expressions of $\pm i$ for $\sqrt{z+1}$

Thus, from the product of $z_1 - 1$ from eq. 6.136a and $z_1 + 1$ from eq. 6.136c we obtain

$$\lim_{z \to i} \sqrt{z^2 - 1} = \sqrt{2}e^{i\pi/2} = i\sqrt{2} \tag{6.137a}$$

From the product of $z_2 - 1$ from eq. 6.136b and $z_2 + 1$ from eq. 6.136d we find

$$\lim_{z \to -i} \sqrt{z^2 - 1} = \sqrt{2}e^{i\pi/2} = i\sqrt{2} \tag{6.137b}$$

Therefore, the sum of the residues of the integrand are

$$a_{-1}(i) + a_{-1}(-i) = \frac{1}{i\sqrt{2}} \tag{6.138}$$

and, from eq. 6.135b, we obtain the expected result

$$\int_1^\infty \frac{x}{(1+x^2)\sqrt{x^2-1}}\, dx = \frac{\pi}{2\sqrt{2}} \tag{6.134}$$
□

Example 6.10: Evaluating an integral using a function with two logarithm branch points

By substituting $x = \tan\phi$, it is straightforward to show that

$$\int_0^1 \frac{1}{(1+x^2)}\, dx = \frac{\pi}{4} \tag{6.139}$$

Because the path of integration is along the real axis between 0 and 1, this integral can be evaluated using a multivalued function that has

- Branch points at 0 and 1
- Cut structure that results in a finite cut between these points
- A discontinuity across that cut that is a constant multiple of $1/(1+x^2)$
  It was shown in ex. 6.5 (and particularly fig. 6.20) that when the cuts of

$$\ell n\left(\frac{z-x_1}{z-x_2}\right)$$

are taken to extend from the two branch points to $+\infty$ along the real axis, the cut structure is equivalent to a finite cut from x_1 to x_2. Therefore, we see that

$$F(z) = \frac{\ell n\left(\dfrac{z-1}{z}\right)}{(1+z^2)} \tag{6.140}$$

- Has branch points at 0 and 1
- Has a finite cut between these points if both associated cuts extend to $+\infty$ along the real axis
- Has a discontinuity across the finite cut given by

$$\Delta(x) = -\frac{2\pi i}{(1+x^2)} \tag{6.141}$$

Figure 6.28 shows the cut structure of $F(z)$ at points along the positive real axis, and a contour appropriate for evaluating the integral of eq. 6.139 by Cauchy's theorem. (A contour that goes around the branch point at +1 and opens along the negative real axis would also be applicable to this problem.)

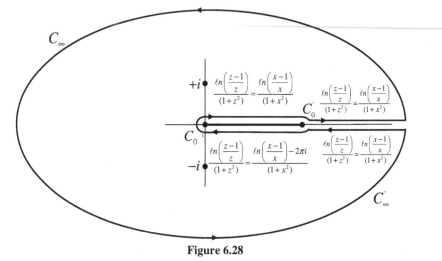

Figure 6.28

Analytic structure and contour for evaluating the integral of eq. 6.139

Writing

$$\oint \frac{\ell n\left(\frac{z-1}{z}\right)}{(z^2+1)}dz = 2\pi i\left[a_{-1}(i)+a_{-1}(-i)\right]$$

$$= \int_0^1 \frac{\ell n\left(\frac{x-1}{x}\right)}{(x^2+1)}dx + \int_1^\infty \frac{\ell n\left(\frac{x-1}{x}\right)}{(x^2+1)}dx \tag{6.142a}$$

$$+ \int_\infty^1 \frac{\ell n\left(\frac{x-1}{x}\right)}{(x^2+1)}dx + \int_1^0 \frac{\left[\ell n\left(\frac{x-1}{x}\right)-2\pi i\right]}{(x^2+1)}dx$$

$$= 2\pi i\int_0^1 \frac{1}{(x^2+1)}dx$$

yields

$$\int_0^1 \frac{1}{(x^2+1)} dx = \left[a_{-1}(i) + a_{-1}(-i)\right] \tag{6.142b}$$

Referring to figs. 6.29, we have

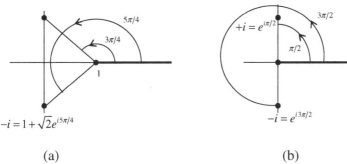

(a)   (b)

Figure 6.29
Expressions of $\pm i$ for (a) $\ell n(z-1)$ and (b) $\ell n(z)$

$$\ell n(z-1)\Big|_{z=+i} = \ell n\left(\sqrt{2}\right) + \frac{3\pi i}{4} \tag{6.143a}$$

$$\ell n(z)\Big|_{z=+i} = \frac{i\pi}{2} \tag{6.143b}$$

$$\ell n(z-1)\Big|_{z=-i} = \ell n\left(\sqrt{2}\right) + \frac{5\pi i}{4} \tag{6.143c}$$

and

$$\ell n(z)\Big|_{z=-i} = \frac{3\pi i}{2} \tag{6.143d}$$

Therefore, eq. 6.143 yields the expected result

$$\int_0^1 \frac{1}{(1+x^2)} dx = \frac{\pi}{4} \tag{6.139}$$

□

Problems

1. Determine the position(s) of the branch point(s) of each function below.

(a) $(z-1)^{5/4}$ (b) $\dfrac{(z+2)^3}{(z-2)^{4/5}}$ (c) $\left[\dfrac{(z+3)}{(z-3)}\right]^{5/2}$

(d) $(z-1)^{3/2}(z+1)^2$ (e) $(z-1)^{1/2}(z-2)^{2/3}$ (f) $\ell n\left(\dfrac{1}{4z+3}\right)$

(g) $\ell n\left(\dfrac{z^2-4z-5}{z^3+z^2-z-1}\right)$ (h) $(z^2-4)^{3/2}\ell n(z)$

2. For each function below, determine the position(s) of the branch point(s) and the number of Riemann sheets required to make the function single-valued.

(a) $\dfrac{(z-2)^3}{(z+2)^{4/7}}$ (b) $(z^2-4)^{3/7}$ (c) $(z+i)^\pi$

(d) $\left[\dfrac{(z+4)}{(z-4)}\right]^{5/3}$ (e) $(z+1)^2(z-1)^{3/2}$ (f) $\left(z^2-4\right)^{3/2}\ell n(z)$

(g) $z^{1/2}+z^2$ (h) $z^{1/6}+z^{2/3}$ (k) $(z^2-1)^3\ell n(z)$

3. Find the value of each complex number below on the principal sheet of a multisheeted complex plane when the principal branch is defined by

(i) $-\pi\le\theta\le\pi$ (ii) $-\pi/2\le\theta\le 3\pi/2$ (iii) $0\le\theta\le 2\pi$

(a) $\ell n(3-4i)$ (b) $\ell n(-3-4i)$ (c) $\ell n\left(ie^i\right)$

(d) $(3+4i)^{1/3}$ (e) $(i-1)^{1/4}$ (f) $\left(-ie^{-i}\right)^{2/5}$

4. For $F(z)=\ell n(z)$, determine the range of $\theta=\arg(z)$ for the

(i) upper   (ii) lower   (iii) right   (iv) left

half-plane when the cut associated with the logarithm branch point is taken to extend from the origin to ∞ along the

(a) positive x (b) negative x (c) positive y (d) negative y

axis.

5. Each function below has one or more branch points at just one value of z. For each function

- Identify that value of z.
- Determine how many sheets are needed for the function to be single valued.
- Extend the cuts from the branch point to $+\infty$ along the real axis and determine the values of the function along the top and along the bottom of the cut.
- Extend the cuts from the branch point to $-\infty$ along the real axis and determine the values of the function along the top and along the bottom of the cut.

(a) $(z-1)^{5/4}$ (b) $z^{1/2}z^{2/3}$ (c) $z^{3/5}\ln(z)$

(d) $(z-1)^{1/6}+(z-1)^{2/3}$ (e) $(z+2)^2+(z+2)^{2/3}$ (f) $z^{1/2}+\ln(z)$

6. With x_1 and x_2 real and with $x_1 < x_2$
   (a) Show that if both cuts associated with the branch points of

   $$F(z)=\left[(z-x_1)(z-x_2)\right]^{1/2}$$

   extend to $-\infty$, the resulting cut structure of $F(z)$ is like that shown in fig. 6.17 with $M = 1$.
   (b) Determine the cut structure for $F(z)$ when the cuts are taken as shown in fig. P6.1.

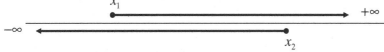

Figure P6.1
A possible orientation of the cuts associated with two branch points

7. (a) Show that if the two cuts associated with the branch points of

$$F(z) = \ell n \left(\frac{z - x_1}{z - x_2} \right)$$

extend to $-\infty$, the resulting cut structure of $F(z)$ is that shown in fig. 6.20.

   (b) Determine the cut structure for $F(z)$ when the cuts are taken as shown in fig. P6.1.

8. Use the properties of a multivalued function to evaluate the following integrals by Cauchy's residue theorem.

   (a) $\displaystyle\int_0^\infty \frac{1}{(x+1)(x^2+x+1)}\,dx$ (b) $\displaystyle\int_{-\infty}^0 \frac{1}{(x-1)(x^2+x+1)}\,dx$

9. For $x_0 > 0$, evaluate

$$\int_{-\infty}^{-x_0} \frac{1}{\left(x_0^2 + x^2\right)}\,dx$$

   (a) By substituting $x = x_0\tan\phi$
   (b) By using the properties of the logarithm function and Cauchy's residue theorem. (Hint: See ex. 6.7.)

10. (a) For the integer $N \geq 2$, and α a positive real constant, use the properties of $\ell n(z)$ and Cauchy's residue theorem to evaluate

$$\int_0^\infty \frac{1}{(\alpha+x)^N}\,dx$$

    (b) By considering

$$\oint \frac{(\ell n(z))^2}{(z+2)^N}\,dz$$

    and using the results of part (a), evaluate

$$\int_0^\infty \frac{\ell n(x)}{(x+2)^N} dx$$

for $N = 4$ and for $N = 5$.

11. For α real and nonzero, consider

$$\oint \frac{(\ell n(z))^2}{(x^2 + \alpha^2)} dx$$

to evaluate

$$\int_0^\infty \frac{\ell n(x)}{(x^2 + \alpha^2)} dx$$

12. With α a real constant, use the properties of a multivalued function and Cauchy's residue theorem to evaluate

$$\int_{-\infty}^{-2} \frac{1}{\left(x^2 + \alpha^2\right)^2} dx$$

Do not transform this integral to another interval (e.g., [0, +∞], [0, −∞] or [+2,+∞]).

13. Use the cut structure of the cube root function and Cauchy's residue theorem to evaluate

$$\int_0^\infty \frac{x^{1/3}}{\left(1 + x^3\right)} dx$$

14. With the integer $N \geq 2$, use the properties of a multivalued function and Cauchy's residue theorem to evaluate

$$\int_2^\infty \frac{1}{(1 + x)^N \sqrt{(x - 2)}} dx$$

Do not transform this integral to another interval (such as [0, +∞]).

15. Using the properties of the cube root function, evaluate the integral below by Cauchy's residue theorem.

$$\int_1^\infty \frac{(x-1)^{2/3}}{(1+x^2)} dx$$

Do not transform the range of integration.

16. Use the properties of a multivalued function and Cauchy's residue theorem to evaluate

$$\int_0^\infty \frac{x^{1/\pi}}{(1+x^2)} dx$$

17. For the integer $N > 0$, α a real constant $0 < \alpha < 1$, and β a real positive constant, use the properties of a multivalued function and Cauchy's residue theorem to evaluate

$$\int_0^\infty \frac{x^{\alpha-1}}{(x+\beta)^N} dx$$

What is the smallest integer N for which this integral is finite?

18. For α and β real constants and $-1 < \alpha < 1$, use the properties of a multivalued function and Cauchy's residue theorem to evaluate

$$\int_0^\infty \frac{x^\alpha}{\left(x^2 + 2x\cos\beta+1\right)} dx$$

19. For α a real nonzero constant, use Cauchy's residue theorem to evaluate

$$\int_0^\infty \frac{x^{1/2}\ell n(x)}{(x^2+\alpha^2)} dx$$

20. Use the properties of a multivalued function to evaluate

$$\int_{-1}^2 \frac{1}{(1+x^2)\sqrt{(2-x)(x+1)}} dx$$

21. For α a real nonzero constant, use the properties of a multivalued function and Cauchy's residue theorem to evaluate

$$\int_{-1}^{2} \frac{1}{(\alpha^2 + x^2)} dx$$

Verify your result by evaluating this integral by substituting $x = \alpha \tan\phi$ or using the method of partial fractions.

22. (a) Use the exponential representation

$$\tanh(w) = \frac{e^w - e^{-w}}{e^w + e^{-w}}$$

to prove that $\tanh^{-1}(z)$ has two branch points and determine the positions of those branch points.

(b) Use the results of part (a) and consider

$$\oint \frac{\left(\tanh^{-1}(z)\right)^2}{(1 + z^2)} dz$$

around an appropriate contour to evaluate

$$\int_{-1}^{1} \frac{\tanh^{-1}(x)}{(1 + x^2)} dx$$

using Cauchy's residue theorem.

Chapter 7

SINGULARITIES OF FUNCTIONS DEFINED BY INTEGRALS

Let $F(z)$ be defined by an integral of the form

$$F(z) = \int_{w_1}^{w_2} G(z, w)\, dw \qquad (7.1)$$

where w_1 and w_2 are complex constants.

Clearly, if the integral can be evaluated in terms of functions with known analytic properties, the singularity structure of $F(z)$ can be determined straightforwardly. It is also possible to determine the analytic structure of $F(z)$ even when the integral of eq. 7.1 cannot be evaluated in such closed form. This can be accomplished from an analysis of the singularities of $G(z,w)$ and the contour taken in the w-plane between w_1 and w_2.

An analysis of the singularity structure of the integral of eq. 7.1 was first presented in the literature by Hadamard, 1898, p. 55. For a more modern treatment, the reader is referred to Eden et al., 1966.

7.1 The Integrand Is Analytic

Let $G(z,w)$ be analytic at all z in a region R of the z-plane and at all w in a region S of the w-plane. As discussed in chapter 3, because $G(z,w)$ is $w = z^3$ analytic everywhere in S, the integral of eq. 7.1 is independent of the path taken between w_1 and w_2. Then, writing $G(z,w)$ as

$$G(z, w) = \frac{\partial}{\partial w} H(z, w) \qquad (7.2)$$

we refer the reader to the discussion presented in chapter 3 to deduce that because $G(z,w)$ is analytic at all z in R and at all w in S, $H(z,w)$ is also analytic at all z in R and at all w in S. Therefore, eq. 7.1 can be written as

$$F(z) = H\left(z, w_2\right) - H\left(z, w_1\right) \tag{7.3}$$

Because $H(z,w)$ is analytic everywhere in R for any value of w, $F(z)$ is analytic at all z in R.

Example 7.1: Integral of an integrand that is analytic

(a) With integer $n \geq 0$, we consider

$$F(z) \equiv \int_{w_1}^{w_2} (z+w)^n \, dw = \frac{(w_2 + z)^{n+1} - (w_1 + z)^{n+1}}{(n+1)} \tag{7.4a}$$

Because the integrand of this integral has no singularities in z or in w, we see that $F(z)$ is analytic at all finite z.

(b) Consider

$$F(z) \equiv \int_{w_1}^{w_2} e^{-wz} \, dw = \frac{1}{z}\left(e^{-w_2 z} - e^{-w_1 z}\right) \tag{7.4b}$$

Because the integrand of the integral is analytic at all finite z and all finite w, $F(z)$ is an entire function. It is straightforward to see that $F(z)$ does not have a pole at $z = 0$ by expanding the exponential functions in their MacLaurin series. □

Thus, if the integrand $G(z,w)$ of eq. 7.1 is analytic at all z in R and all w in S, $F(z)$ is analytic at all z in R. Therefore, if $F(z)$ is to be singular at some point z_0 in R, $G(z,w)$ must be singular at one or more points in the z-plane and/or one or more points in the w-plane.

7.2 The Integrand is Singular

Let $G(z,w)$ be analytic everywhere in a region R of the z-plane, and let $G(z,w)$ be singular at one or more points in the w-plane. This analysis is independent of whether the singularities are poles or branch points. To introduce this analysis, we let $G(z,w)$ have one singularity in the w-plane at the point w_0. If w_0 is independent of z, it is called a *fixed singularity*. If w_0

depends on z, the position of w_0 changes as z is varied. Such a singularity is referred to as a *movable* or *variable singularity*.

Fixed singularity of the integrand

Let w_0 be a fixed singularity in the w-plane. It was discussed in chapter 5 that as long as w_0 is not on the contour, the integral of eq. 7.1 is not singular.

Example 7.2: Integral of an integrand with a fixed singularity

The integrand of

$$F(z) \equiv \int_{-1}^{1} \frac{(w-z)}{(w-i)} dw \tag{7.5}$$

has a simple pole at $w = i$. We take the contour to extend along the real axis as shown in fig. 7.1.

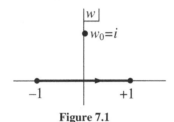

Figure 7.1

Contour and singularity of the integral of eq. 7.5

Because $w_0 = i$ is not on the contour, $F(z)$ should be analytic at all finite z. This is easily verified by evaluating the integral in closed form. We obtain

$$F(z) = 2 - (z-i) \; \ell n \left(\frac{(1-i)}{(-1-i)} \right) \tag{7.6}$$

which is a linear function of z. It is therefore analytic at all finite z as predicted. □

Movable singularity of the integrand

It was also discussed in chapter 5 that if $w_0(z)$ appears to be on the contour, the contour must be deformed away from the singularity if a finite result is to be obtained. An example of this is presented in the discussion resulting in eqs. 5.67. If the contour cannot be deformed away from $w_0(z)$, the integral is singular.

Endpoint and pinch singularities

Let $G(z,w)$ be singular in the w-plane at just one point $w_0(z)$ that varies as z is varied. Because $G(z,w)$ has no other singularities, then at any z, $G(z,w)$ is analytic in a region around w_0. If z has a value that causes w_0 to move to the contour, the contour can be deformed away from the singularity without changing the value of the integral.

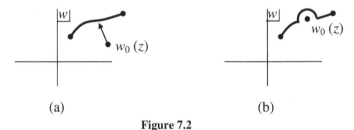

(a)   (b)

Figure 7.2
Migration of a singularity to the contour and
deformation of the contour away from the singularity

Therefore, simply having $w_0(z)$ move toward the contour will not generate a singularity of the integral.

Because the endpoints of the integral are fixed points in the w-plane, if z causes $w_0(z)$ to move to an endpoint, the contour cannot be deformed away from the singularity. Then the integral is singular at that value of z. Such a singularity is called an *endpoint singularity*.

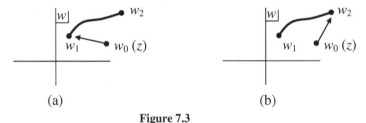

(a)   (b)

Figure 7.3
Migration of a singularity to the endpoints of the contour

These endpoint singularities occur at those values of z for which

$$w_0(z) = w_1 \qquad\qquad (7.7a)$$

and

$$w_0(z) = w_2 \tag{7.7b}$$

Example 7.3: Endpoint singularities of the integral of an integrand with one movable singluarity

The integrand of

$$F(z) = \int_{w_1}^{w_2} \frac{1}{(w-z)} dw \tag{7.8}$$

has a simple pole at

$$w_0(z) = z \tag{7.9}$$

Therefore, $F(z)$ has endpoint singularities at

$$z = \pm w_1 \tag{7.10a}$$

and

$$z = \pm w_2 \tag{7.10b}$$

This can be verified easily inasmuch as the integral of eq. 7.8 can be written in closed form as

$$F(z) = \ell n \left(\frac{w_2 - z}{w_1 - z} \right) \tag{7.11}$$

which has logarithm branch points at the values of z given in eqs. 7.10. □

Let an integrand have two singularities in the w-plane denoted by $w_0(z)$ and $w_0'(z)$. In order for these singularities to coincide, at least one of them must be a movable singularity.

For a given z, either both singularities lie on the same side of the contour, or they lie on opposite sides of the contour. When z is varied, any singularity that depends on z can migrate toward the contour.

If both singularities migrate to the same point from the same side of the contour, the contour can be deformed away from that point as shown in fig. 7.4a. Then, a coincidence of w_0 and w_0' does not give rise to a singularity of the integral.

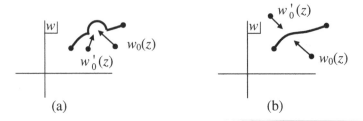

(a)   (b)

Figure 7.4
Two singularities on (a) opposite sides of the contour and (b) the same side of the contour

If w_0 and w_0' migrate to the same point from opposite sides of the contour, the contour cannot be deformed away from the singularities and is therefore pinched by the coincidence of the singularities. Then, the integral is singular at the value of z that makes the two singularities pinch the contour as shown in fig. 7.4b. Such a singularity is called a *pinch singularity*.

$w_0(z)$ and $w_0'(z)$ can migrate from opposite sides of the contour to a common point that is outside the range of integration as shown in fig. 7.4c. In that case the contour will not be pinched, and the integral will not have a singularity due to the coincidence of w_0 and w_0'.

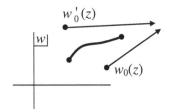

Figure 7.4c
Coincidence of two singularities from opposite sides of the contour that do not pinch the contour

Example 7.4: Pinch singularity of the integral of an integrand with one fixed and one movable singluarity

The singularities of the integrand of

$$F(z) \equiv \int_{-1}^{1} \frac{1}{(w-i)(w-z)}\, dw \tag{7.12}$$

are a fixed pole at

$$w_0 = i \qquad (7.13a)$$

and a variable pole at

$$w_0'(z) = z \qquad (7.13b)$$

Clearly, w_0 cannot migrate to the endpoints but $w_0'(z)$ can. Thus, $F(z)$ has endpoint singularities at

$$w_0'(z) = z = \pm 1 \qquad (7.14)$$

Taking the contour along the real axis of the w-plane, we see that if $w_0'(z)$ is on the opposite side of the contour from w_0, and if they coincide somewhere between the endpoints, $F(z)$ will also have a pinch singularity. Taking $y > 0$ places both poles on the same side of the contour. Then the contour will be not be pinched; $F(z)$ will have no singularity.

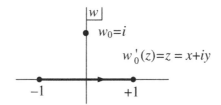

Figure 7.5

The two poles of the integrand on the same side of the
contour result in no pinch singularity

If we take $y < 0$ and then vary y to a value of $+1$, the coincidence of the two poles generates a pinch singularity of $F(z)$.

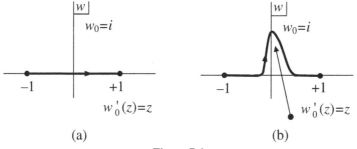

(a)   (b)

Figure 7.6

The two poles of the integrand on opposite sides of the contour
result in a pinch singularity of the integral

To understand this analysis, we note that the integral of eq. 7.12 can be expressed in closed form as

$$F(z) = \frac{1}{(z-i)}\left[\ell n\left(\frac{z-1}{z+1}\right) - \ell n\left(\frac{i-1}{i+1}\right) \right] \qquad (7.15)$$

We see that $\ell n\,[(z-1)/(z+1)]$ has two logarithm branch points at $z = \pm 1$. These are the endpoint singularities of $F(z)$ predicted above.

It was shown in ex. 6.5 that if the cuts associated with the branch points of $\ell n\,[(z-1)/(z+1)]$ are both taken to extend to $+\infty$ (or to $-\infty$) along the real axis, the resulting structure is a finite cut extending from -1 to $+1$ along the real axis, with points along the real axis above the cut given by

$$\left[\ell n\left(\frac{z-1}{z+1}\right)\right]_{above} = \ell n\left|\frac{z-1}{z+1}\right| + i\big[\arg(z-1) - \arg(z+1)\big] \qquad (7.16a)$$

Below the cut,

$$\left[\ell n\left(\frac{z-1}{z+1}\right)\right]_{below} = \ell n\left|\frac{z-1}{z+1}\right| + i\big[\arg(z-1) - \arg(z+1)\big] + 2\pi i \quad (7.16b)$$

Let the value of $\ell n\,[(i-1)/(i+1)]$ define the principal sheet of the z-plane for $F(z)$ of eq. 7.15. This can be done by expressing $i \pm 1$ as

$$i+1 = \sqrt{2}\,e^{i\pi/4} \qquad (7.17a)$$

and

$$i-1 = \sqrt{2}\,e^{i3\pi/4} \qquad (7.17b)$$

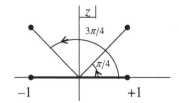

Figure 7.7

$i \pm 1$ on the principal sheet of the logarithm function

If z is chosen with $y > 0$, then when $z \to i$, it does so from above the real axis. As shown in fig. 7.8a, this causes z to approach i on the principal sheet of the logarithm. Then,

$$\lim_{z \to i} \ell n\left(\frac{z-1}{z+1}\right) - \ell n\left(\frac{i-1}{i+1}\right) = 0 \tag{7.18}$$

Thus, using l'Hopital's rule, when $z \to i$, we see from eq. 7.15 that

$$\lim_{z \to i} F(z) = \lim_{z \to i} \frac{2}{(z^2 - 1)} = -1 \tag{7.19}$$

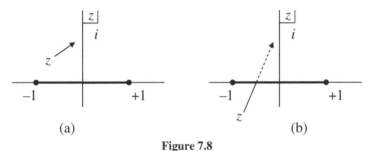

Figure 7.8

$z \to i$ on the (a) principal sheet (b) second sheet

If $z \to i$ starting from a point with $y < 0$, z must migrate onto the second sheet to access i as shown in fig. 7.8b. Therefore, because the number

$$\ell n\left(\frac{i-1}{i+1}\right)$$

is on the principal sheet,

$$\lim_{z \to i} \ell n\left(\frac{z-1}{z+1}\right) - \ell n\left(\frac{i-1}{i+1}\right) = -2\pi i \tag{7.20}$$

Then, at points z near i

$$\lim_{z \to i} F(z) = \frac{-2\pi i}{(z-i)} \tag{7.21}$$

Thus, $F(z)$ has a simple pole at $z = i$. The singularity that results by starting at a point below the real axis causes the contour to be pinched when $z \rightarrow i.$ □

This example illustrates that the choice of the "starting" point of a movable singularity can determine whether a contour can be pinched by singularities of the integrand. In some cases, contours can be pinched no matter how one chooses the "starting" position of z.

Example 7.5: Analyticity of an integral when two movable singularities of the integrand do not pinch the contour

We consider

$$F(z) = \int_{-1}^{1} \frac{1}{(w - z + 1)(4w - 3z)} dw \qquad (7.22)$$

taking the contour along the real axis of the w-plane.
   The singularities of the integrand at

$$w_0(z) = z - 1 \qquad (7.23a)$$

and

$$w_0'(z) = \tfrac{3}{4} z \qquad (7.23b)$$

generate endpoint singularities when

$$z - 1 = \pm 1 \Rightarrow z = 0, 2 \qquad (7.24a)$$

and

$$\tfrac{3}{4} z = \pm 1 \Rightarrow z = \pm \tfrac{4}{3} \qquad (7.24b)$$

To determine whether the contour can be pinched by w_0 and w_0', we write $z = x + iy$. With $y > 0$, both w_0 and w_0' are above the real axis in the w-plane. If $y < 0$, both singularities are below the real axis. Therefore, even when $w_0 = w_0'$, which occurs for $z = 4$, the contour cannot be pinched by these singularities.

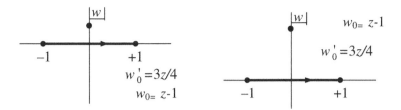

Figure 7.9

Contour and possible positions of the poles for the integral of eq. 7.22

The fact that the integral of eq. 7.22 does not contain pinch singularities can be seen from the closed form of $F(z)$, which we express as

$$F(z) = \frac{1}{(z-4)} \ell n \left[\frac{(2-z)(4+3z)}{z(4-3z)} \right]$$

$$= \frac{1}{(z-4)} \left[\ell n(2-z) + \ell n(4+3z) - \ell n(z) - \ell n(4-3z) \right]$$

(7.25)

The logarithm branch points at $z = 0$, 2, and $\pm 4/3$ arise from the endpoint coincidences predicted above. To demonstrate that there is no singularity at $z = 4$ arising from the coincidence of w_0 and w_0', let us approach the limit $z \to 4$ from above the x-axis. Then, using the analysis of eqs. 6.48 to 6.51 and fig. 6.12a (or equivalently eqs. 6.52 to 6.55 and fig. 6.12b), we have

$$\lim_{z \to 4+i\varepsilon} \ell n(2-z) = \ell n(-2-i\varepsilon) = \ell n(2) + i\pi \qquad (7.26a)$$

$$\lim_{z \to 4+i\varepsilon} \ell n(4+3z) = \ell n(16+i\varepsilon) = \ell n(16) \qquad (7.26b)$$

$$\lim_{z \to 4+i\varepsilon} \ell n(z) = \ell n(4+i\varepsilon) = \ell n(4) \qquad (7.26c)$$

and

$$\lim_{z \to 4+i\varepsilon} \ell n(4-3z) = \ell n(-8-i\varepsilon) = \ell n(8) + i\pi \qquad (7.26d)$$

Substituting eqs. 7.26 into the bracket of logarithms in eq. 7.25, we obtain

$$\lim_{z \to 4+i\varepsilon} \left[\ell n(2-z) + \ell n(4+3z) - \ell n(z) - \ell n(4-3z) \right] = 0 \qquad (7.27)$$

Then, using l'Hopital's rule, it is straightforward to show that

$$F(4) = \lim_{z \to 4+i\varepsilon} \frac{1}{(z-4)} \ell n \left[\frac{(2-z)(4+3z)}{z(4-3z)} \right] = \frac{1}{16} \tag{7.28}$$

Thus the integral of eq. 7.22 is analytic at $z = 4$ as predicted. □

Example 7.6: Singularity of an integral when two movable singularities of the integrand pinch the contour

Changing the sign of z in one of the factors in the denominator of eq. 7.22, we consider

$$F(z) = \int_{-1}^{1} \frac{1}{(w-z+1)(4w+3z)} dw \tag{7.29}$$

The path of integration is again taken along the real axis in the w-plane.

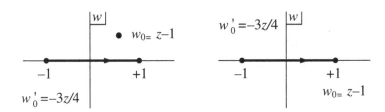

Figure 7.10
Contour and possible positions of the poles for the integral of eq. 7.29

The singularities of the integrand at

$$w_0(z) = z - 1 \tag{7.30a}$$

and

$$w_0'(z) = -\tfrac{3}{4} z \tag{7.30b}$$

generate endpoint singularities at

$$z - 1 = \pm 1 \Rightarrow z = 0, 2 \tag{7.31a}$$

and

$$-\tfrac{3}{4}z = \pm 1 \Rightarrow z = \pm\tfrac{4}{3} \tag{7.31b}$$

the same points at which the function defined in eq. 7.22 is singular.

Writing the singularities of the integrand as

$$w_0(x, y) = (x-1) + iy \tag{7.32a}$$

and

$$w_0'(x, y) = -\tfrac{3}{4}x - \tfrac{3}{4}iy \tag{7.32b}$$

we see that the imaginary parts of w_0 and w_0' have opposite signs. There-fore, they must be on opposite sides of the contour, and the sign of y is not important. If the coincidence of w_0 and w_0' occurs at a point z for which $-1 < \text{Re}(w_0) < 1$, $F(z)$ has a pinch singularity at that value of z. We find that $w_0 = w_0'$ at $z = x = 4/7$ from which $w_0 = w_0' = -3/7$. Because $-1 < -3/7 < 1$, the contour is pinched by the coincidence of w_0 and w_0'.

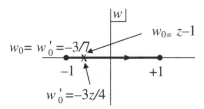

Figure 7.11
Pinch of the contour of the integral of eq. 7.29

This result is verified by analyzing the closed form of $F(z)$;

$$F(z) = \frac{1}{(7z-4)} \ell n \left[\frac{(2-z)(4-3z)}{z(4+3z)} \right]$$

$$= \frac{1}{(7z-4)} \left[\ell n (2-z) + \ell n (4-3z) - \ell n(z) - \ell n(4+3z) \right] \tag{7.33}$$

The endpoint singularities predicted in eqs. 7.30 are the branch points of the individual logarithms. Again, referring to eqs. 6.48 to 6.51 and fig. 6.12a, we have

$$\lim_{z \to 4/7+i\varepsilon} \ell n(2-z) = \ell n\left(\frac{10}{7} - i\varepsilon\right) = \ell n\left(\frac{10}{7}\right) + 2\pi i \qquad (7.34a)$$

$$\lim_{z \to 4/7+i\varepsilon} \ell n(4-3z) = \ell n\left(\frac{16}{7} - i\varepsilon\right) = \ell n\left(\frac{16}{7}\right) + 2\pi i \qquad (7.34b)$$

$$\lim_{z \to 4/7+i\varepsilon} \ell n(z) = \ell n\left(\frac{4}{7} + i\varepsilon\right) = \ell n\left(\frac{4}{7}\right) \qquad (7.34c)$$

and

$$\lim_{z \to 4/7+i\varepsilon} \ell n(4+3z) = \ell n\left(\frac{40}{7} + i\varepsilon\right) = \ell n\left(\frac{40}{7}\right) \qquad (7.34d)$$

from which

$$\lim_{z \to 4/7+i\varepsilon} \left[\ell n(2-z) + \ell n(4-3z) - \ell n(z) - \ell n(4+3z)\right] = 4\pi i \qquad (7.35)$$

Therefore, in a small neighborhood around $z = 4/7$,

$$F(z) \sim \frac{4\pi i}{7z - 4} \qquad (7.36)$$

So, the predicted singularity of $F(z)$ arising from a pinch of the contour is a simple pole at $z = 4/7$. □

Example 7.7: Analyticity of an integral when two movable singularities of the integrand do not pinch the contour

Let

$$F(z) = \int_{-1}^{1} \frac{1}{(w - z + 2)(w + 2z)} \, dw \qquad (7.37)$$

taking the contour to be along the real axis in the w-plane from -1 to 1.

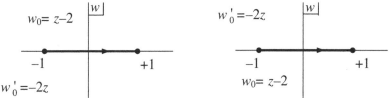

Figure 7.12
Contour and possible positions of the poles for the integral of eq. 7.37

The poles of the integrand coincide with the endpoints at

$$w_0(z) = z - 2 = \pm 1 \Rightarrow z = 1, 3 \tag{7.38a}$$

and

$$w_0'(z) = -2z = \pm 1 \Rightarrow z = \pm \tfrac{1}{2} \tag{7.38b}$$

We note that the imaginary parts of w_0 and w_0' have opposite signs and therefore are on opposite sides of the contour. Setting $w_0 = w_0'$, we find that these singularities coincide at $z = 2/3$. At this value of z, $w_0 = w_0' = -4/3$ which is outside the range of integration.

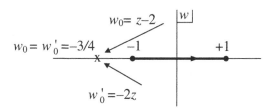

Figure 7.13
Coincidence of the poles does not pinch the contour

Therefore, this coincidence of singularities does not pinch the contour, and the integral of eq. 7.37 is not singular at $z = 2/3$.

To see this, we evaluate the integral of eq. 7.37 in closed form; we have

$$F(z) = \frac{1}{(2-3z)} \ell n \left[\frac{(2z+1)(1-z)}{(2z-1)(3-z)} \right]$$

$$= \frac{1}{(2-3z)} \left[\ell n(2z+1) + \ell n(1-z) - \ell n(2z-1) - \ell n(3-z) \right]$$

$$(7.39)$$

As in the previous examples, we see that the endpoint singularities are reflected in the branch points of the individual logarithms. Again following the analysis of eqs. 6.48 to 6.51 and fig. 6.12a, we have

$$\lim_{z \to 2/3+i\varepsilon} \ell n(2z+1) = \ell n \left(\frac{7}{3} + i\varepsilon \right) = \ell n \left(\frac{7}{3} \right) \qquad (7.40a)$$

$$\lim_{z \to 2/3+i\varepsilon} \ell n(1-z) = \ell n \left(\frac{1}{3} - i\varepsilon \right) = \ell n \left(\frac{1}{3} \right) + 2\pi i \qquad (7.40b)$$

$$\lim_{z \to 2/3+i\varepsilon} \ell n(2z-1) = \ell n \left(\frac{1}{3} + i\varepsilon \right) = \ell n \left(\frac{1}{3} \right) \qquad (7.40c)$$

and

$$\lim_{z \to 2/3+i\varepsilon} \ell n(3-z) = \ell n \left(\frac{7}{3} - i\varepsilon \right) = \ell n \left(\frac{7}{3} \right) + 2\pi i \qquad (7.40d)$$

Therefore,

$$\lim_{z \to 2/3} \left[\ell n(2z+1) + \ell n(1-z) - \ell n(2z-1) - \ell n(3-z) \right] = 0 \qquad (7.41)$$

Then, using l'Hopital's rule, it is straightforward to obtain

$$F\left(\tfrac{2}{3} \right) = \lim_{z \to 2/3} \frac{1}{(2-3z)} \ell n \left[\frac{(2z+1)(1-z)}{(2z-1)(3-z)} \right] = \frac{18}{7} \qquad (7.42)$$

Thus, as predicted, $F(z)$ is analytic at $z = 2/3$. □

7.3 Limits of the Integral are Variable

Limits are analytic functions of z

The discussion of section 7.2 involves the singularities of integrals with fixed limits. This analysis can be extended to limits that are analytic functions of z. When $w_1(z)$ and $w_2(z)$ are analytic functions of z and the integrand of

$$F(z) = \int_{w_1(z)}^{w_2(z)} G(z,w)dw \qquad (7.43)$$

has one or more singularities that coincide with endpoints at a certain value of z, then $F(z)$ has endpoint singularities at those values of z. If $G(z,w)$ has two or more singularities that can pinch the contour at values of z in the range between $w_1(z)$ and $w_2(z)$, then $F(z)$ can have pinch singularities.

Example 7.8: Singularities of an integral with variable endpoints

Let

$$F(z) \equiv \int_{z-1}^{z+1} \frac{1}{w^2 - z^2} dw \qquad (7.44)$$

with two possible contours as shown in fig. 7.14.

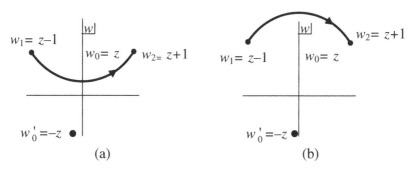

Figure 7.14
Poles and possible contours for the integral of eq. 7.44

The integrand has poles at

$$w_0(z) = z \qquad (7.45a)$$

and

$$w_0'(z) = -z \qquad\qquad (7.45b)$$

Because there is no value of z for which $z = z \pm 1$, $F(z)$ does not have singularities generated by endpoint coincidences with $w_0(z)$. However, endpoint coincidences with $w_0'(z)$ do generate singularities of $F(z)$. These singularities are at values of z given by

$$-z = z - 1 \Rightarrow z = \tfrac{1}{2} \qquad\qquad (7.46a)$$

and

$$-z = z + 1 \Rightarrow z = -\tfrac{1}{2} \qquad\qquad (7.46b)$$

Referring to fig. 7.14, it is clear that the existence of a pinch singularity depends on the choice of contour. If the contour is that shown in fig. 7.14a, the poles are on opposite sides of the contour and coincide at $z = 0$. Thus, $F(z)$ has a pinch singularity at $z = 0$. If the pinch singularity structure involves one or more branch points, then $w_0(z) \to 0$ on the first sheet of the multivalued $F(z)$ and $w_0'(z) \to 0$ on the second sheet.

This can be verified for the current example because the integral of eq. 7.44 can be evaluated in closed form. We obtain

$$F(z) = \frac{1}{2z} \ell n \left(\frac{1 - 2z}{1 + 2z} \right) \qquad\qquad (7.47)$$

The branch points at $z = \pm 1/2$ are the endpoint coincidences predicted in eqs. 7.46. Referring to ex. 6.5, both associated cuts can be taken to extend to $+\infty$ (or $-\infty$) resulting in a finite cut between $z = -1/2$ and $z = 1/2$.

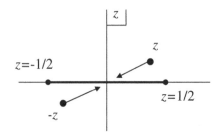

Figure 7.15
Value of $\ell n \, [(1 - 2z)/(1 + 2z)]$ when $z \to 0$

Letting $z \to 0$ from above the cut,

$$\lim_{z \to 0 + i\varepsilon} \ell n(1 - 2z) = \ell n(1 - i\varepsilon) = 2\pi i \qquad (7.48a)$$

and

$$\lim_{z \to 0 + i\varepsilon} \ell n(1 + 2z) = \ell n(1 + i\varepsilon) = 0 \qquad (7.48b)$$

Therefore, in the neighborhood of the origin,

$$F(z) \sim \frac{i\pi}{z} \qquad (7.49a)$$

so that the singularity of $F(z)$ predicted above is a pole at $z = 0$.

If the contour is that shown in fig. 7.14b, the two poles cannot pinch the contour and $F(z)$ is analytic at $z = 0$. In that case,

$$\lim_{z \to 0} F(z) = \lim_{z \to 0} \frac{1}{2z} \ell n\left(\frac{1 - 2z}{1 + 2z}\right) = -2 \qquad (7.49b)$$

so $F(z)$ is analytic at the origin as predicted. $\square$

Integrals with limits that have singularities

When the limits of an integral contain singularities, the singularity structure of the integral cannot be predicted by the techniques of this chapter. This is illustrated by two examples.

Example 7.9: Analyticity of an integral with a singular limit and an integrand that is analytic

For an integer $N > 0$ we see that the function

$$F(z) = \int_0^{\sqrt{z}} (zw)^{2N} dw = \frac{z^{3N}\sqrt{z}}{2N + 1} \qquad (7.50a)$$

has a square root branch point at $z = 0$, whereas

$$F(z) = \int_0^{\sqrt{z}} (zw)^{2N+1} dw = \frac{z^{3N+1}}{2N + 2} \qquad (7.50b)$$

is an entire function. Thus, the singularities of the integrand and limits are not predictors of the singularity structure of the integral. ☐

Example 7.10: Analyticity of an integral with a singular limit and an integrand that is singular

For integer $N > 0$, the integral

$$F(z) = \int_1^{e^{1/z}} \frac{z^{N+1}}{w} dw = z^N \tag{7.51}$$

has a singular integrand and a singular limit. However, the integral is an entire function. ☐

We conclude this material by noting that the analysis we have presented for predicting endpoint and pinch singularities only determines the location of the singularities of the integral. Unless the integral can be evaluated in closed form (making this analysis unnecessary), the type of singularity (poles or branch points) cannot be determined by the techniques presented in this chapter.

Problems

For problems 1 and 2, predict the values of z at which each integral has endpoint singularities (if any). Then evaluate each integral and verify your prediction.

1. $F(z) = \int_0^1 \sqrt{z + w}\, dw$

2. $F(z) = \int_0^1 \ell n(z + w)\, dw$

For prob. 3 through 8, predict the values of z at which each integral has pinch singularities (if any) over a contour along the real axis in the w-plane. For the integrals of prob. 3 and 4, evaluate each integral in closed form and verify your prediction.

3. $F(z) = \int_{-1}^1 \frac{1}{\left(w^2 - iz^2\right)} dw$

4. $\quad F(z) = \int_0^1 \dfrac{1}{(w-iz)^2} \, dw$

5. $\quad F(z) = \int_{-1}^1 \dfrac{\ell n(w-z)}{(w-3z)^2} \, dw$

6. $\quad F(z) = \int_{-1}^1 \dfrac{\ell n(w-z)}{(w+3z)^2} \, dw$

7. $\quad F(z) = \int_1^2 \dfrac{\ell n(w-z)}{(w+3z)^2} \, dw$

8. $\quad F(z) = \int_{-1}^1 \dfrac{\ell n(w-z+3)}{(w+3z-3)} \, dw$

In problems 9, 10, and 11, the path of integration is along the real axis of the w-plane.

9. Determine the pinch singularities of

$\quad$ (a) $\quad F(z) = \int_{-5}^5 \dfrac{\ell n(w+3z-4)}{(w-3z-1)(w+4z+6)} \, dw$

$\quad$ (b) $\quad F(z) = \int_0^5 \dfrac{\ell n(w+3z-4)}{(w-3z-1)(w+4z+6)} \, dw$

10. (a) What is the smallest possible value of the real positive number α such that

$$ F(z) = \int_{-\alpha}^\alpha \dfrac{(w-2z+1)^{1/4}}{(w+z-5)} \, dw $$

$\quad$ has a pinch singularity?

(b) For the minimum value of α found in part (a), what are the endpoint singularities of

$$F(z) = \int_{-2\alpha}^{2\alpha} \frac{(w-2z+1)^{1/4}}{(w+z-5)} \, dw$$

11. (a) For what range of values of the real parameter β does

$$F(z) = \int_0^1 \frac{(w+z-\beta)^{1/3}}{(2z-w-5\beta)^2} \, dw$$

(b) For the value of β at the midpoint of the range found in part (a), what are the endpoint singularities of $F(z)$?

(c) For the value of β at the midpoint of the range found in part (a), at what z does $F(z)$ have a pinch singularity?

For each integral in prob. 12 through 17, b and c are real positive constants, $x = \mathrm{Re}(z)$, and the contour of integration is along the real axis of the w-plane. ε is the infinitesimal positive real quantity. (See the discussion of principal value integrals in chapter 5.) For each integral, determine

- The value(s) of x at which the integral is singular and identify whether each singularity is an endpoint singularity or a pinch singularity
- If there is any range of the ratio b/c for which the integral does not have a pinch singularity

12. $F(x) = \int_{-c}^{c} \dfrac{1}{(w-x+b-i\varepsilon)(w+x+b+i\varepsilon)} \, dx$

13. $F(x) = \int_{-c}^{c} \dfrac{1}{(w-x+b-i\varepsilon)(w+x+b-i\varepsilon)} \, dx$

14. $F(x) = \int_{-c}^{c} \dfrac{1}{(w-x+b-i\varepsilon)(w+x-b+i\varepsilon)} \, dx$

15. $F(x) = \int_{-c}^{c} \dfrac{1}{(w-x+b+i\varepsilon)(w+x-b+i\varepsilon)} \, dx$

16. $F(x) = \int_{-c}^{c} \frac{1}{(w+x+b-i\varepsilon)(w+x-b+i\varepsilon)} dx$

17. $F(x) = \int_{-c}^{c} \frac{1}{(w+x+b-i\varepsilon)(w+x-b-i\varepsilon)} dx$

18. Determine all points z at which

$$F(z) = \int_{2z-1}^{2z+1} \frac{\ell n(w-z)}{(w+3z+2)} dw$$

is singular. Identify each singularity as either an endpoint or pinch singularity.

19. The incomplete elliptic functions of the first and second kind are defined as

$$K_i(z,\phi_0) = \int_0^{\phi_0} \frac{1}{\sqrt{(1-z^2 \sin^2 \phi)}} d\phi$$

and

$$E_i(z,\phi_0) = \int_0^{\phi_0} \sqrt{(1-z^2 \sin^2 \phi)} d\phi$$

respectively. For $\phi_0 = \pi/6$, and setting $z \sin\phi \equiv w$, these can be written as

$$K_i(z,\pi/6) = \int_0^{z/2} \frac{1}{\sqrt{(z^2-w^2)(1-w^2)}} dw$$

and

$$E_i(z,\pi/6) = \int_0^{z/2} \frac{(1-w^2)}{\sqrt{(z^2-w^2)}} dw$$

Taking the path of integration to be along the real axis of the w-plane, determine the values of z at which these elliptic functions are singular.

Chapter 8

CONFORMAL MAPPING

8.1 Properties of a Mapping

Let R be a region in the z-plane defined by points $x + iy$, and let S be a region of the w-plane defined by points $u + iv$. If

$$w = f(z) \tag{8.1}$$

defines a set of points in S for each point in R, eq. 8.1 is a *mapping* or a *transformation* from R to S. This mapping can also be written as two real transformation equations

$$u = u(x, y) \tag{8.2a}$$

and

$$v = v(x, y) \tag{8.2b}$$

Each point w that is obtained by mapping the point z by eq. 8.1 is called an *image of z*. The point z that maps to the image w is the *source of w*. The set of all source points is called the *domain* of the mapping.

If each point in R has only one image in S and each point in S has only one source in R, the mapping is said to be *one-to-one*. If eq. 8.1 generates two or more images in S for a each source in R, the mapping is called *one-to-many*. If two or more sources in R have the same image in S, the mapping is called *many-to-one*.

Let w_1 and w_2 be the images of z_1 and z_2, respectively. If $f(z)$ is a one-to-one mapping for all z in R, then

$$w_1 = w_2 \Rightarrow z_1 = z_2 \tag{8.3a}$$

and

$$z_1 = z_2 \Rightarrow w_1 = w_2 \tag{8.3b}$$

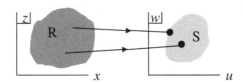

Figure 8.1a

A one-to-one mapping

If $f(z)$ is a many-to-one mapping

$$z_1 = z_2 \Rightarrow w_1 = w_2 \tag{8.4a}$$

but

$$w_1 = w_2 \not\Rightarrow z_1 = z_2 \tag{8.4b}$$

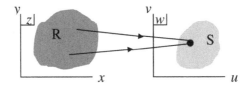

Figure 8.1b

A two-to-one mapping

If $f(z)$ is a one-to-many mapping

$$w_1 = w_2 \Rightarrow z_1 = z_2 \tag{8.5a}$$

but

$$z_1 = z_2 \not\Rightarrow w_1 = w_2 \tag{8.5b}$$

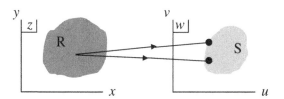

Figure 8.1c

A one-to-two mapping

The number of sources that map to one image, or the number of images obtained from one source is referred to as the *multiplicity* of the mapping .

Example 8.1: Mappings of various multiplicities

In the following three examples, where applicable, we take $N \geq 2$ to be an integer and α and β to be constants with $\alpha \neq 0$.

(a) The general form of the *linear transformation*

$$w = \alpha z + \beta \tag{8.6}$$

with $\alpha \neq 0$. This is discussed in detail in section 8.2. Because

$$w_2 - w_1 = \alpha(z_2 - z_1) \tag{8.7}$$

we see that if $w_1 = w_2$, this requires that $z_1 = z_2$. Conversely $z_1 = z_2$ requires $w_1 = w_2$. That is,

$$w_1 = w_2 \iff z_1 = z_2 \tag{8.8}$$

Therefore, the linear mapping is one-to-one.

(b) For the mapping

$$w = \alpha z^N + \beta \tag{8.9}$$

there are N different sources in the form

$$z = re^{i\phi/N} e^{i2\pi k/N} \quad k = 0,1,\ldots,(N-1) \tag{8.10}$$

that will map to one value of w, which we can express as

$$w = \alpha r^N e^{i\phi} + \beta \tag{8.11}$$

For example, let

$$z_1 = r e^{i\phi/N} \tag{8.12a}$$

and

$$z_2 = r e^{i\phi/N} e^{i2\pi/N} \neq z_1 \tag{8.12b}$$

Then

$$w_1 = \alpha r^N e^{i\phi} + \beta \tag{8.13a}$$

and

$$w_2 = \alpha r^N e^{i\phi} e^{i2\pi} + \beta = \alpha r^N e^{i\phi} + \beta = w_1 \tag{8.13b}$$

Therefore, because both z_1 and z_2 map into one point

$$z_1 = z_2 \Rightarrow w_1 = w_2 \tag{8.4a}$$

but

$$w_1 = w_2 \not\Rightarrow z_1 = z_2 \tag{8.4b}$$

Thus, N different values of z have the same image, so the mapping of eq. 8.9 is an N-to-one mapping.
(c) Consider the mapping

$$w = \alpha z^{1/N} + \beta \tag{8.14}$$

Let

$$z_1 = r e^{i\theta} \tag{8.15a}$$

and

$$z_2 = re^{i\theta}e^{i2\pi} = z_1 \tag{8.15b}$$

Then

$$w_1 = r^{1/N}e^{i\theta/N} \tag{8.16a}$$

and

$$w_2 = r^{1/N}e^{i\theta/N}e^{i2\pi/N} \neq w_1 \tag{8.16b}$$

Therefore,

$$w_1 = w_2 \Rightarrow z_1 = z_2 \tag{8.5a}$$

but

$$z_1 = z_2 \not\Rightarrow w_1 = w_2 \tag{8.5b}$$

Because there are N different images of each source, the mapping of eq. 8.14 is one-to-N. □

Into and onto mappings

Let R be a region in the z-plane and S be a region in the w-plane. If every point in S is the image of one or more points in R under some mapping, then the transformation is said to be a mapping of R *onto* S. If there is at least one point in S that is not an image of some point in R, the transformation is a mapping of R *into* S. This is illustrated in figs. 8.2.

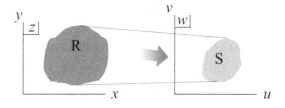

Figure 8.2a
Mapping of R onto S

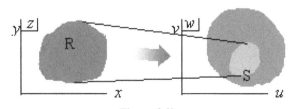

Figure 8.2b

Mapping of R into S

Example 8.2: Into and onto mappings

Let z_0 be a point in the upper half of the z-plane. It is shown in ex. 8.9 that for $\text{Im}(z_0) \neq 0$, the mapping

$$w = \frac{z - z_0}{z - z_0^*} \tag{8.17}$$

maps the entire upper half of the z-plane, including the real axis, to the unit circle centered at the origin of the w-plane.

Let the region R be the upper half of the z-plane, including the x-axis. Taking S to be the unit circle centered at the origin of the w-plane, the mapping of eq. 8.17 maps R onto S. If S is defined to be any region containing the unit circle plus at least one additional point outside the unit circle, eq. 8.17 maps R into S. □

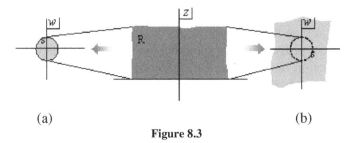

(a)   (b)

Figure 8.3

Mapping by the transformation of eq. 8.17, of the upper half of the z-plane
(a) onto the unit circle and (b) into a region within the entire w-plane

Invariant points

If, for a given point z, a mapping $f(z)$ has the same value as z, the point is called an *invariant point* or a *fixed point* of the mapping. Such points satisfy

$$w = f(z) = z \tag{8.18}$$

Example 8.3: Invariant points of a mapping

The mapping

$$w = z^3 \tag{8.19}$$

has three invariant points found by solving

$$z = z^3 \tag{8.20}$$

Clearly these invariant points are $z = 0, \pm 1$, and they map to the points $w = 0, \pm 1$. □

Mapping of curves

Let λ represent a set of parameters that are independent of x and y (and are therefore independent of u and v). For a specific set of values of λ, a curve in the z-plane is a set of points constrained by an equation of the form

$$G(x, y; \lambda) = 0 \tag{8.21}$$

The function G is specific to the type of curve one is describing. For example, a family of straight lines is specified by

$$G(x, y; \{a,b\}) = y - ax - b = 0 \tag{8.22a}$$

with specific values of the parameters a and b defining a particular line in the family. The family of circles centered at a point (a,b) and of radius R, is described by

$$G(x, y; \{a,b,R\}) = (x-a)^2 + (y-b)^2 - R^2 = 0 \tag{8.22b}$$

Assigning specific values to the set of parameters a, b, and R defines one of the circles in that family.

Because x and y can be expressed in terms of z and z^*, the equation of a curve given in eq. 8.21 can also be expressed as

$$G(z, z^*; \lambda) = 0 \tag{8.23}$$

For example, the equation of a straight line of slope m passing through the origin can be expressed as

$$G(z, z^*; \{m,0\}) = (i - m)z^* - (i + m)z = 0 \tag{8.24a}$$

and the equation of a circle centered at the origin, of radius R is

$$G(z, z^*; \{0, 0, R\}) = zz^* - R^2 = 0 \qquad (8.24b)$$

Inverting eqs. 8.2 to express x and y as functions of u and v, the equation of a curve can also be expressed in the w-plane as

$$G(x(u, v), y(u, v); \lambda) = 0 \qquad (8.25a)$$

which can be renamed as a function H with

$$H(u, v; \lambda) = H(w, w^*; \lambda) = 0 \qquad (8.25b)$$

Equation 8.25b describes a curve Γ in the w-plane that is the image of the curve C in the z-plane described in eq. 8.25a.

Example 8.4: Mapping a straight line

In the z-plane, the equation of a straight line of slope 1 passing through the origin is given by

$$y - x = 0 \qquad (8.26a)$$

This is expressed in the form of eq. 8.24a as

$$G(z, z^*; \{1, 0\}) = (i - 1)z^* - (i + 1)z = 0 \qquad (8.26b)$$

The mapping

$$w = z^2 \qquad (8.27)$$

transforms the straight line to

$$H(w, w^*; \{1, 0\}) = (i - 1)\sqrt{w^*} - (i + 1)\sqrt{w} = 0 \qquad (8.28a)$$

which can be simplified as

$$w + w^* = 2u = 0 \qquad (8.28b)$$

This is the equation of the entire imaginary axis of the w-plane.

Writing eq. 8.27 in terms of its real and imaginary parts, we have

$$u = x^2 - y^2 \tag{8.29a}$$

and

$$v = 2xy \tag{8.29b}$$

Then, using eq. 8.26a, these become

$$u = 0 \tag{8.30a}$$

and

$$v = 2x^2 = 2y^2 \tag{8.30b}$$

which restricts the image of the straight line to the origin and the positive imaginary axis of the *w*-plane.

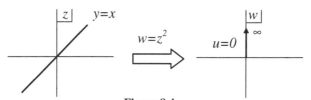

Figure 8.4
Transformation of the line $y = x$ onto the origin and positive
imaginary axis of the *w*-plane

Points on the line $y = x$ are of the form (x, x) or equivalently (y, y). We note that both (x, x) and $(-x, -x)$ map to the point $(u,v) = (0, 2x^2)$. Therefore, the transformation of eq. 8.27 is a two-to-one mapping.

To determine if there are any invariant points, we set $w = z$ in eq. 8.27. Then from

$$z = z^2 \tag{8.31}$$

we see from this that the mapping has two invariant points at $z = 0, 1$. □

Tangent to a curve

Let C be a curve in the z-plane and let Γ be the image of C in the w-plane. A differential segment dw along Γ is related to dz, a differential segment along C, by

$$dw = \frac{df}{dz}dz = f'(z)dz \tag{8.32}$$

As noted in eqs. 8.25, even though C and Γ are described by different equations, both curves are defined by the same constants and are therefore parametrized by the same λ. Let w_0 be a point on Γ that is the image of z_0 on C. Then from eq. 8.32, $\tau(w_0)$, the tangent to Γ at w_0 is related to $t(z_0)$, the tangent to C at z_0 by

$$\tau(w_0) \equiv \frac{dw}{d\lambda}\bigg)_{w_0} = f'(z_0)\frac{dz}{d\lambda}\bigg)_{z_0} \equiv f'(z_0)t(z_0) \tag{8.33}$$

In order for eqs. 8.32 and 8.33 to have meaning, $f'(z_0)$ must be defined. This requires that $f'(z)$ be analytic everywhere in a neighborhood containing z_0. But if $f'(z_0) = 0$, the differential line element along Γ at w_0 cannot be related to the differential line element along C at z_0. When $f'(z_0) = 0$, z_0 is called a *critical point* on a curve. We see from eq. 8.33 that z_0 cannot be a critical point if the tangent to C at z_0 is related to the tangent to Γ at w_0. In the discussion that follows, we require that z_0 not be a critical point.

Expressing the three factors in eq. 8.33 in polar form, we write

$$\tau(w_0) \equiv |\tau(w_0)|e^{i\psi(w_0)} \tag{8.34a}$$

$$f'(z_0) \equiv |f'(z_0)|e^{i\phi(z_0)} \tag{8.34b}$$

and

$$t(z_0) \equiv |t(z_0)|e^{i\theta(z_0)} \tag{8.34c}$$

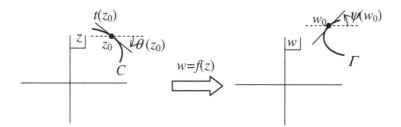

Figure 8.5

Tangent to a curve in the z-plane and its image in the w-plane

under the mapping $w = f(z)$

Equation 8.34a shows that $\tau(w_0)$ is oriented at an angle $\psi(w_0)$ to the u-axis and from eq. 8.34c, we see that $t(z_0)$ makes an angle $\theta(z_0)$ with the x-axis. Then eq. 8.33 becomes

$$\left|\tau(w_0)\right|e^{i\psi(w_0)} = \left|f'(z_0)\right|\left|t(z_0)\right|e^{i[\phi(z_0)+\theta(z_0)]} \tag{8.35}$$

Thus, the magnitude of $\tau(w_0)$ is related to the magnitude of $t(z_0)$ by

$$\left|\tau(w_0)\right| = \left|f'(z_0)\right|\left|t(z_0)\right| \tag{8.36a}$$

and the angle $\tau(w_0)$ makes with the u-axis is related to the angle between $t(z_0)$ and the x-axis by

$$\psi(w_0) = \phi(z_0) + \theta(z_0) \tag{8.36b}$$

From this, we learn that because z_0 is not a critical point, if $f(z)$ is analytic at z_0, the mapping $f(z)$ affects the tangent to a curve C at z_0 in two ways:

- $\left|t(z_0)\right|$ is modified by the multiplicative factor $\left|f'(z_0)\right|$, called the *magnification factor* (or simply the *magnification*) of the mapping. It is the factor by which the magnitude of the tangent is enlarged or shrunk by the mapping. Because $\left|f'(z_0)\right|$ depends on z_0, the magnification varies from point to point on C.
- The angle between the tangent to C and the x-axis at z_0 differs from the angle the tangent to Γ makes with the u-axis at w_0 by an additive amount $\phi(z_0)$, called the *argument* of the mapping. Because ϕ depends on z_0, it too varies from point to point on C.

Example 8.5: Magnification and argument of a mapping

(a)   In ex. 8.4 we considered the transformation of a straight line of unit slope under the mapping

$$w = z^2 \tag{8.27}$$

Because

$$f'(z_0) = 2z_0 \tag{8.37}$$

we see that the magnification of the mapping is

$$\left| f'(z_0) \right| = 2 \left| z_0 \right| \tag{8.38a}$$

For the tangent to the line of slope 1, the argument of every point z_0 is

$$\phi = \pi/4 \tag{8.38b}$$

Therefore, each point on the line in the z-plane can be expressed as

$$z = re^{i\pi/4} \tag{8.39a}$$

and the image of each point on the line can be written as

$$w = r^2 e^{i\pi/2} = ir^2 \tag{8.39b}$$

That is, the image of each point on the line lies on the positive imaginary axis in the w-plane as discussed in ex. 8.4.

(b) Let us consider the transformation of the straight line of unit slope under the mapping

$$w = e^z = e^x e^{iy} \tag{8.40a}$$

from which

$$f'(z_0) = e^{x_0} e^{iy_0} \tag{8.40b}$$

Therefore, the magnification is

$$\left| f'(z_0) \right| = e^{x_0} \tag{8.41a}$$

and

$$\phi(z_0) = y_0 \tag{8.41b}$$

Let z_0 and z_1 be two points on the line. The distance between them

$$D_z = \left| z_1 - z_0 \right| \tag{8.42a}$$

is magnified by the mapping so that the distance between the images of z_0 and z_1 is

$$D_w = e^{x_0} \left| (z_1 - z_0) \right| \tag{8.42b}$$

and the angle between the tangent to the image curve at w_0 and the u-axis is given by

$$\psi = y_0 + \pi/4 \tag{8.43}$$

$\square$

Conformal and isogonal mappings

Let C_1 and C_2 be two curves in the z-plane that intersect at z_0 and let Γ_1 and Γ_2 be the images of C_1 and C_2 under the mapping $f(z)$. Then Γ_1 and Γ_2 intersect at w_0, the image of z_0.

We require $f(z)$ to be analytic at all points on C_1 and C_2 and that z_0 is not a critical point. We take λ_1 to represent the set of parameters that define C_1 and Γ_1. Likewise, λ_2 parametrizes C_2 and Γ_2. Then, from eqs. 8.34 the tangents to the two curves at the point of intersection are given by

$$\tau_1(w_0) = f'(z_0) \frac{dz}{d\lambda_1} \bigg)_{z_0} = f'(z_0) t_1(z_0) \tag{8.44a}$$

and

$$\tau_2(w_0) = f'(z_0)\frac{dz}{d\lambda_2}\bigg)_{z_0} = f'(z_0)t_2(z_0) \tag{8.44b}$$

Referring to eqs. 8.35 and 8.36, the polar forms of these equations yield

$$\left|\tau_1(w_0)\right| = \left|f'(z_0)\right|\left|t_1(z_0)\right| \tag{8.45a}$$

and

$$\left|\tau_2(w_0)\right| = \left|f'(z_0)\right|\left|t_2(z_0)\right| \tag{8.45b}$$

with

$$\psi_1(w_0) = \theta_1(z_0) + \phi(z_0) \tag{8.46a}$$

and

$$\psi_2(w_0) = \theta_2(z_0) + \phi(z_0) \tag{8.46b}$$

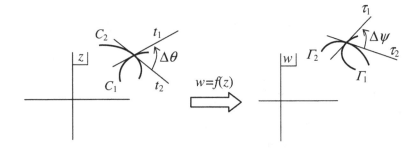

Figure 8.6

The tangents to two curves under the mapping $w = f(z)$

Because both $\left|f'(z_0)\right|$ and ϕ are the same for the transformation of both curves, eqs. 8.45 yield

$$\frac{\left|\tau_1(w_0)\right|}{\left|\tau_2(w_0)\right|} = \frac{\left|t_1(z_0)\right|}{\left|t_2(z_0)\right|} \tag{8.47}$$

and from eqs. 8.46 we obtain

$$\Delta\psi(w_0) \equiv \psi_1(w_0) - \psi_2(w_0) = \theta_1(z_0) - \theta_2(z_0) \equiv \Delta\theta(z_0) \qquad (8.48)$$

Referring to fig. 8.6, eq. 8.48 states that the magnitude and sign of the angle subtended by the tangents to the two curves in the z-plane is the same as the magnitude and sign of the angle subtended by the tangents to the images of those curves in the w-plane. A mapping under which the magnitude and sign of the angle between the tangents to two curves is unchanged is called a *conformal mapping*. As discussed above, eqs. 8.47 and 8.48 result only from a mapping by a function that is analytic at z_0 and for which z_0 is not a critical point. Thus, an equivalent definition of a conformal mapping is one that is generated by a function that is analytic in a region surrounding z_0 and for which z_0 is not a critical point.

The sign of the difference between the angles of eq. 8.48 is called the *sense* of the angle subtended by the tangents. A transformation for which

$$\left|\psi_1(w_0) - \psi_2(w_0)\right| = \left|\theta_1(z_0) - \theta_2(z_0)\right| \qquad (8.49)$$

but for which the sense of the angle may not be preserved is called an *isogonal transformation*. As such, a conformal transformation is isogonal but an isogonal transformation is not necessarily conformal.

Inverse mapping

Let

$$w = f(z) \qquad (8.1)$$

be a conformal mapping at all z in a region R. Then $f(z)$ is analytic everywhere in R and R does not contain a critical point of $f(z)$.

An *inverse mapping* is defined by solving eq. 8.1 for z to find the function $g(w)$, defined by

$$z = g(w) \qquad (8.50)$$

which maps points in a region S of the w-plane to points in R of the z-plane. Substituting eq. 8.50 into eq. 8.1 we obtain

$$w = f\left[g(w)\right] \qquad (8.51a)$$

and substituting eq. 8.1 into eq. 8.50 yields

$$z = g[f(z)] \tag{8.51b}$$

Equations 8.51 are the defining equations of functions that are inverses of each other.

Taking the differentials of eqs. 8.1 and 8.50 we have

$$dw = f'(z)dz \tag{8.32}$$

and

$$dz = g'(w)dw \tag{8.52}$$

From these, we see that functions that are inverses of each other satisfy

$$g'(w) = \frac{1}{f'(z)}\bigg)_{z=g(w)} \tag{8.53a}$$

or equivalently

$$f'(z) = \frac{1}{g'(w)}\bigg)_{w=f(z)} \tag{8.53b}$$

Because $f(z)$ is analytic everywhere in R, $f'(z)$ is defined at all z in R. Because R contains no critical points, $f'(z) \neq 0$ everywhere in R. Therefore, $g'(w)$ is defined at all w in S and so is analytic everywhere in S. Thus $g(w)$ is analytic at all w in S.

The results expressed in eqs. 8.53 allow us to develop other relations between the derivatives of $f(z)$ and $g(w)$. With

$$w_0 \equiv f(z_0) \tag{8.54}$$

the Taylor series expansion of $f(z)$ around z_0 is given by

$$w - w_0 = f'(z_0)(z - z_0) + \frac{1}{2!}f''(z_0)(z - z_0)^2 + \cdots \tag{8.55}$$

Because $g(w)$ is analytic in the region S containing w_0, then it has a Taylor expansion around w_0 given by

$$z - z_0 = g'(w_0)(w - w_0) + \frac{1}{2!}g''(w_0)(w - w_0)^2 + \cdots \tag{8.56}$$

where

$$z_0 = g(w_0) \tag{8.57}$$

Substituting $z - z_0$ as given in eq. 8.56 into eq. 8.55, we obtain

$$\begin{aligned}
w - w_0 &= f'(z_0)\left[g'(w_0)(w - w_0) + \frac{1}{2!}g''(w_0)(w - w_0)^2 + \cdots \right] \\
&\quad + \frac{1}{2!}f''(z_0)\left[g'(w_0)(w - w_0) + \frac{1}{2!}g''(w_0)(w - w_0)^2 + \cdots \right]^2 + \cdots \\
&= f'(z_0)g'(w_0)(w - w_0) \\
&\quad + \frac{1}{2!}\left[f'(z_0)g''(w_0) + f''(z_0)\left(g'(w_0)\right)^2 \right](w - w_0)^2 + \cdots
\end{aligned} \tag{8.58}$$

Equating the coefficients of $(w - w_0)$, we obtain

$$f'(z_0)g'(w_0) = 1 \tag{8.59a}$$

which is eq. 8.53 evaluated at $z = z_0$ and $w = w_0$. The left-hand side of eq. 8.58 has no terms in $(w - w_0)^2$, therefore the corresponding coefficient on the right-hand side of eq. 8.58 is zero. Thus

$$f'(z_0)g''(w_0) + f''(z_0)\left(g'(w_0)\right)^2 = 0 \tag{8.59b}$$

In probl. 4, the reader will show that

$$\frac{1}{3!}\left[f'(z_0)g'''(w_0) + f'''(z_0)\left(g'(w_0)\right)^3 \right] + \frac{1}{2!}f''(z_0)g''(w_0)g'(w_0) = 0 \tag{8.59c}$$

Relations such as eqs. 8.59 are derived by substituting $(w - w_0)$ from eq. 8.55 into eq. 8.56. Those equations can also be obtained by interchanging the corresponding derivatives of $f(z_0)$ and $g(w_0)$. Thus, for example, from eqs. 8.59b and 8.59c, we have

$$g'(w_0)f''(z_0) + g''(w_0)(f'(z_0))^2 = 0 \qquad (8.60a)$$

and

$$\frac{1}{3!}\left[g'(w_0)f'''(z_0) + g'''(w_0)(f'(z_0))^3 \right] + \frac{1}{2!}g''(w_0)f''(z_0)f'(z_0) = 0$$
$$(8.60b)$$

Example 8.6: The inverse of a mapping

In order to ensure that all derivatives are nonzero, we take α to be a noninteger constant and consider the mapping

$$w = f(z) = z^\alpha \qquad (8.61a)$$

where $f(z)$ is analytic in any region R of a multisheeted complex z-plane that does not contain the origin. The inverse of the mapping, which is also analytic in a multisheeted w-plane in a region S that does not contain the origin, is given by

$$z = g(w) = w^{1/\alpha} \qquad (8.61b)$$

so that

$$z = g[f(z)] = \left[z^\alpha \right]^{1/\alpha} = z \qquad (8.62a)$$

and

$$w = f[g(w)] = \left[w^{1/\alpha} \right]^\alpha = w \qquad (8.62b)$$

We note that

$$f'(z) = \alpha z^{\alpha-1} \qquad (8.63a)$$

and

$$g'(w) = \frac{1}{\alpha} w^{\frac{1}{\alpha}-1} = \frac{1}{\alpha} \left(z^\alpha\right)^{\frac{1}{\alpha}-1} = \frac{1}{\alpha z^{\alpha-1}}$$

(8.63b)

Therefore, in any regions R and S for which z and w are not zero,

$$f'(z)g'(w) = 1$$

(8.59a)

In prob. 5, the reader will demonstrate that eqs. 8.59b and 8.59c are also satisfied by the mapping $w = z^\alpha$ and its inverse. $\square$

Conformal mapping of a differential area

Let $f(z)$ be a conformal mapping that transforms a region R of the z-plane onto a region S of the w-plane. A differential line element in R from the point (x, y) to the point $(x + dx, y + dy)$ is described by

$$dz = dx + idy$$

(8.64a)

Similarly the differential line element in S from (u,v) to $(u + du, v + dv)$ is

$$dw = du + idv$$

(8.64b)

Referring to fig. 8.7, we define a rectangular differential area element in R with sides parallel to the x-axis and the y-axis. These sides are given by

$$dz_1 = dx$$

(8.65a)

and

$$dz_2 = idy$$

(8.65b)

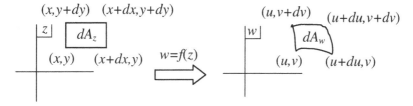

Figure 8.7

Differential area elements in the z- and w-planes

The images of dz_1 and dz_2 are differential curves in the w-plane given by

$$dw_1 = du_1 + i\,dv_1 \tag{8.66a}$$

and

$$dw_2 = du_2 + i\,dv_2 \tag{8.66b}$$

In prob. 22 of chapter 2, the reader was asked to prove that the area of a parallelogram of sides z_1 and z_2 is given by

$$A = \left| \mathrm{Im}(z_1^* z_2) \right| \tag{8.67}$$

Using this result, the differential area element of the rectangle in the z-plane is the expected result

$$dA_z = dx\,dy \tag{8.68a}$$

Substituting eqs. 8.66 into eq. 8.67, the differential area element of the parallelogram in the w-pane is

$$dA_w = du_1\,dv_2 - du_2\,dv_1 \tag{8.68b}$$

Because dz_1 is along a line defined by $y = $ constant, and dz_2 is along a line described by $x = $ constant, the images of these line elements in the w-plane can be written

$$dw_1 = \left(\left. \frac{df}{dz} \right|_{y=constant} \right) * dz_1 = \frac{\partial f}{\partial x} dx \tag{8.69a}$$

and

$$dw_2 = \left(\left. \frac{df}{dz} \right|_{x=constant} \right) * dz_2 = \frac{1}{i}\frac{\partial f}{\partial y} i\,dy = \frac{\partial f}{\partial y} dy \tag{8.69b}$$

With

$$w = f(z) = u(x,\,y) + iv(x,\,y) \tag{8.70}$$

the partial derivatives of eqs. 8.69 are given by

$$\frac{\partial f}{\partial x} = \frac{\partial u}{\partial x} + i\frac{\partial v}{\partial x} \tag{8.71a}$$

and

$$\frac{\partial f}{\partial y} = \frac{\partial u}{\partial y} + i\frac{\partial v}{\partial y} \tag{8.71b}$$

Then

$$dw_1 = \left(\frac{\partial u}{\partial x} + i\frac{\partial v}{\partial x}\right)dx \tag{8.72a}$$

and

$$dw_2 = \left(\frac{\partial u}{\partial y} + i\frac{\partial v}{\partial y}\right)dy \tag{8.72b}$$

Therefore, the differential area element in the *w*-plane is given by

$$dA_w = \left|\text{Im}(dw_1^* dw_2)\right| = \left|\frac{\partial u}{\partial x}\frac{\partial v}{\partial y} - \frac{\partial u}{\partial y}\frac{\partial v}{\partial x}\right|dx\,dy$$
$$= \left|\frac{\partial u}{\partial x}\frac{\partial v}{\partial y} - \frac{\partial u}{\partial y}\frac{\partial v}{\partial x}\right|dA_z \tag{8.73}$$

It is expected that the reader recognizes that this can be expressed in terms of a 2 x 2 determinant

$$\left|\frac{\partial u}{\partial x}\frac{\partial v}{\partial y} - \frac{\partial u}{\partial y}\frac{\partial v}{\partial x}\right| = \begin{vmatrix} \dfrac{\partial u}{\partial x} & \dfrac{\partial u}{\partial y} \\ \dfrac{\partial v}{\partial x} & \dfrac{\partial v}{\partial y} \end{vmatrix} \tag{8.74}$$

This 2 x 2 determinant, called the *Jacobian determinant of the transformation*, is also denoted by

$$\begin{vmatrix} \dfrac{\partial u}{\partial x} & \dfrac{\partial u}{\partial y} \\[2mm] \dfrac{\partial v}{\partial x} & \dfrac{\partial v}{\partial y} \end{vmatrix} \equiv \frac{\partial(u,v)}{\partial(x,y)} \equiv J(u,v;x,y) \equiv J(w;z) \tag{8.75}$$

Because $f(z)$ is analytic, $u(x, y)$ and $v(x, y)$ satisfy the CR conditions

$$\frac{\partial u}{\partial x} = \frac{\partial v}{\partial y} \tag{3.7a}$$

and

$$\frac{\partial u}{\partial y} = -\frac{\partial v}{\partial x} \tag{3.7b}$$

at all z in R. Therefore, the Jacobian determinant can be written

$$J(w;z) = \left(\frac{\partial u}{\partial x}\right)^2 + \left(\frac{\partial v}{\partial x}\right)^2 = \left(\frac{\partial u}{\partial y}\right)^2 + \left(\frac{\partial v}{\partial y}\right)^2 \tag{8.76}$$

which is positive for all x and y. It was shown in chapter 3 that when $f(z)$ is analytic in a neighborhood containing z, then

$$f'(z) = \frac{\partial u}{\partial x} + i\frac{\partial v}{\partial x} \tag{3.11a}$$

or equivalently

$$f'(z) = \frac{\partial v}{\partial y} - i\frac{\partial u}{\partial y} \tag{3.11b}$$

Substituting either of these expressions into eq. 8.76 yields

$$J(w;z) = \left|f'(z)\right|^2 = \left|\frac{dw}{dz}\right|^2 \tag{8.77}$$

8.2 Linear and Bilinear Transformations

The linear transformation

As noted earlier, the general form of the *linear transformation* is

$$w = \alpha z + \beta \tag{8.6}$$

where α and β are constants and $\alpha \neq 0$. Clearly, the inverse of this mapping is

$$z = \frac{1}{\alpha} w - \frac{\beta}{\alpha} \tag{8.78}$$

which is a linear mapping from the w-plane to the z-plane.

By setting $w = z$ in eq. 8.6, we see that when $\alpha \neq 1$, the linear transformation has one invariant point at

$$z = \frac{\beta}{1 - \alpha} \tag{8.79}$$

Translation

If $\alpha = 1$, eq. 8.6 becomes

$$w = z + \beta \tag{8.80}$$

which describes a *translation* by the constant β of the points being mapped. Setting $w = z$ in eq. 8.80, we see that for $\beta \neq 0$ there is no invariant point for a translation. That is, when a region in the z-plane undergoes a translation, all points in the region are moved by the same nonzero amount. No point remains at its original position.

Let z_1 and z_2 be points in the z-plane and let w_1 and w_2 be their images in the w-plane. The lines between these points are described by $z_1 - z_2$ and $w_1 - w_2$. If the mapping is the translation of eq. 8.80, we see that

$$w_1 - w_2 = z_1 - z_2 \tag{8.81}$$

That is, a translation does not modify the length of a line or its orientation, but only changes its position with respect to the coordinate axes.

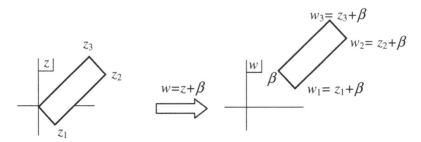

Figure 8.8

Translation of a polygon by the mapping of eq. 8.80

Because a polygon is constructed from three or more lines, the size and orientation of a polygon are not affected by a translation; only the position of the polygon is changed.

Magnification and rotation

When $\beta = 0$ and α is expressed in polar form, the linear mapping becomes

$$w = |\alpha| e^{i\gamma} z \qquad\qquad (8.82)$$

Then, a line between two points transforms as

$$w_1 - w_2 = |\alpha|(z_1 - z_2) e^{i\gamma} \qquad\qquad (8.83)$$

Therefore, the length of a line in the z-plane, $|z_1 - z_2|$ becomes magnified by a factor $|\alpha|$ and the line is rotated by an angle γ. Thus, the lengths of the sides of a polygon and the orientation of the polygon with respect to the axes will be modified, but its shape will remain unchanged by the linear transformation with $\beta = 0$.

Example 8.7: Transformation of a polygon under the mapping $w = \alpha z$

A polygon with vertices at $z_0 = 0$, $z_1 = 1 + i$, $z_2 = 2$ and $z_3 = 1 - i$ is the square with sides of length $\sqrt{2}$ shown on the left side of fig. 8.9. Under the mapping

$$w = 2e^{i\pi/4} z \qquad\qquad (8.84)$$

the vertices are mapped onto the points

$$w_0 = 0 \tag{8.85a}$$

$$w_1 = 2\sqrt{2}e^{i\pi/2} \tag{8.85b}$$

$$w_2 = 4e^{i\pi/4} \tag{8.85c}$$

and

$$w_3 = 2\sqrt{2} \tag{8.85d}$$

These four points form the vertices of a square of side $2^{3/2}$ shown on the right side of fig. 8.9. □

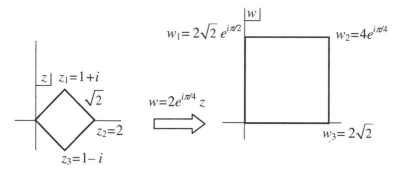

Figure 8.9

Magnification and rotation of a square

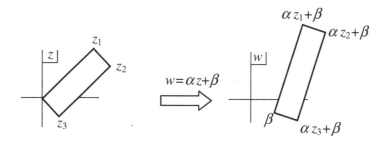

Figure 8.10

Translation, magnification, and rotation

of a rectangle under a general linear mapping

From the discussions above, we see that the general linear mapping generates a translation plus a magnification and rotation of a polygon, but does not affect the shape of a figure as shown in fig. 8.10. Thus, for example, under a linear mapping, a line will map to a line, a rectangle maps to a rectangle, a circle maps to a circle, and so on. The argument of α is the angle through which the figure is rotated.

We have seen that if one knows the values of α and β, it is straightforward to determine the image of a polygon in the z-plane under a given linear mapping. If one knows the coordinates of two points on the original polygon in the z-plane, and the images of those two points in the w-plane, one can determine α and β and thus the linear mapping. Because it is straightforward to determine the inverse of a linear mapping, if one knows the coordinates of the vertices of the image of a polygon in the w-plane, one can easily determine the coordinates of the vertices of the source polygon in the z-plane under a given linear mapping.

Example 8.8: Transformation of a triangle under a linear mapping

An isosceles triangle with base of length 6 and with sides of length 5 is shown in fig. 8.11a.

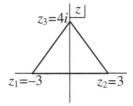

Figure 8.11a
Isosceles triangle to be mapped

(a) The image of this triangle under the linear mapping

$$w = (1+i)z + (1-i) \tag{8.86}$$

is found by determining that the images of the vertices of the triangle are

$$w_1 = (1+i)z_1 + (1-i) = -2 - 4i \tag{8.87a}$$

$$w_2 = (1+i)z_2 + (1-i) = 4 + 2i \tag{8.87b}$$

and

$$w_3 = (1+i)z_3 + (1-i) = -3 + 3i \tag{8.87c}$$

The lengths of the sides of the triangle in the w-plane are given by

$$|w_1 - w_3| = |1 - 7i| = 5\sqrt{2} \tag{8.88a}$$

$$|w_2 - w_3| = |7 - i| = 5\sqrt{2} \tag{8.88b}$$

and

$$|w_1 - w_2| = |6 + 6i| = 6\sqrt{2} \tag{8.88c}$$

Therefore, the image of the isosceles triangle of fig. 8.11a is the isosceles triangle shown in fig. 8.11b. All the points on the source triangle have been translated by $(1 - i)$, all lines have been rotated by $\pi/4$, and the lengths of the sides have been magnified by $\sqrt{2}$.

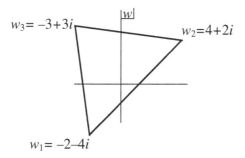

Figure 8.11b

Image of the triangle of fig. 8.11a under the mapping of eq. 8.86

(b) We next determine the linear mapping that transforms the isosceles triangle of fig. 8.11a to the isosceles triangle of fig. 8.11c. To find the mapping that transforms the triangle of fig. 8.11a to that of fig. 8.11c, we determine α and β for the mapping

$$w = \alpha z + \beta \tag{8.6}$$

by substituting the coordinates of any two vertices of the source triangle and the image triangle into eq. 8.6. It is a straightforward exercise to show that the mapping is

$$w = -2iz \qquad\qquad (8.89)$$

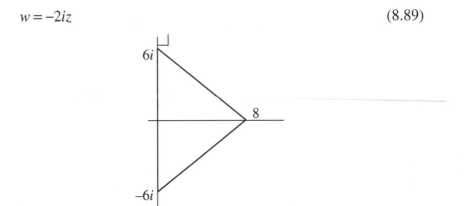

Figure 8.11c
Image of the triangle of fig. 8.11a under a linear mapping to be determined

(c) Let us determine the source triangle that transforms into the image shown in fig. 8.11c under the mapping

$$w = -2i(z + 3) \qquad\qquad (8.90)$$

We do this by finding the sources of $w_1 = 8$, $w_2 = 6i$, and $w_3 = -6i$ from the inverse mapping

$$z = \frac{i}{2}w - 3 \qquad\qquad (8.91)$$

We obtain $z_1 = -3 + 4i$, $z_2 = -6$, and $z_3 = 0$. The triangle with vertices at these coordinates is shown in fig. 8.11d. □

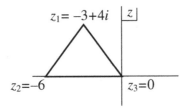

Figure 8.11d
Source of the triangle of fig. 8.11c under the mapping of eq. 8.90

The bilinear transformation

The general form of the *bilinear* (also called the *fractional* or *Moebius*) transformation is

$$w = \frac{\alpha z + \beta}{\gamma z + \delta} \tag{8.92a}$$

which can also be written in the form

$$w = \frac{1}{\gamma}\left[\alpha + \frac{\beta\gamma - \alpha\delta}{\gamma z + \delta}\right] \tag{8.92b}$$

If $\gamma = 0$, the bilinear mapping reduces to a linear mapping, which has already been discussed. Thus, we require that $\gamma \neq 0$. Then, the bilinear mapping has a simple pole at $z = -\delta/\gamma$.
  Clearly, if $\beta\gamma - \alpha\delta = 0$, eq. 8.92b becomes

$$w = \frac{\alpha}{\gamma} \tag{8.93}$$

which is a transformation of all z to one value of w. Such a transformation is an infinite-to-one mapping. For this reason, we also require $\beta\gamma - \alpha\delta \neq 0$. Then, the mapping has no zero at any finite z. With these properties, we see that the bilinear mapping is a conformal mapping at all finite z except at $z = -\delta/\gamma$.
  If $\alpha = 0$, the bilinear mapping becomes

$$w = \frac{\beta}{\gamma z + \delta} \tag{8.94}$$

which is called an *inversion*. This can be written in the form

$$w = \frac{\beta/\gamma}{z + \delta/\gamma} \equiv \frac{\beta'}{z + \delta'} \tag{8.95}$$

Using the form given in eq. 8.92b, it is straightforward to see that the mapping of the tangent to a curve by a bilinear mapping is given by

$$\frac{dw}{d\lambda} = \frac{(\alpha\delta - \beta\gamma)}{(\gamma z + \delta)^2}\frac{dz}{d\lambda} \tag{8.96a}$$

or

$$\tau(w) = \frac{(\alpha\delta - \beta\gamma)}{(\gamma z + \delta)^2} t(z) \tag{8.96b}$$

Setting $w = z$, the two invariant points are found to be at

$$z = \frac{(\alpha - \delta) \pm \sqrt{(\alpha - \delta)^2 - 4\beta\gamma}}{2\gamma} \tag{8.97}$$

Solving eq. 8.92a for z, we see that the inverse mapping

$$z = \frac{\delta w - \beta}{-\gamma w + \alpha} \tag{8.98}$$

is also a bilinear transformation.

The multiplicity of the bilinear mapping is found by letting w_1 and w_2 be the images of z_1 and z_2, respectively. Then

$$w_1 = \frac{\alpha z_1 + \beta}{\gamma z_1 + \delta} \tag{8.99a}$$

and

$$w_2 = \frac{\alpha z_2 + \beta}{\gamma z_2 + \delta} \tag{8.99b}$$

If $w_1 = w_2$, then

$$\frac{\alpha z_1 + \beta}{\gamma z_1 + \delta} = \frac{\alpha z_2 + \beta}{\gamma z_2 + \delta} \tag{8.100a}$$

from which

$$z_1(\beta\gamma - \alpha\delta) = z_2(\beta\gamma - \alpha\delta) \tag{8.100b}$$

Because $\beta\gamma - \alpha\delta \neq 0$, we see that for a bilinear mapping

$$z_1 = z_2 \Rightarrow w_1 = w_2 \tag{8.101a}$$

Because the inverse mapping of a bilinear transformation is also bilinear

$$w_1 = w_2 \Rightarrow z_1 = z_2 \tag{8.101b}$$

Therefore, the bilinear mapping is one-to-one.

Example 8.9: Image of the real axis and upper half-plane under the bilinear mapping $w = (z - z_0)/(z - z_0{}^*)$

In ex. 8.2, it was stated without proof that if $\mathrm{Im}(z_0) \neq 0$, the transformation

$$w = f(z) = \frac{z - z_0}{z - z_0^*} \tag{8.17}$$

maps the upper half of the z-plane, including the x-axis, onto the unit circle centered at the origin of the w-plane. We demonstrate this in the current example.

We point out the following about this mapping.

- If z_0 is real, all points z are mapped onto the single point $w = 1$. Therefore, we require that $\mathrm{Im}(z_0) \neq 0$. Because we are mapping the upper half of the z-plane, we take $\mathrm{Im}(z_0) > 0$.
- Because $w(z_0) = 0$, the image of z_0 is the origin of the w-plane.
- From

$$f'(z) = \frac{2\,\mathrm{Im}(z_0)}{(z - z_0^*)^2} \tag{8.102}$$

  we see that the mapping has no critical points.
- Because we are mapping the upper half of the z-plane, $\mathrm{Im}(z) > 0$ for all z in eq. 8.17. Because $\mathrm{Im}(z_0^*) < 0$, it is not one of the points being mapped. Therefore, the denominator of eq. 8.17 is never zero and the mapping does not have a pole in the region being mapped.

Because the mapping is analytic everywhere in the upper half of the z-plane and has no singularities in that region, the mapping is a conformal mapping.

Setting $z = x$, we see that the images of points on the x-axis are given by

$$w = \frac{x - z_0}{x - z_0^*} \tag{8.103a}$$

The square of the magnitude of w for any finite x is

$$|w|^2 = \left(\frac{x-z_0}{x-z_0^*}\right)\left(\frac{x-z_0}{x-z_0^*}\right)^* = \left(\frac{x-z_0}{x-z_0^*}\right)\left(\frac{x-z_0^*}{x-z_0}\right) = 1 \qquad (8.103b)$$

That is, the image of the x-axis is the circumference of the unit circle.

To determine the image of a point off the x-axis in the upper half of the z-plane, we express z and z_0 in polar form. Then

$$|w|^2 = \frac{\left(re^{i\theta} - r_0 e^{i\theta_0}\right)\left(re^{-i\theta} - r_0 e^{-i\theta_0}\right)}{\left(re^{i\theta} - r_0 e^{-i\theta_0}\right)\left(re^{-i\theta} - r_0 e^{i\theta_0}\right)}$$

$$= \frac{\left(r^2 + r_0^2 - 2rr_0\cos\theta\cos\theta_0\right) - \left(2rr_0\sin\theta\sin\theta_0\right)}{\left(r^2 + r_0^2 - 2rr_0\cos\theta\cos\theta_0\right) + \left(2rr_0\sin\theta\sin\theta_0\right)} \qquad (8.104)$$

We note that because $-1 \le \cos\theta\cos\theta_0 \le 1$,

$$(r-r_0)^2 \le r^2 + r_0^2 - 2rr_0\cos\theta\cos\theta_0 \le (r+r_0)^2 \qquad (8.105)$$

In addition, Because z and z_0 are in the upper half-plane, both

$$\sin\theta > 0 \qquad (8.106a)$$

and

$$\sin\theta_0 > 0 \qquad (8.106b)$$

Therefore, the two expressions in parentheses that appear in both the numerator and in the denominator of eq. 8.104 are positive; $|w|^2$ is the ratio of the difference to the sum of positive numbers. Thus, the images of points in the upper half of the z-plane map into points in the w-plane that satisfy

$$|w|^2 < 1 \qquad (8.107)$$

These are points in the interior of the unit circle. ◻

It was shown above that if two points on a polygon in the z-plane and their images in the w-plane are known, one can determine the two constants defining the linear mapping that transforms the polygon in the z-plane to a polygon in the w-plane. Similarly, because $\gamma \neq 0$, we can write

$$w = \frac{\dfrac{\alpha}{\gamma}z + \dfrac{\beta}{\gamma}}{z + \dfrac{\delta}{\gamma}} \equiv \frac{\alpha' z + \beta'}{z + \delta'} \tag{8.108}$$

That is, a bilinear transformation is defined by three constants. As such, one can determine the bilinear mapping that transforms any polygon in the z-plane to a particular polygon in the w-plane if three source points and their images are known.

Example 8.10: Determination of a bilinear mapping for a given source and image

Let us determine the bilinear mapping that transforms the square of side 2, centered at the origin of the z-plane, to the parallelogram shown in fig. 8.12. The coordinates of the vertices of the square and their images, the vertices of the parallelogram, are given in the figure.

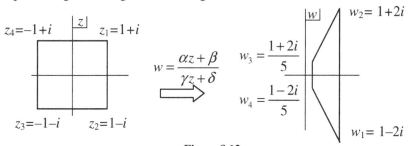

Figure 8.12

Square in the z-plane and its parallelogram image in the
w-plane under a bilinear mapping

Substituting the values of z of any three of the vertices of the square and their corresponding images into eq. 8.108, we obtain

$$\alpha' = \frac{\alpha}{\gamma} = 1 \tag{8.109a}$$

$$\beta' = \frac{\beta}{\gamma} = 1 \tag{8.109b}$$

and

$$\delta' = \frac{\delta}{\gamma} = -1 \qquad\qquad (8.109c)$$

Thus, the parallelogram in fig. 8.12 is the image of the square under the mapping

$$w = \frac{z+1}{z-1} \qquad\qquad (8.110)$$
$$\square$$

8.3 Schwarz–Christoffel Transformation

A *Schwarz–Christoffel* (SC) transformation is derived from the mapping of the differential line element tangent to a curve C in the z-plane onto the differential line element tangent to Γ, the image of that curve in the w-plane. This family of mappings was discovered by E. Schwarz and independently by H. Christoffel (see Schwarz, 1869, pp. 121–136, and Christoffel, 1870, pp. 359–369).

As noted earlier, the mapping of the differential line element along C onto the differential line element along Γ is given by

$$dw = f'(z)\,dz \qquad\qquad (8.32)$$

so the tangent is also mapped by $f'(z)$ as

$$\tau(w) = f'(z)\,t(z) \qquad\qquad (8.33)$$

The derivative of the SC mapping is defined by

$$f'(z) \equiv \alpha(z - z_1)^{-k_1}(z - z_2)^{-k_2}\cdots(z - z_N)^{-k_N} \qquad\qquad (8.111)$$

where $k_1, k_2, \ldots, k_N$ are N distinct real nonzero constants, $z_1, z_2, \ldots, z_N$ are N distinct fixed points in the z-plane, and α is a complex constant. Integrating eq. 8.111, we obtain the general form of the SC mapping

$$w = f(z) = \beta + \alpha\int(z - z_1)^{-k_1}(z - z_2)^{-k_2}\cdots(z - z_N)^{-k_N}\,dz \qquad (8.112)$$

The integral of the mapping describes the shape of the image under the SC transformation. As with the linear mapping, α gives rise to a constant magnification of that image by a factor $|\alpha|$ and a rotation of that image by an angle $\arg(\alpha)$. The constant β generates a translation of the magnified and rotated image.

Inverse of the SC mapping

The inverse of the SC mapping, given by

$$z = g(w)$$

(8.50)

can be found, in principle, by solving the first-order, nonlinear differential equation obtained from eq. 8.53,

$$g'(w) = \frac{1}{f'(z)}\Bigg)_{z=g(w)} = \frac{1}{\alpha}\big(g(w) - z_1\big)^{k_1}\big(g(w) - z_2\big)^{k_2}\cdots\big(g(w) - z_N\big)^{k_N}$$

(8.113)

When $g'(w)$ contains a single factor, eq. 8.113 becomes

$$g'(w) = \frac{1}{\alpha}\big(g(w) - z_1\big)^{k_1}$$

(8.114a)

Dividing by

$$\big(g - z_1\big)^{k_1}$$

this is integrated straightforwardly to obtain

$$g(w) = \begin{cases} z_1 + \left[\dfrac{1-k_1}{\alpha}\right]^{1/(1-k_1)} w^{1/(1-k_1)} & k_1 \neq 1 \\[4mm] g(w) = z_1 + e^{w/\alpha} & k_1 = 1 \end{cases}$$

(8.114b)

If $g'(w)$ contains two or more factors, eq. 8.113 becomes a difficult, and more likely an impossible, nonlinear differential equation to solve in closed form.

Mapping the real axis

An important application of the SC transformation is its mapping of the real axis and either the upper or lower half of the z-plane.

Let each $z_n = x_n$ be a point on the real axis. Then the mapping of eq. 8.111 becomes

$$f'(z) = \alpha(z - x_1)^{-k_1}(z - x_2)^{-k_2} \cdots (z - x_N)^{-k_N} \tag{8.115}$$

where we order the factors so that $-\infty < x_1 < x_2 < \cdots < x_N < \infty$. For those k_n that are not integers, $f(z)$ has a branch point at each z_n. We restrict z to the principal sheet of the multisheeted complex plane. Then $f(z)$ is analytic and has a nonzero derivative at all points on this sheet, and is therefore a conformal mapping everywhere except at each branch point and at all points on its associated cut.

With $f'(z)$ given by eq. 8.115, we see from eq. 8.33 that the SC transformation of the tangent $t(z)$ to the tangent $\tau(w)$ is given by

$$\tau(w) = \alpha(z - x_1)^{-k_1}(z - x_2)^{-k_2} \cdots (z - x_N)^{-k_N} t(z) \tag{8.116}$$

Defining the argument of a factor in the transformation as

$$\arg(z - x_n) \equiv \theta_n \tag{8.117}$$

the argument of $\tau(w)$ is found from

$$|\tau|e^{i\psi} = |\alpha||z - x_1|^{-k_1}|z - x_2|^{-k_2} \cdots |z - x_N|^{-k_N} e^{i[\phi - (k_1\theta_1 + k_2\theta_2 + \cdots + k_N\theta_N)]} \tag{8.118}$$

Following the analysis of a function with one or more branch points presented in chapter 6, the real axis is divided into regions that are defined by the branch points. The analytic structure of the function in a given region is defined by the way the cuts are taken to extend into that region. For the remainder of this chapter, we take each cut to extend from the associated branch point to $+\infty$ along the real axis. Then, as shown in many examples of chapter 6, the cut structure of $f(z)$ is that shown in fig. 8.13.

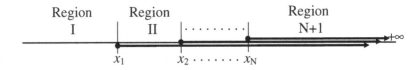

Figure 8.13
Cut structure for a function with N branch points

Because the real axis is a straight line, the shape of its image can be deduced directly from $f'(z)$ and the mapping of the tangent to the real axis. Clearly, $t_n(x)$, the tangent at a point x in the nth region is along the real axis and so is real. We take $t_n(x)$ to point in the direction of increasing x.

- Region I is defined by $-\infty < x < x_1$.
  Referring to fig. 8.13, none of the cuts extends into this region. Thus,

$$\theta_n = \arg(x - x_n) = \pi \quad 1 \le n \le N \tag{8.119}$$

Then, in region I,

$$\tau_1(w) = \left|\tau_1\right|e^{i\psi_1}$$
$$= \left|\alpha\right|\left|x - x_1\right|^{-k_1}\left|x - x_2\right|^{-k_2}\cdots\left|x - x_N\right|^{-k_N} e^{i\left[\phi-(k_1+k_2+\cdots+k_N)\pi\right]}t_1(x) \tag{8.120}$$

so that

$$\psi_1 = \arg(\tau_1) = \phi - \sum_{\ell=1}^{N} k_\ell \pi \tag{8.121}$$

- Region II is defined by $x_1 < x < x_2$.
  There is only one cut in this region, extending from x_1 to $+\infty$. Points in this region are to the left of the branch points $x_2,\ldots, x_N$. Therefore, the arguments of the factors containing these branch points are

$$\theta_n = \arg(x - x_n) = \pi \quad 2 \le n \le N \tag{8.122}$$

If points on the real axis are accessed along the top of the cut in region II, then

$$\theta_1 = \arg(x - x_1) = 0 \tag{8.123a}$$

If points are accessed along the bottom of the cut,

$$\theta_1 = \arg(x - x_1) = 2\pi \tag{8.123b}$$

Therefore, along the top of the cut in this region,

$$\psi_{\mathrm{II}} = \arg(\tau_{\mathrm{II}}) = \phi - \sum_{n=2}^{N} k_n \pi = \psi_{1} + k_1 \pi \qquad (8.124a)$$

and along the bottom of that cut

$$\psi_{\mathrm{II}} = \arg(\tau_{\mathrm{II}}) = \phi - 2k_1 \pi - \sum_{\ell=2}^{N} k_\ell \pi = \psi_{1} - k_1 \pi \qquad (8.124b)$$

- Region III is defined by $x_2 < x < x_3$.
  The two cuts associated with x_1 and x_2 extend into this region. Because x is to the left of $x_3, x_4, \ldots, x_N$

$$\theta_n = \arg(x - x_n) = \pi \quad 3 \le n \le N \qquad (8.125)$$

When points are accessed along the tops of both cuts in this region,

$$\theta_1 = \theta_2 = 0 \qquad (8.126a)$$

If points are accessed along the top of the cut associated with x_1 and along the bottom of the cut associated with x_2,

$$\theta_1 = 0, \qquad \theta_2 = 2\pi \qquad (8.126b)$$

and when points are accessed along the bottom of the cut associated with x_1 and along the top of the cut associated with x_2

$$\theta_1 = 2\pi, \qquad \theta_2 = 0 \qquad (8.126c)$$

When points are accessed along the bottom of both cuts,

$$\theta_1 = \theta_2 = 2\pi \qquad (8.126d)$$

Referring to eq. 8.124a, when points are accessed along the top of both cuts

$$\psi_{\mathrm{III}} = \arg(\tau_{\mathrm{III}}) = \phi - \sum_{n=3}^{N} k_n \pi = \psi_{1} + (k_1 + k_2)\pi = \psi_{\mathrm{II}} + k_2 \pi \quad (8.127a)$$

If points are accessed along the top of the cut associated with x_1 and along the bottom of the cut associated with x_2

$$\psi_{\mathrm{III}} = \arg(\tau_{\mathrm{III}}) = \phi - 2\pi k_2 - \sum_{n=3}^{N} k_n \pi = \psi_{\mathrm{I}} + (k_1 - k_2)\pi = \psi_{\mathrm{II}} - k_2 \pi$$

$$(8.127b)$$

Referring to eq. 8.124b, when points are accessed along the bottom of the cut associated with x_1 and along the top of the cut associated with x_2

$$\psi_{\mathrm{III}} = \arg(\tau_{\mathrm{III}}) = \phi - 2\pi k_1 - \sum_{n=3}^{N} k_n \pi = \psi_{\mathrm{I}} - (k_1 - k_2)\pi = \psi_{\mathrm{II}} + k_2 \pi$$

$$(8.127c)$$

and when points are accessed along the bottom of both cuts

$$\psi_{\mathrm{III}} = \arg(\tau_{\mathrm{III}}) = \phi - 2\pi(k_1 + k_2) - \sum_{n=3}^{N} k_n \pi = \psi_{\mathrm{I}} - (k_1 + k_2)\pi$$
$$= \psi_{\mathrm{II}} - k_2 \pi$$

$$(8.127d)$$

The pattern we have developed for finding ψ_n, the argument of the image of the tangent at any point in the nth region of the x-axis, is straightforward. However, the determination of that angle quickly becomes cumbersome. Many authors simplify the process somewhat by restricting access to points along the tops of all cuts, thus restricting each θ_n to 0 (for $x > x_n$) or π (for $x < x_n$). With or without this simplification it is possible, using the analysis above, to deduce the argument of the tangent to the curve in the w-plane that is the image of the real axis in the z-plane.

Referring to eqs. 8.121, 8.124, and 8.127, we see that the image of each region of the x-axis is a curve in the w-plane, the tangent to which makes a constant angle to the u-axis. A curve with a tangent of constant orientation is a straight line. Thus, the image of each region of the x-axis is a straight line in the w-plane. These tangents and therefore, the lines they define, change orientation abruptly at the images of the branch points. Such lines define the sides and vertices of a polygon. The length of the $(n + 1)^{\text{th}}$ side of the polygon, the image of the $(n + 1)^{\text{th}}$ region of the x-axis (between x_n and x_{n+1}), is

$$L_n = f(x_{n+1}) - f(x_n)$$

$$(8.128)$$

To get a sense of the way the polygon in the w-plane is formed, we denote w_n as the image of x_n, we take all $k_n > 0$, and, in all regions in which there are cuts, we access points along the tops of the cuts. Then, in regions II and III, for example, the change in orientation of the tangents in the w-plane at w_1 and w_2, are found from eqs. 8.124a and 8.127a to be

$$\psi_{\mathrm{II}} - \psi_{\mathrm{I}} = k_1 \pi \qquad (8.129a)$$

and

$$\psi_{\mathrm{III}} - \psi_{\mathrm{II}} = k_2 \pi \qquad (8.129b)$$

Referring to eqs. 8.129a and 8.129b, we see that each $\psi_{n+1} - \psi_n$ is positive. Therefore, when points in regions n and $n + 1$ are accessed along the tops of cuts, the angles between the tangents in two successive regions are positive and are thus taken in the counterclockwise direction.

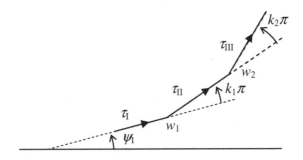

Figure 8.14a

SC mapping of regions I, II, and III of the x-axis

when points are accessed along the tops of all cuts

From eqs. 8.124b and 8.127d, we see that if points in regions II and III are accessed along the bottoms of the cuts in these regions,

$$\psi_{\mathrm{II}} - \psi_{\mathrm{I}} = -k_1 \pi \qquad (8.130a)$$

and

$$\psi_{\mathrm{III}} - \psi_{\mathrm{II}} = -k_2 \pi \qquad (8.130b)$$

Because these angles are negative, they represent angles between tangents in adjacent regions taken in the clockwise direction.

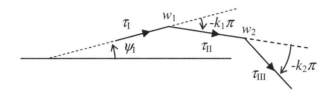

Figure 8.14b
SC mapping of regions I, II, and III of the *x*-axis
when points are accessed along the bottoms of all cuts

A generalization of the results of eqs. 8.129 and 8.130 is that the images of any two adjacent regions of the *x*-axis are two adjacent sides of a polygon in the *w*-plane, the tangents to which subtend an angle $k_n \pi$. If we constrain each k_n to the ranges $-1 \leq k_n < 1$, the angle subtended by two adjacent tangents is between $-\pi$ and π. From now on, we restrict each constant to the range $0 < k_n \leq 1$ and explicitly write either a positive or negative sign in the exponent of each factor $(x - x_n)$.

In prob. 38, the reader is asked to analyze the changes in orientation of tangents in the *w*-plane when points are accessed along one side of the cut associated with x_1 and along the opposite side of the cut associated with x_2. This analysis indicates to the reader that taking all points to be accessed along the same side (e.g., the top) of all cuts leads to the following simplified analysis of the SC mapping of the real axis. With the constants k_n defined as in eq. 8.115, and with all k_n positive,

- Accessing points along the tops (bottoms) of all cuts results in angles between adjacent sides of the polygon in the *w*-plane being taken in the counterclockwise (clockwise) direction.

With all k_n negative,

- Accessing points along the tops (bottoms) of all cuts results in angles between adjacent sides of the polygon in the *w*-plane being taken in the clockwise (counterclockwise) direction.

Mapping of points that are not on the real axis

Let all factors in $f'(x)$ be of the form

$$(x - x_n)^{-k_n}$$

and let x be a point in the nth region of the x-axis. Referring to fig. 8.14a, the nth side of the polygon makes an angle $k_n - {}_1\pi$ with the $(n - 1)^{th}$ side taken in the counterclockwise direction. A point z in the upper half-plane is accessed by a counterclockwise rotation by an angle γ relative to the x-axis. Because the SC mapping is conformal except at the branch points and their associate cuts, the magnitude and sense of an angle are preserved by the mapping. Therefore, the image of z in the w-plane is accessed by a counterclockwise rotation by γ relative to the nth side of the polygon. Therefore, because all $k_n > 0$, the image of the upper half of the z-plane is the interior of the polygon shown in fig. 8.15a.

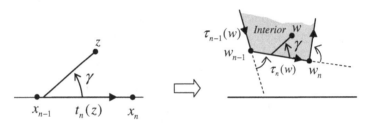

Figure 8.15a

Image of a point in the upper half-plane onto the interior of a polygon
when all factors are of the form $(x - x_n)^{-k}$

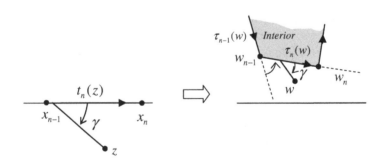

Figure 8.15b

Image of a point in the lower half-plane onto the exterior of a polygon
when all factors are of the form $(x - x_n)^{-k}$

A point z in the lower half-plane is accessed by a clockwise rotation from a point x in the nth region through an angle γ relative to the x-axis. Again, because the SC mapping is conformal, the image of that point must also be accessed by a clockwise rotation by γ from the image of that n^{th} region. This places the image of z outside the polygon image of the x-axis.

By identical reasoning, if all k parameters are negative, the lower half of the z-plane maps onto the interior of the polygon image of the x-axis and the image of the upper half of the z-plane is the exterior of that polygon.

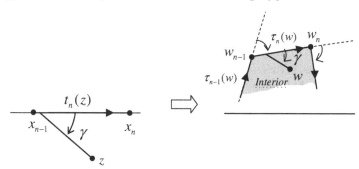

Figure 8.15c

Image of a point in the lower half-plane onto the interior of a polygon when all factors are of the form $(x - x_n)^{-k}$

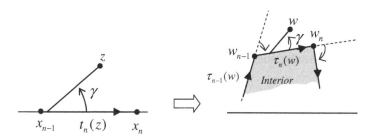

Figure 8.15d

Image of a point in the upper half-plane onto the exterior of a polygon when all factors are of the form $(x - x_n)^{-k}$

Closed form of the SC mapping

Unless the integral in eq. 8.112 can be evaluated in closed form, the mapping $f(z)$ cannot be expressed in terms of elementary functions. Then, the positions of the vertices of the polygon image of the real axis cannot be determined. Without a knowledge of the positions of the vertices, the lengths of the sides of the polygon cannot be found. There are only a few SC mappings that can be evaluated as closed expressions.

Example 8.11: The image of the real axis and upper half-plane under a SC mapping with one branch point

The simplest SC transformation is generated by the function with derivative given by

$$f'(z) = \alpha(z - x_1)^{-k} \tag{8.131}$$

With $0 < k < 1$, $f'(z)$ has one branch point on the real axis. This separates the real axis at x_1 into two regions. Therefore, the image of the real axis is a polygon with one vertex and two sides. Such a configuration is an open wedge with sides that extend to infinity.

If points with $x > x_1$ are accessed along the top of the cut associated with x_1, then referring to eqs. 8.121 and 8.124a, we see that the angles relative to the u-axis made by τ_I and τ_{II}, the tangents to the two sides of the wedge are given by

$$\psi_I = \phi - k\pi \tag{8.132a}$$

and

$$\psi_{II} = \phi \tag{8.132b}$$

Therefore, the angle subtended by the two tangents is

$$\psi_{II} - \psi_I = k\pi \tag{8.133a}$$

Referring to figs. 8.14a and 8.15a, we see that this results in a wedge with interior angle given by

$$\gamma = \pi - k\pi \tag{8.133b}$$

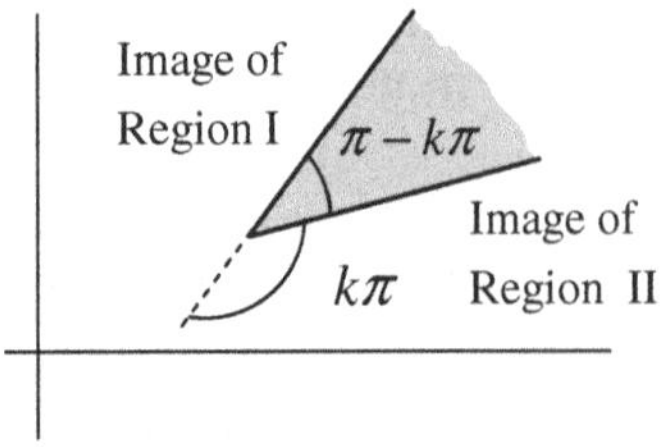

Figure 8.16
Image of the real axis under the SC mapping of eq. 8.131

In prob. 37, the reader is asked to analyze this problem when points with $x > x_1$ are accessed along the bottom of the cut.

The SC transformation with one branch point is one of the few mappings for which $f'(z)$ can be integrated in closed form. With $k < 1$ the result is

$$w = f(z) = \beta + \alpha \int (z - x_1)^{-k} dz = \beta + \frac{\alpha}{1-k}(z - x_1)^{1-k} \tag{8.134}$$

This algebraic expression of $f(z)$ allows us to analyze the mapping of the two regions on the x-axis directly and compare the results to those obtained from the analysis of the mapping of the tangents.

- Region I is defined by $-\infty < x < x_1$.
  Because points in this region are to the left of the branch point at x_1,

  $$\arg(x - x_1) = \pi \tag{8.135}$$

  Therefore,

  $$w = \beta + \frac{\alpha}{1-k}|x - x_1|^{1-k} e^{i(1-k)\pi} = \beta + \frac{|\alpha|}{1-k}(x_1 - x)^{1-k} e^{i[\phi + (1-k)\pi]} \tag{8.136a}$$

  from which we see that the angle relative to the u-axis made by the image of region I is given by

  $$\psi_I \equiv \arg(w - \beta) = \phi + \pi(1 - k) \tag{8.136b}$$

- Region II is defined by $x_1 < x < \infty$.
  Taking points in this region to be accessed along the top of the cut associated with x_1,

  $$\arg(x - x_1) = 0 \tag{8.137}$$

  With this,

  $$w = \beta + \frac{\alpha}{1-k}|x - x_1|^{1-k} = \beta + \frac{|\alpha|}{1-k}(x - x_1)^{1-k} e^{i\phi} \tag{8.138}$$

so, the image of region II makes an angle with the u-axis of

$$\psi_{\text{II}} = \arg(w - \beta) = \phi \tag{8.139}$$

Therefore, as shown in fig. 8.16, the angle interior to the wedge is given by

$$\gamma = |\psi_{\text{II}} - \psi_{\text{I}}| = \pi(1 - k) \tag{8.140}$$

and the interior of the wedge is the image of the upper half of the z-plane.
    The inverse of the mapping of eq. 8.134 is

$$z = g(w) = x_1 + \left[\frac{(1 - k)}{\alpha} (w - \beta) \right]^{1/(1-k)} \tag{8.141}$$

which maps the wedge of interior angle $\pi(1 - k)$ in the w-plane onto the real axis of the z-plane. This is the result that we obtained in eq. 8.114c for a point $z_1 = x_1$ on the real axis. $\square$

Example 8.12: The image of the real axis and upper half-plane under a SC mapping with two branch points

A second SC transformation that can be expressed in closed form is obtained from

$$f'(z) = \alpha(z - x_1)^{-\frac{1}{2}}(z + x_1)^{-\frac{1}{2}} = \alpha(z^2 - x_1^2)^{-\frac{1}{2}} \tag{8.142}$$

which has square root branch points at $z = \pm x_1$. The mapping is given by

$$w = f(z) = \beta + \alpha \int \left(z^2 - x_1^2 \right)^{-1/2} dz \tag{8.143}$$

Substituting

$$z = x_1 \cosh \mu \tag{8.144}$$

we obtain

$$w = f(z) = \beta + \alpha \cosh^{-1} \left(\frac{z}{x_1} \right) \tag{8.145a}$$

Referring to problem 11a in chapter 2, if z is restricted to the principal branches of both the logarithm and square root functions, eq. 8.145a can be written in the form

$$w = f(z) = \beta + \alpha \ln \left[\frac{z + \sqrt{(z^2 - x_1^2)}}{x_1} \right] \tag{8.145b}$$

from which we see that $f(z)$ has branch points at $z = \pm x_1$. Again, both associated cuts are taken to extend from the branch points to $+\infty$.

To determine the image of the x-axis under this mapping, we first note that the sign of x_1 and $\phi = \arg(\alpha)$ do not affect the results. Therefore, we simplify the problem by taking x_1 to be positive and set $\phi = 0$, making α real and positive. Then, setting $z = x$, the mapping of each region of the real axis is as follows:

- Region I is defined by $-\infty < x < -x_1$.
  We see that because points in this region are to the left of both branch points,

$$\theta_1 = \theta_2 = \arg(x - x_1) = \arg(x + x_1) = \pi \tag{8.146}$$

Thus, the image of the tangent to the real axis of region I is

$$\begin{aligned} \tau_1(w) = \left| \tau_1(w) \right| e^{i\psi_1} &= \alpha \left| (x^2 - x_1^2)^{-1/2} \right| e^{-i\frac{1}{2}(\theta_1 + \theta_2)} \left| t_1(x) \right| \\ &= \alpha (x^2 - x_1^2)^{-1/2} e^{-i\pi} \left| t_1(x) \right| \end{aligned} \tag{8.147a}$$

from which

$$\psi_{\mathrm{I}} = -\pi \tag{8.147b}$$

The image of $-\infty$ is found from eq. 8.145b by setting

$$z = |x| e^{i\pi} \tag{8.148}$$

so that with $|x| \gg x_1$,

$$w_{-\infty} = \beta + \alpha \lim_{|x| \to \infty} \ln \left[\frac{2|x| e^{i\pi}}{x_1} \right] = \beta + \infty + i\pi\alpha \tag{8.149a}$$

We can ignore the finite $\text{Re}(\beta)$ relative to ∞, but because $\text{Im}(\beta)$ is multiplied by i, it must not be compared to the real ∞ and therefore cannot be ignored. Thus, eq. 8.149a becomes

$$w_{-\infty} = \infty + i\left(\text{Im}(\beta) + \pi\alpha\right) \tag{8.149b}$$

For $x = -x_1 = x_1 e^{i\pi}$, eq. 8.145b becomes

$$w_{-x_1} = \beta + i\pi\alpha = \text{Re}(\beta) + i\left(\text{Im}(\beta) + \pi\alpha\right) \tag{8.150}$$

Therefore, we see from eqs. 8.149 and 8.150 that the image of region I is a line from $+\infty$ to $\text{Re}(\beta)$ parallel to the u-axis, a distance $\text{Im}(\beta) + \pi\alpha$ above (or below) the u-axis.

- Region II is defined by $-x_1 < x < x_1$.
  Only the cut associated with $-x_1$ extends into this region. We take points in this region to be accessed along the top of this cut. Then we have

$$\theta_1 = \arg(x - x_1) = \pi \tag{8.151a}$$

and

$$\theta_2 = \arg(x + x_1) = 0 \tag{8.151b}$$

Therefore,

$$\begin{aligned}\tau_{\text{II}}(w) &= \left|\tau_{\text{II}}(w)\right| e^{i\psi_{\text{II}}} = \alpha\left|(x^2 - x_1^2)^{-1/2}\right| e^{-i\frac{1}{2}(\theta_1 + \theta_2)} \left|t_{\text{II}}(x)\right| \\ &= (x_1^2 - x^2)^{-1/2} e^{-i\pi/2} \left|t_{\text{II}}(x)\right|\end{aligned} \tag{8.152a}$$

from which

$$\psi_{\text{II}} = -\pi/2 \tag{8.152b}$$

Again referring to eq. 8.145b, we see that with $z = -x_1 = x_1 e^{i\pi}$,

$$w_{-x_1} = \beta + i\pi\alpha = \text{Re}(\beta) + i\left(\text{Im}(\beta) + \pi\alpha\right) \tag{8.153a}$$

Setting $z = x_1$, eq. 8.145b yields

$$w_{+x_1} = \beta = \text{Re}(\beta) + i\,\text{Im}(\beta) \tag{8.153b}$$

Thus, the image of region II is a line with a constant real part $\text{Re}(\beta)$ and an imaginary part that varies from $\text{Im}(\beta) + \pi\alpha$ to $\text{Im}(\beta)$.

- Region III is defined by $x_1 < x < \infty$.
  Both cuts extend into region III. We take points in this region to be accessed along the tops of both cuts, so that

$$\theta_1 = \arg(x - x_1) = \theta_2 = \arg(x + x_1) = 0 \tag{8.154}$$

Therefore,

$$\psi_{\text{III}} = 0 \tag{8.155}$$

With $z = x_1$, eq. 8.145b yields

$$w_{+x_1} = \beta = \text{Re}(\beta) + i\,\text{Im}(\beta) \tag{8.156a}$$

The image of $+\infty$ is given by eq. 8.145b to be

$$w_{+\infty} = \beta + \infty = \infty + i\,\text{Im}(\beta) \tag{8.156b}$$

Thus, the image of region III is a line from $\text{Re}(\beta)$ to $+\infty$ at a distance $\text{Im}(\beta)$ from the u-axis.

The image of the real axis, and the upper half of the z-plane is the open three-sided polygon shown in fig. 8.17.

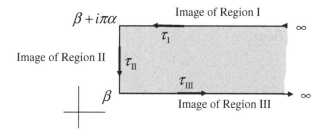

Figure 8.17
Image of the real axis and upper half of the z-plane
under the mapping of eqs. 8.145

The inverse mapping, which transforms the open three-sided polygon of fig. 8.17 onto the real axis is found straightforwardly from eq. 8.145a to be

$$z = g(w) = x_1 \cosh\left(\frac{w - \beta}{\alpha}\right) \tag{8.157}$$

□

Open and closed polygons

We have seen that if $f'(z)$ has N branch points on the real axis, these points separate the axis into $N + 1$ regions. Each region maps onto one side of a polygon with $N + 1$ sides. The orientation of each side is different from the orientation of the sides adjacent to it.

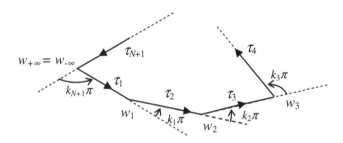

Figure 8.18
A closed polygon with $N + 1$ sides

A polygon is closed if at every vertex, each side is joined to an adjacent side. Therefore, in order to map the real axis in the z-plane onto a closed polygon in the w-plane, the images of $x = \pm\infty$ must satisfy

$$w_{x=-\infty} = w_{x=+\infty} \tag{8.158}$$

with $w_{x=-\infty}$ and $w_{x=+\infty}$ finite.

Let

$$f'(z) = \alpha(z - x_1)^{-k_1}(z - x_2)^{-k_2}\cdots(z - x_N)^{-k_N} \tag{8.159}$$

be the derivative of the function that maps the x-axis onto a closed polygon in the w-plane. Because the polygon is closed, there exists a nonzero angle between the tangents to the sides of the polygon that are the images of regions I and $N + 1$. Referring to fig. 8.19, we define this angle in terms of an additional k-number denoted by k_{N+1}.

A minimum of three sides is required to form a closed polygon. If none of the sides of a closed polygon crosses, it is straightforward to demonstrate that the sum of internal angles for closed polygons with three, four, five, and six sides are π, 2π, 3π, and 4π, respectively.

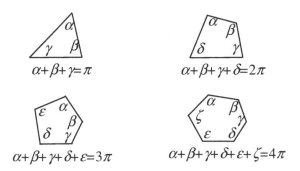

$$\alpha + \beta + \gamma = \pi \qquad\qquad \alpha + \beta + \gamma + \delta = 2\pi$$

$$\alpha + \beta + \gamma + \delta + \varepsilon = 3\pi \qquad\qquad \alpha + \beta + \gamma + \delta + \varepsilon + \zeta = 4\pi$$

Figure 8.19
Sums of internal angles of closed polygons

From this we deduce that for $N \geq 2$, the sum of internal angles in a closed polygon with $N + 1$ uncrossed sides is $(N - 1)\pi$.

When points on the x-axis are accessed along the tops of all cuts, the external angle between two adjacent sides of a closed polygon that join at the point w_n is $k_n\pi$ (see fig. 8.14a). Therefore, the internal angle between such adjacent sides is $\pi - k_n\pi$. Having defined the angle $k_{N+1}\pi$ as the angle between the sides that are the images of regions I and $N + 1$, the sum of internal angles of a closed polygon with $N + 1$ sides is given by

$$(N-1)\pi = \sum_{n=1}^{N+1} (\pi - k_n\pi) = (N+1)\pi - \pi\sum_{n=1}^{N+1} k_n \tag{8.160a}$$

Therefore, the condition for a polygon to be closed and have no crossed lines is

$$\sum_{n=1}^{N+1} k_n = 2 \tag{8.160b}$$

From this we see that the sum of external angles of any closed polygon with uncrossed sides is 2π.

The sides of a polygon can only cross if the polygon has four or more sides. For such a polygon, eq. 8.160b will not be valid.

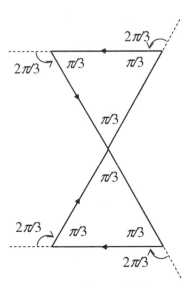

Figure 8.20
A four sided polygon with crossed sides

Example 8.13: A SC mapping that yields a polygon with crossing sides

Consider the four-sided polygon of fig. 8.20. To simplify the analysis, we take all external angles between adjacent sides to be the same $2\pi/3$. Then, we see from fig. 8.20 that

$$k_1 = k_2 = k_3 = k_4 = 2/3 \tag{8.161}$$

Then eq. 8.160b for this polygon becomes

$$\sum_{\ell=1}^{4} k_\ell = 8/3 \neq 2 \tag{8.162}$$
□

We note that this four-sided polygon can be viewed as two equilateral triangles with a common vertex and no sides in common. In two dimensions, any closed polygon which has sides that cross can be viewed as a collection of closed polygons, each with uncrossed sides and one vertex in common with an adjacent polygon. From now on, we only consider polygons with uncrossed sides so that eq. 8.160b holds.

If it is known that a polygon is closed, then k_{N+1} is defined from the angle between the images of the tangents of regions I and $(N + 1)$. From the geometry of the polygon, as indicated in fig. 8.18, or from eq. 8.160b,

$$k_{N+1} = 2 - \sum_{\ell=1}^{N} k_\ell \qquad (8.163)$$

To determine if the polygon is closed, one must determine if

$$w_{x=-\infty} = w_{x=+\infty} \qquad (8.158)$$

Even when $f'(x)$ cannot be integrated in closed form, it is possible to determine if the image of the x-axis is a closed polygon by testing whether $w_{+\infty} = w_{-\infty}$. When $z = x \to \pm\infty$, the finite quantities x_1, x_2, ... ,x_N can be ignored in each factor of $f'(x)$. Then, with $z = x$, eq. 8.159 can be approximated as

$$f'(x) \simeq \alpha x^{-(k_1+k_2+\cdots+k_N)} \qquad (8.164)$$

If $\Sigma k_n = 1$, then for large $|x|$,

$$w = f(x) \simeq \beta + \alpha \, \ell n(x) \qquad (8.165)$$

Then, when $x \to -\infty$, so that $\arg(x) \equiv \theta_{-\infty} = \pi$,

$$w_{-\infty} = \beta + \alpha \lim_{|x|\to\infty} \ell n|x| + i\pi\alpha \qquad (8.166a)$$

and when $x \to +\infty$, and $\arg(x) \equiv \theta_{+\infty} = 0$,

$$w_{+\infty} = \beta + \alpha \lim_{|x|\to\infty} \ell n|x| \neq w_{-\infty} \qquad (8.166b)$$

Therefore, when $\Sigma k_n = 1$, the resulting polygon is open.
When $\Sigma k_n \neq 1$ and $|x|$ is large,

$$w = f(x) \simeq \beta + \frac{\alpha}{(1-k_1-k_2-\cdots-k_N)} x^{(1-k_1-k_2-\cdots-k_N)} \qquad (8.167)$$

Thus, with $\theta_{-\infty} = \pi$ and $\theta_{+\infty} = 0$,

$$w_{-\infty} = \beta + \alpha e^{i\pi(1-k_1-k_2-\cdots-k_N)} \lim_{x\to-\infty} \frac{|x|^{(1-k_1-k_2-\cdots-k_N)}}{(1-k_1-k_2-\cdots-k_N)} \qquad (8.168a)$$

and

$$w_{+\infty} = \beta + \alpha \lim_{x \to +\infty} \frac{|x|^{(1-k_1-k_2-\cdots-k_N)}}{(1-k_1-k_2-\cdots-k_N)} \tag{8.168b}$$

We see that if $\Sigma k_n > 1$, the limit in eqs. 8.168 is zero and $w_{+\infty} = w_{-\infty}$. Then the polygon is closed and, from eqs. 8.160b and 8.163

$$\sum_{n=1}^{N} k_n > 1 \Rightarrow k_{N+1} < 1 \tag{8.169}$$

When $\Sigma k_n < 1$, the limit in eqs. 8.168 are infinite. In addition, for $0 < \Sigma k_n < 1$, we see that $\pi(1 - \Sigma k_n)$ cannot be an integer multiple of 2π. Then, comparing eqs. 8.168a and 8.168b, we see that

$$\alpha e^{i\pi(1-)} \neq \alpha \tag{8.170}$$

Thus, $w_{+\infty} \neq w_{-\infty}$ and the polygon is open.

Example 8.14: A SC mapping of the real axis and upper half-plane onto an open polygon

Referring to fig. 8.17, it was shown in ex. 8.12 that setting $k_1 = k_2 = 1/2$, the x-axis with branch points at $\pm x_1$ is mapped onto the open three-sided polygon by

$$w = \beta + \alpha \ln\left[\frac{x + \sqrt{x^2 - x_1^2}}{x_1}\right] \tag{8.171}$$

with α real. To show that the polygon is open (because $k_1 + k_2 = 1$), we see from eq. 8.170 that when $x \to \pm\infty$,

$$w_{\pm\infty} = \beta + \alpha \lim_{|x| \to \infty} \ln\left(\frac{2|x|}{x_1}\right) + i\alpha \arg(x) \tag{8.172}$$

When $x \to +\infty$, $\arg(x) = 0$. When $x \to -\infty$, $\arg(x) = \pi$. Therefore,

$$w_{+\infty} = \beta + \alpha \lim_{|x| \to \infty} \ell n \left(\frac{2|x|}{x_1} \right) \tag{8.173a}$$

and

$$w_{-\infty} = \beta + \alpha \lim_{|x| \to \infty} \ell n \left(\frac{2|x|}{x_1} \right) + i\alpha\pi \tag{8.173b}$$

Because $w_{+\infty} \neq w_{-\infty}$, the three-sided polygon image of the x-axis under this mapping is not a closed triangle. □

Determining the SC mapping

When discussing the linear and bilinear mappings, it was shown in exs. 8.8b and 8.10 that from a knowledge of a sufficient number of source points and their corresponding images, one can determine the parameters that define the mapping. The same is applicable in determining the SC mapping of the real axis onto a specified polygon.

Example 8.15: Determining the SC mapping that transforms the real axis onto a rectangle

Let us find the SC transformation that maps the real axis onto the rectangle of fig. 8.21, where a and b are real and positive.

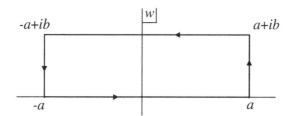

Figure 8.21

A rectangular image of the x-axis

(See, for example, Churchill and Brown, 1990, p. 290; Carrier et al, 1966, p. 139; Ahlfors, 1976, p. 238 for other discussions of this problem.)

Because the rectangle is a four-sided polygon, it must be the image of the x-axis divided into four regions by three branch points. Therefore, the derivative of the mapping is given by

$$f'(x) = \alpha(x - x_1)^{-k_1}(x - x_2)^{-k_2}(x - x_3)^{-k_3} \qquad (8.174a)$$

and the mapping we seek is

$$w = f(x) = \beta + \alpha \int (x - x_1)^{-k_1}(x - x_2)^{-k_2}(x - x_3)^{-k_3}\, dx \qquad (8.174b)$$

To completely specify this mapping requires that we find the values of eight parameters, k_1, k_2, k_3, x_1, x_2, x_3, α, and β.

It is clear that a rectangle has external (and internal) angles of $\pi/2$. Therefore, $k_1 = k_2 = k_3 \equiv k = 1/2$, and from eq. 8.163, $k_4 = 1/2$. With these k-numbers, the mapping becomes

$$w = \beta + \alpha \int \frac{1}{\sqrt{(x - x_1)(x - x_2)(x - x_3)}}\, dx \qquad (8.175)$$

Referring to fig. 8.21, we see that the vertices of the rectangle are at $w = -a$, $w = +a$, $w = -a + ib$, and $w = a + ib$. Therefore, the rectangle is symmetric about the imaginary w-axis. Because each side of the rectangle is either parallel or perpendicular to the u-axis, we can assign

$$\phi = \arg(\alpha) = 0 \qquad (8.176a)$$

or

$$\phi = \arg(\alpha) = \pi/2 \qquad (8.176b)$$

As shown in the following the choice of ϕ simply determines which branch point on the x-axis maps onto which vertex of the rectangle. Because the value of ϕ cannot be specified, we take it to be zero to simplify the arithmetic. Then α is real.

Because the vertices of the rectangle are the images of the points $-\infty$, x_1, x_2, x_3, and $+\infty$, we obtain from eq. 8.175

$$w_{-\infty} = \beta + \alpha \int_{x=-\infty} \frac{1}{\sqrt{(x - x_1)(x - x_2)(x - x_3)}}\, dx \qquad (8.177a)$$

$$w_1 = \beta + \alpha \int_{x=x_1} \frac{1}{\sqrt{(x-x_1)(x-x_2)(x-x_3)}} dx \qquad (8.177b)$$

$$w_2 = \beta + \alpha \int_{x=x_2} \frac{1}{\sqrt{(x-x_1)(x-x_2)(x-x_3)}} dx \qquad (8.177c)$$

$$w_3 = \beta + \alpha \int_{x=x_3} \frac{1}{\sqrt{(x-x_1)(x-x_2)(x-x_3)}} dx \qquad (8.177d)$$

and

$$w_\infty = \beta + \alpha \int_{x=+\infty} \frac{1}{\sqrt{(x-x_1)(x-x_2)(x-x_3)}} dx \qquad (8.177e)$$

To simplify the arithmetic, we take all points in regions II, III, and IV to be accessed along the tops of all cuts. Therefore, for each factor,

$$(x-x_n)^{-1/2} = e^{-i\pi/2}(x_n-x)^{-1/2} \qquad x < x_n \qquad (8.178a)$$

and

$$(x-x_n)^{-1/2} = e^{i0}(x-x_n)^{-1/2} \qquad x > x_n \qquad (8.178b)$$

Then, using eqs. 8.177 and 8.178, the sides of the rectangle are given by

$$w_1 - w_\infty$$

$$= \alpha \left[\int_{x=x_1} \frac{1}{\sqrt{(x-x_1)(x-x_2)(x-x_3)}} dx - \int_{x=-\infty} \frac{1}{\sqrt{(x-x_1)(x-x_2)(x-x_3)}} dx \right]$$

$$= \alpha e^{-i3\pi/2} \int_{-\infty}^{x_1} \frac{1}{\sqrt{(x_1-x)(x_2-x)(x_3-x)}} dx$$

$$= i\alpha \int_{-\infty}^{x_1} \frac{1}{\sqrt{(x_1-x)(x_2-x)(x_3-x)}} dx$$

$$(8.179a)$$

Similarly,

$$w_2 - w_1 = \alpha e^{-i\pi} \int_{x_1}^{x_2} \frac{1}{\sqrt{(x-x_1)(x_2-x)(x_3-x)}} dx$$

$$= -\alpha \int_{x_1}^{x_2} \frac{1}{\sqrt{(x-x_1)(x_2-x)(x_3-x)}} dx \tag{8.179b}$$

$$w_3 - w_2 = \alpha e^{-i\pi/2} \int_{x_2}^{x_3} \frac{1}{\sqrt{(x-x_1)(x-x_2)(x_3-x)}} dx$$

$$= -i\alpha \int_{x_2}^{x_3} \frac{1}{\sqrt{(x-x_1)(x-x_2)(x_3-x)}} dx \tag{8.179c}$$

and

$$w_{+\infty} - w_3 = \alpha \int_{x_3}^{\infty} \frac{1}{\sqrt{(x-x_1)(x-x_2)(x-x_3)}} dx \tag{8.179d}$$

These expressions for the sides of the rectangle cannot be chosen arbitrarily. They must conform to the expressions given in eqs. 8.179. We note that all the integrals of eqs. 8.179 are real and positive. Thus, with α real and positive, we see from eq. 8.179a, for example, that $w_1 - w_{-\infty}$ must be a positive imaginary number. Therefore, referring to the parameters in fig. 8.21, this side must be given by

$$w_1 - w_{-\infty} = ib \tag{8.180a}$$

so that

$$b = \alpha \int_{-\infty}^{x_1} \frac{1}{\sqrt{(x_1-x)(x_2-x)(x_3-x)}} dx \tag{8.180b}$$

We take the interior of the rectangle to be the image of the upper half of the z-plane. Therefore, the external angles between tangents to the adjacent sides of the rectangle must be taken in the counterclockwise sense. Thus, the tangents must be oriented as shown in fig. 8.21. With these constraints, eq. 8.180a can only be satisfied by taking

$$w_1 = a + ib \tag{8.181a}$$

and

$$w_{-\infty} = a \tag{8.181b}$$

The remaining vertices are assigned by traversing the rectangle in the counterclockwise direction. That is,

$$w_2 = -a + ib \tag{8.182a}$$

$$w_3 = -a \tag{8.182b}$$

and

$$w_{+\infty} = a = w_{-\infty} \tag{8.182c}$$

These assignments, along with eqs. 8.179, yield

$$\alpha \int_{x_1}^{x_2} \frac{1}{\sqrt{(x - x_1)(x_2 - x)(x_3 - x)}} dx = 2a \tag{8.183a}$$

$$\alpha \int_{x_2}^{x_3} \frac{1}{\sqrt{(x - x_1)(x - x_2)(x_3 - x)}} dx = b \tag{8.183b}$$

and

$$\alpha \int_{x_3}^{\infty} \frac{1}{\sqrt{(x - x_1)(x - x_2)(x - x_3)}} dx = 2a \tag{8.183c}$$

The reader should convince himself or herself that the above assignment of vertices is the only one that results in the integrals having the correct sign and the correct factor of i.

Comparing eqs. 8.180b and 8.183b, the integrals must satisfy

$$\int_{-\infty}^{x_1} \frac{1}{\sqrt{(x_1 - x)(x_2 - x)(x_3 - x)}} dx = \int_{x_2}^{x_3} \frac{1}{\sqrt{(x - x_1)(x - x_2)(x_3 - x)}} dx \tag{8.184a}$$

and from eqs. 8.183a and 8.183c, we must have

$$\int_{x_1}^{x_2} \frac{1}{\sqrt{(x-x_1)(x_2-x)(x_3-x)}} dx = \int_{x_3}^{\infty} \frac{1}{\sqrt{(x-x_1)(x-x_2)(x-x_3)}} dx$$

(8.184b)

To demonstrate that eq. 8.184a is valid, we make the substitution

$$x = x_1 - (x_3 - x_1)\tan^2 \mu$$
(8.185a)

in the left side integral and substitute

$$x = x_2 + (x_3 - x_2)\sin^2 \mu$$
(8.185b)

in the integral on the right side. Then eq. 8.184a becomes

$$\int_0^{\pi/2} \frac{\sec \mu}{\sqrt{\left((x_2 - x_1) + (x_3 - x_1)\tan^2 \mu\right)}} d\mu$$
$$= \int_0^{\pi/2} \frac{1}{\sqrt{\left((x_2 - x_1) + (x_3 - x_2)\sin^2 \mu\right)}} d\mu$$

(8.186)

It is a straightforward algebraic exercise to show that the integrands of these integrals are equal for any finite x_1, x_2, and x_3, as are those of the integrals of eqs. 8.184.

Because of the equalities of these integrals, there are only two independent equations from which we can determine α, x_1, x_2, and x_3. These are, for example,

$$\alpha \int_{x_1}^{x_2} \frac{1}{\sqrt{(x-x_1)(x_2-x)(x_3-x)}} dx = 2a$$
(8.183a)

and

$$\alpha \int_{x_1}^{x_2} \frac{1}{\sqrt{(x-x_1)(x-x_2)(x_3-x)}} dx = b$$
(8.183b)

Substituting

$$x = x_1 + (x_2 - x_1)\sin^2 \mu \qquad (8.187)$$

in eq. 8.183a we obtain

$$a = \frac{\alpha}{\sqrt{(x_3 - x_1)}} \int_0^{\pi/2} \frac{1}{\sqrt{1 - p^2 \sin^2 \mu}} d\mu \qquad (8.188)$$

where

$$p^2 = \frac{(x_2 - x_1)}{(x_3 - x_1)} \qquad (8.189)$$

Because $x_1 < x_2 < x_3$, we see that $0 < p^2 < 1$.

The integral in eq. 8.188 is called the *complete elliptic integral of the first kind* and is commonly denoted by $K(p^2)$ (see prob. 19 of chapter 7). It cannot be evaluated exactly, but many of its properties, including values obtained by numerical integration methods, exist in the literature. The reader is referred, for example, to Cohen, 1992, pp.274–278 and pp. 686–699.

To evaluate the integral of eq. 8.183b, we let

$$x = x_2 + (x_3 - x_2)\sin^2 \mu \qquad (8.190)$$

We obtain

$$b = \frac{2\alpha}{\sqrt{(x_2 - x_1)}} \int_0^{\pi/2} \frac{1}{\sqrt{1 + q^2 \sin^2 \mu}} d\mu \equiv \frac{2\alpha}{\sqrt{(x_2 - x_1)}} L(q^2) \qquad (8.191)$$

where

$$q^2 = \frac{(x_3 - x_2)}{(x_2 - x_1)} = \frac{(x_3 - x_1)}{(x_2 - x_1)} - 1 = \frac{1}{p^2} - 1 \qquad (8.192)$$

Because $0 < p^2 < 1$, we see that $0 < q^2 < \infty$.

The properties of the function $L(q^2)$ defined in eq. 8.191 have not been discussed in the literature and the integral has not been given a name. Like the elliptic integral, its values must be obtained numerically.

To determine the values of α, x_1, x_2, and x_3, we combine eqs. 8.188 and 8.191 to obtain

$$pK(p^2) = \frac{2a}{b} L\left(\frac{1}{p^2} - 1\right) \tag{8.193}$$

One must assign a given value to a/b, then solve this equation numerically for p. For example, when $2a/b = 1$,

$$pK(p^2) = L\left(\frac{1}{p^2} - 1\right) \tag{8.194}$$

is satisfied by $p^2 = 1/2$. This yields

$$x_3 = 2x_2 - x_1 \tag{8.195}$$

which can be satisfied by $x_2 = 0$ and $x_1 = -x_3$. This result reflects the fact that when $2a/b = 1$ and the rectangle is symmetric around the imaginary axis in the w-plane, the branch points can be taken to be symmetric around the origin of the z-plane.

If, for example, $2a/b = 1.25$, then

$$pK(p^2) = 1.25L\left(\frac{1}{p^2} - 1\right) \tag{8.196}$$

is satisfied by $p^2 \cong 0.7295$, a value for which the branch points on the x-axis cannot be taken to be symmetric around the origin.

For a given value of p^2, the value of α is found from either eq. 8.188 or 8.191. For $2a/b = 1$ and therefore $p^2 = 1/2$, we refer to a table of elliptic functions such as that found in Cohen, 1992, p. 278 to determine that $pK(p^2) = 1.3110$. Therefore, with $x_3 = -x_1$ and $x_2 = 0$, eq. 8.188 yields

$$\alpha = \frac{a\sqrt{2x_3}}{1.3110} \tag{8.197}$$

Then, β is found in terms of α or x_3 using one of eqs. 8.177. For example, referring to eq. 8.177e,

$$\beta = a - \alpha \lim_{x \to \infty} \int \frac{1}{\sqrt{x}\sqrt{x^2 - x_3^2}} dx \qquad (8.198)$$

This indefinite integral cannot be expressed in terms of elementary functions. However, because x is large (and ultimately infinite), we can ignore x_3 relative to x to obtain

$$\beta = a - \alpha \lim_{x \to \infty} \int x^{-3/2} dx = a \qquad (8.199)$$

Because the integral of eq. 8.175 cannot be evaluated in closed form, the inverse mapping, which transforms the rectangle in the w-plane onto the x-axis, cannot be expressed in terms of elementary functions. □

8.4 Applications

There is a certain class of applied problems that can be solved using conformal mappings. Because such mappings are applied to geometric configurations in two dimensions, if a problem is to be solved by conformal mapping, it must be possible to describe the problem in two dimensions.

Let a problem be described by coordinates (X,Y,Z). We temporarily denote the usual Cartesian coordinates in three dimensions by capital letters in order to distinguish them from the lowercase letters in $z = x + iy$ that denote points in the complex plane. If a problem in three dimensions can be treated as a two-dimensional problem, the system must have geometric symmetry in one of the dimensions.

Example 8.16: A three-dimensional configuration that can be treated as two-dimensional

An infinitely long cylinder, positioned along the Z-axis, with positive charge uniformly distributed over its surface, is shown in fig. 8.22. Finding the electrostatic properties of this system is a three-dimensional problem that can be treated in two dimensions.

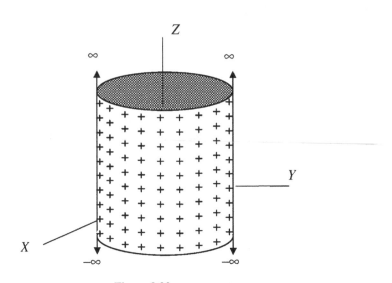

Figure 8.22
An infinitely long uniformly charged cylinder along the Z-axis

Because the cylinder is infinitely long and charged uniformly along its length, the values of functions that describe the electrostatic properties in the space around and on the cylinder at a given X, Y, and Z are the same at all values of Z. Thus, these functions are independent of Z and thus depend only on X and Y. Therefore, determining such a function becomes a problem in two dimensions, and it may be possible to view it as a problem in the complex. □

Laplace's equation

If a problem is described by a function $\Phi(x, y)$ which satisfies Laplace's equation in two dimensions

$$\nabla^2_{x,y} \Phi(x, y) = \frac{\partial^2 \Phi}{\partial x^2} + \frac{\partial^2 \Phi}{\partial y^2} = 0 \qquad (8.200)$$

the problem may be soluble by conformal mapping techniques. The reason for this is that if $\Phi(z)$ satisfies Laplace's equation, and w is the image of z under the conformal mapping $f(z)$, then

$$\Phi(w) = \Phi\big(f(z)\big) \qquad (8.201)$$

satisfies Laplace's equation

$$\nabla^2_{u,v} \Phi = \frac{\partial^2 \Phi}{\partial u^2} + \frac{\partial^2 \Phi}{\partial v^2} = 0 \qquad (8.202)$$

in the w-plane. The proof of this is given in appendix 6.

Boundary conditions

The solution to an ordinary differential equation always involves one or more constants of integration. The values of these constants are determined by having known values of the dependent variable $\Phi(x)$ and/or its derivatives, at specified values of the independent variable x. The values of $\Phi(x)$ and its derivatives at various points x_k are referred to as *boundary values* and the set of equations specifying those boundary values is called *boundary conditions*.

As an example, one could find four constants of integration from a knowledge of four independent constraint equations of the form

$$\Phi(x_1) = a \qquad (8.203a)$$

$$\Phi(x_2) = b \qquad (8.203b)$$

$$\Phi'(x_1) = c \qquad (8.203c)$$

and

$$\Phi'(x_2) = d \qquad (8.203d)$$

where a, b, c, and d are constants.

The set of points at which the boundary values are given define the boundaries of the problem. For example, if the values of $\Phi(x)$ and $\Phi'(x)$ are specified at $x_1 = 0$ and $x_2 = 1$ in eqs. 8.203, this implies that $0 \leq x \leq 1$.

For a solution to be completely defined, the number of independent boundary conditions must be the same as the *order* of the differential equation (the highest derivative of Φ). If the number of independent boundary conditions is less than the order of the differential equation the solution is said to be *underconstrained*. The solution to an underconstrained differential equation contains one or more undetermined constants of integration. A differential equation for which the number of independent boundary conditions is greater than the order of the equation is *overconstrained*. If a differential equation is overconstrained, a solution does not exist.

Because Laplace's equation in two dimensions is a second-order differential equation in each of two variables, the solution is completely specified by four independent boundary conditions, two in each variable.

Many applied problems that are described by Laplace's equation in two dimensions are constrained by boundary conditions of the form

$$\left[\alpha\Phi(x, y)+\beta\frac{\partial\Phi}{\partial n}\right]_{x_k,y_m} = \gamma_{km} \qquad (8.204)$$

where α, β, and γ_{km} are constants. x_k and y_m are points on the boundary, and the quantity n is the magnitude of a vector that is perpendicular (normal) to the boundary.

For example, if a problem is defined at points inside and on a circle, the circumference of the circle is the boundary of the problem. The normal to that boundary is a vector $\vec{r}$ pointing outward along the radius of the circle, and the magnitude of the normal to that boundary is

$$n = \left|\vec{r}\right| = \sqrt{\left(x^2 + y^2\right)} \qquad (8.205)$$

Then with

$$x = r\cos\theta \qquad (2.21a)$$

and

$$y = r\sin\theta \qquad (2.21b)$$

we have

$$\frac{\partial\Phi}{\partial n} = \frac{\partial\Phi}{\partial r} = \frac{\partial\Phi}{\partial x}\frac{\partial x}{\partial r} + \frac{\partial\Phi}{\partial y}\frac{\partial y}{\partial r} = \cos\theta\frac{\partial\Phi}{\partial x} + \sin\theta\frac{\partial\Phi}{\partial y} \qquad (8.206)$$

When $\alpha \neq 0$, $\beta = 0$, eq. 8.204 is know as the *Dirichlet boundary condition*

$$\Phi\left(z_{boundary}\right) = \frac{\gamma}{\alpha} = constant \qquad \text{(Dirichlet)} \qquad (8.207a)$$

For $\alpha = 0$, $\beta \neq 0$, the condition is the *Neumann boundary condition*

$$\left.\frac{\partial \Phi}{\partial n}\right|_{boundary} = \frac{\gamma}{\beta} = constant \qquad \text{(Neumann)} \qquad (8.207b)$$

The *Gauss* or *mixed boundary condition* is eq. 8.204 with $\alpha \neq 0$ and $\beta \neq 0$. That is,

$$\alpha \Phi\left(z_{boundary}\right) + \beta \left.\frac{\partial \Phi}{\partial n}\right|_{boundary} = constant \quad \text{(Gauss/mixed)} \quad (8.207c)$$

Let $f(z)$ be a conformal transformation that maps a region R in the z-plane onto a region S in the w-plane. Referring to eq. 8.207a, let the boundary of R be defined by a constant λ given by

$$\Phi\left(z_{boundary}\right) = \lambda \qquad (8.208a)$$

The boundary of S is then defined by

$$\Phi\left(z_{boundary}\right) = \Phi\left(g(w_{boundary})\right) = \Phi(w_{boundary}) = \lambda \qquad (8.208b)$$

Thus, the value of λ that defines the boundary condition in the z-plane also defines the boundary condition in the w-plane. That is, the value of Φ at the boundary of R is the same as the value of Φ at the boundary of S, even though the dependence of Φ on z in R is not the same as the dependence of Φ on w in S under the mapping. The terminology used to describe this is that the Dirichlet boundary condition is *invariant under a conformal mapping*.

Example 8.17: Invariance of the Dirichlet boundary condition under a conformal mapping

The circumference of a circle of radius λ, centered at the origin is defined by

$$\Phi(x, y) = x^2 + y^2 = \lambda^2 \qquad (8.209)$$

The mapping

$$w = e^z = e^x(\cos y + i \sin y) \qquad (8.210)$$

can also be expressed as

$$u(x, y) = e^x \cos y \qquad\qquad (8.211\text{a})$$

and

$$v(x, y) = e^x \sin y \qquad\qquad (8.211\text{b})$$

Then

$$x = \tfrac{1}{2} \ln\left(u^2 + v^2\right) \qquad\qquad (8.212\text{a})$$

and

$$y = \tan^{-1}\left(\frac{v}{u}\right) \qquad\qquad (8.212\text{b})$$

Therefore, eq. 8.209 becomes

$$\left[\tfrac{1}{2}\ln\left(u^2 + v^2\right)\right]^2 + \left[\tan^{-1}\left(\frac{v}{u}\right)\right]^2 = \Phi(u, v) = \lambda^2 \qquad (8.213)$$

So, as stated, the dependence of Φ on x and y in the z-plane is different from its dependence on u and v in the w-plane. As such, the shape of the boundary changes under the mapping but the value of λ that defines that boundary is unchanged by the mapping. □

It is shown in appendix 7 that

$$\left(\frac{\partial \Phi}{\partial n}\right)_{z_{boundary}} = \left(\nabla_{xy}\Phi\right)_{z_{boundary}} = \left(\frac{\partial \Phi}{\partial x} + i\frac{\partial \Phi}{\partial y}\right)_{z_{boundary}} \qquad (\text{A7.11})$$

The equivalent expression for the normal derivative at the boundary of S in the w-plane is

$$\left(\frac{\partial \Phi}{\partial n}\right)_{w_{boundary}} = \left(\nabla_{uv}\Phi\right)_{w_{boundary}} = \left(\frac{\partial \Phi}{\partial u} + i\frac{\partial \Phi}{\partial v}\right)_{w_{boundary}} \qquad (8.214)$$

In addition, the mapping of the gradient of a function is shown to be

$$\nabla_{xy}\Phi = \left(\nabla_{uv}\Phi\right)\left(\frac{\partial v}{\partial y} + i\frac{\partial u}{\partial y}\right) \qquad (A7.14a)$$

or

$$\nabla_{xy}\Phi = \left(\nabla_{uv}\Phi\right)\left(\frac{\partial u}{\partial x} - i\frac{\partial v}{\partial x}\right) \qquad (A7.14b)$$

The inverses of eqs. A7.14 are

$$\nabla_{uv}\Phi = \left(\nabla_{xy}\Phi\right)\left(\frac{\partial y}{\partial v} + i\frac{\partial x}{\partial v}\right) \qquad (A7.15a)$$

or

$$\nabla_{uv}\Phi = \left(\nabla_{xy}\Phi\right)\left(\frac{\partial x}{\partial u} - i\frac{\partial y}{\partial u}\right) \qquad (A7.15b)$$

If $u(x, y)$ and $v(x, y)$ are not constants,

$$\frac{\partial v}{\partial y} + i\frac{\partial u}{\partial y} \neq 0 \qquad (8.215a)$$

and

$$\frac{\partial u}{\partial x} - i\frac{\partial v}{\partial x} \neq 0 \qquad (8.215b)$$

Therefore, we see from eqs. A7.14 and A7.15 that

$$\nabla_{xy}\Phi = 0 \Longleftrightarrow \nabla_{uv}\Phi = 0 \qquad (8.216a)$$

and

$$\nabla_{xy}\Phi \neq 0 \Longleftrightarrow \nabla_{uv}\Phi \neq 0 \qquad (8.216b)$$

We see from eqs. A7.11 and 8.214 that the normal derivative of a function is its gradient. Therefore, from eq. 8.216a, we see that if $\partial\Phi/\partial n$ is zero in the z-plane, it must also be zero in the w-plane under a conformal mapping. That is, the Neumann boundary condition

$$\frac{\partial\Phi}{\partial n}\bigg|_{boundary} = 0 \qquad\qquad (8.217)$$

is invariant under a conformal mapping. If the normal derivative is not zero at the boundary, its transformation is given by eqs. A7.14 or A7.15.

Applications of conformal mapping to problems in electrostatics

There are many physical systems that are described by Laplace's equation subject to Dirichlet or to invariant Neumann boundary conditions. These include heat conduction in a medium that has uniform thermal properties, the nonturbulent flow of a fluid, and certain static electrical systems (systems in which charge does not move). The methods used to solve Laplace's equation subject to such boundary conditions using conformal mappings are the same for all such physical systems. We apply the conformal mapping techniques to problems in electrostatics to illustrate the method of solution, understanding that these techniques are applicable to problems describing other physical systems. The reader can find solutions to problems involving various physical systems by conformal mapping methods in texts such as Spiegel, 1964; Saff and Snider, 1976; Churchill and Brown, 1990 and Marsden and Hoffman, 1999.

It is important for the reader to have a basic knowledge of the properties of electrostatics. These include knowing that the potential energy of a system of charged particles is described by a function called the *electrostatic potential* (also called the *potential*) which satisfies Laplace's equation. The reader should also be familiar with the idea that an *equipotential surface* (or in two dimensions an *equipotential curve*) is defined as a surface (curve) on which the potential has the same value at all points. The equation for the equipotential surface is found by setting the potential to a constant. Such concepts are treated in most introductory physics texts. See, for example, Serway and Jewitt, 2004, pp. 765–766.

The general approach for determining the electrostatic potential by conformal mapping methods is to transform a complicated geometry of the distribution of charges in the z-plane (defined as one for which one cannot easily solve Laplace's equation) to a simple geometry in the w-plane (for which a solution to Laplace's equation is easily found). After solving the problem with the simpler geometry, one applies the inverse transformation back to the *z*-plane to find the potential with the original geometry. We illustrate this approach with several examples.

Example 8.18: Potential within the space of a wedge in a conducting block

We consider an infinitely large block of metal with a wedge of angle γ cut out of it as shown in fig. 8.23. The region within the wedge is a vacuum. The block extends to $\pm\infty$ in the dimension perpendicular to the page. Therefore, this geometry can be viewed as two-dimensional.

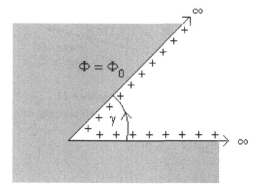

Figure 8.23
Conducting block with a wedge of angle γ cut out of the block

Because charge moves freely within a metal, all charge placed in the conductor is distributed in such a way that the potential is the same at all points along these edges. We denote this potential by Φ_0.

The potential $\Phi(z)$ at points in the vacuum region within the wedge (defined by $0 \leq \arg(z) \leq \gamma$), satisfies Laplace's equation. It is this potential that we determine by methods of conformal mapping.

The one boundary condition satisfied by $\Phi(z)$ is

$$\Phi(r, \theta = 0) = \Phi(r, \theta = \gamma) = \Phi_0 \tag{8.218}$$

Because this Dirichlet boundary condition is the only constraint applied to the system, the system is underconstrained. We use physical properties of this system to specify a second constant of integration.

It was shown in ex. 8.11 and fig. 8.16 that the SC mapping with derivative given by

$$f'(z) = \alpha(z - x_1)^{-k} \quad 0 < k < 1 \tag{8.131}$$

transforms the real axis in the z-plane onto the wedge shown in fig. 8.24.

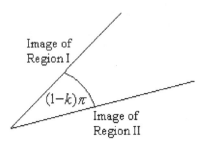

Image of
Region I

$(1-k)\pi$

Image of
Region II

Figure 8.24
Image of the real axis under a SC mapping with one branch point

Therefore, the inverse mapping

$$z = x_1 + \left[\frac{(1-k)}{\alpha} (w - \beta) \right]^{\frac{1}{(1-k)}} \tag{8.141}$$

transforms the wedge in the w-plane with internal angle $(1 - k)\pi$ onto the real axis of the z-plane. By interchanging z and w in eq. 8.141, we obtain the mapping

$$w = u_1 + \left[\frac{(1-k)}{\alpha} (z - \beta) \right]^{\frac{1}{(1-k)}} \tag{8.219}$$

which transforms the wedge in the z-plane onto the real axis of the w-plane. To apply this to the configuration of fig. 8.23, we set

$$\gamma = \pi(1 - k) \tag{8.220a}$$

$$u_1 = \beta = 0 \tag{8.220b}$$

and

$$\frac{(1-k)}{\alpha} = 1 \tag{8.220c}$$

(so α is real.) Then, the mapping of eq. 8.219 becomes

$$w = z^{\pi/\gamma} \tag{8.221}$$

To see that this transformation maps the wedge of fig. 8.23 onto the real axis of the w-plane, we note that points along the horizontal part of the wedge are described by

$$z = re^{i0} = r \tag{8.222}$$

These points map onto

$$w = r^{\pi/\gamma} \tag{8.223}$$

Because $0 \le r \le \infty$, this describes points along the u-axis with $0 \le u \le \infty$. Points along the slanted edge of the wedge are described by

$$z = re^{i\gamma} \tag{8.224}$$

the images of which are given by

$$w = r^{\pi/\gamma} e^{i\gamma(\pi/\gamma)} = -r^{\pi/\gamma} \tag{8.225}$$

Again, with $0 \le r \le \infty$, eq. 8.225 describes points along the u-axis in the region $-\infty \le u \le 0$. Therefore, the mapping of eq. 8.221 transforms the edges of the wedge onto the entire u-axis.

Points within the wedge satisfy $0 \le \arg(z) = \theta \le \gamma$. The mapping of eq. 8.221 transforms these points onto

$$w = r^{\pi/\gamma} e^{i\theta\pi/\gamma} \tag{8.226}$$

where $0 \le \arg(w) = \theta\pi/\gamma \le \pi$. Therefore, the space within the wedge maps onto the upper half of the w-plane.

The boundary of the image of the wedge is defined by $v = 0$. Because the Dirichlet boundary condition is invariant under the conformal mapping, the boundary condition of eq. 8.218 is mapped to

$$\Phi(u, v = 0) = \Phi_0 \tag{8.227}$$

Therefore, the mapping of eq. 8.221 transforms the problem of finding the potential in the region within the wedge of fig. 8.23 to the problem of finding the potential in the upper half of the *w*-plane due to a flat metal surface, extending along the entire *u*-axis and maintained at a potential Φ_0 by a uniformly distributed charge.

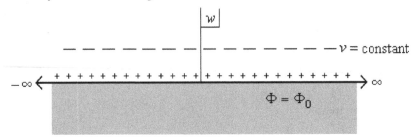

Figure 8.25

Image of the charged wedge of fig. 8.23 under the mapping of eq. 8.221

Because all points on the surface of the flat plate are at the same potential, it is clear that the potential has the same value at all points (u,v) that are the same distance *v* above the plate. Thus, the potential at any point must be independent of the value of *u* and can therefore be written

$$\Phi(u,v) = \Phi(v) \tag{8.228}$$

Then Laplace's equation in the *w*-plane becomes

$$\frac{d^2\Phi}{dv^2} = 0 \tag{8.229}$$

Two integrations of this differential equation yield

$$\Phi(v) = a_0 + a_1 v \tag{8.230}$$

for $v \geq 0$.   Then, applying the Dirichlet boundary condition, we obtain

$$\Phi(0) = \Phi_0 = a_0 \tag{8.231a}$$

The constant a_1 is obtained from the property that the *electrostatic field*, which is the derivative of the potential, is a constant for a charged flat plate. As with Φ_0, this constant field E_0 depends on how much charge is distributed over a given area of the plate. Referring to eq. 8.230 and to Serway and Jewett, 2004, p. 749, for example,

$$\frac{\partial \Phi}{\partial v} = a_1 = -E_0 \tag{8.231b}$$

Then

$$\Phi(v) = \Phi_0 - E_0 v \tag{8.232}$$

To complete the analysis, the potential must be expressed in terms of coordinates in the z-plane. Writing

$$v = \text{Im}(w) = \text{Im}\left(r^{\pi/\gamma} e^{i\theta\pi/\gamma}\right) = r^{\pi/\gamma} \sin\left(\frac{\theta\pi}{\gamma}\right) \tag{8.233}$$

the potential is given by

$$\Phi(r,\theta) = \Phi_0 - E_0 r^{\pi/\gamma} \sin\left(\frac{\theta\pi}{\gamma}\right) \tag{2.234a}$$

or equivalently

$$\Phi(x,y) = \Phi_0 - E_0 \left(x^2 + y^2\right)^{\pi/2\gamma} \sin\left[\frac{\pi}{\gamma} \tan^{-1}\left(\frac{y}{x}\right)\right] \tag{2.234b}$$

The equation for the family of equipotential curves is obtained by setting the potential to a constant. We see from eq. 8.234a that $\Phi(r,\theta)$ is constant when

$$r^{\pi/\gamma} \sin\left(\frac{\theta\pi}{\gamma}\right) = \Lambda = constant \tag{8.235}$$

This equation for the family of equipotential curves can be written as

$$r = \left[\Lambda \csc\left(\frac{\theta\pi}{\gamma}\right)\right]^{\gamma/\pi} \tag{8.236a}$$

or

$$\theta = \frac{\gamma}{\pi} \sin^{-1}\left(\Lambda r^{-\pi/\gamma}\right)$$
(8.236b)
□

Example 8.19: Potential within the space of a wedge in a conducting block with an insulator

We consider the problem of a wedge of angle γ cut from an infinite metal block (such as that in ex. 8.18) b. In this example, the block is separated electrically into two sections by an insulating strip as shown in fig. 8.26.

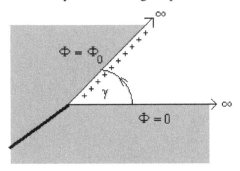

Figure 8.26

Wedge in a metal block that is separated into two sections
by an insulating strip

Because of the insulating strip the surfaces of the two sections can be charged differently and can therefore be maintained at different potentials. We take the potentials to be $\Phi = 0$ and $\Phi = \Phi_0 > 0$ as shown in fig. 8.26.

It was shown in ex. 8.18 that the mapping

$$w = z^{\pi/\gamma}$$
(8.221)

transforms the geometry of fig. 8.26 to the flat plate of fig. 8.27.

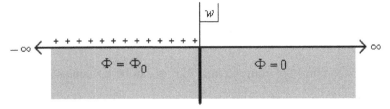

Figure 8.27

Image of a wedge with insulating strip under the mapping of eq. 8.221

In ex. 8.18, the symmetry in the u variable allows Laplace's equation to be expressed as an easily solved ordinary second-order differential equation. In this example, the region defined by $u < 0$ (the image of the slanted part of the wedge) is maintained at a potential $\Phi = \Phi_0$ and the region $u > 0$ (the image of the flat side of the wedge) is at a potential $\Phi = 0$. Because the two halves of the plate are at different potentials, the symmetry in u does not exist. Therefore, the potential in the upper half of the w-plane depends on u and v and Laplace's equation does not reduce to an ordinary differential equation with a straightforward solution. Thus, the mapping of eq. 8.221 is not appropriate for this problem.

Instead, we consider the transformation

$$w = \ell n(z) \tag{8.237}$$

which is a SC mapping obtained by integrating

$$f'(z) = z^{-1} \tag{8.238}$$

with the constant of integration taken to be zero. Under this mapping, points on the side of the wedge, defined by $\theta = \arg(z) = 0$, are of the form

$$z = re^{i0} = r \tag{8.239}$$

The images of these points are given by

$$w = \ell n(r) \tag{8.240}$$

Because $0 \leq r \leq \infty$ the points described by eq. 8.240 lie on the u-axis with $-\infty \leq u \leq +\infty$. That is, the side of the wedge defined by $\theta = 0$ maps onto the entire u-axis.

The argument of all points on the slanted side of the wedge is $\theta = \gamma$. These points are described in polar form as

$$z = re^{i\gamma} \tag{8.241}$$

which map onto points given by

$$w = \ell n(r) + i\gamma \tag{8.242}$$

This describes points on a line parallel to the u-axis defined by $-\infty \leq u \leq +\infty$ and $v = \gamma$.

Points in the region within the wedge, defined by $0 < \theta < \gamma$, are transformed onto points given by

$$w = \ln(r) + i\theta \tag{8.243}$$

which describes points with $-\infty \le u \le +\infty$ and $0 < v < \gamma$.

Thus, under the mapping of eq. 8.237, the slanted wedge of angle γ with its two halves electrically separated is transformed to the configuration of a parallel plate capacitor with a plate separation γ as shown in fig. 8.28.

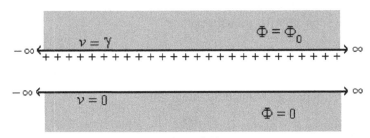

Figure 8.28

Image of the wedge of fig. 8.26 under the logarithm mapping of eq. 8.237

We see that the logarithm mapping of eq. 8.237 results in an image that is symmetric in the u variable. Therefore, the potential is independent of u and Laplace's equation becomes

$$\frac{d^2\Phi}{dv^2} = 0 \tag{8.229}$$

with solution

$$\Phi(v) = a_0 + a_1 v \tag{8.230}$$

The Dirichlet boundary conditions

$$\Phi(0) = 0 \tag{8.244a}$$

and

$$\Phi(\gamma) = \Phi_0 \tag{8.244b}$$

lead straightforwardly to the solution

$$\Phi(v) = \frac{\Phi_0}{\gamma} v = \frac{\Phi_0}{\gamma} \operatorname{Im}(w) = \frac{\Phi_0}{\gamma} \theta = \frac{\Phi_0}{\gamma} \tan^{-1}\left(\frac{y}{x}\right) \tag{8.245}$$

The equation for the family of equipotential curves, found by setting the potential to a constant, is given by

$$\theta = \Lambda = constant \tag{8.246a}$$

or equivalently,

$$y = x \tan \Lambda \tag{8.246b}$$

That is, the equipotential curves are straight lines of slope $\tan \Lambda$ passing through the origin. □

Example 8.20: Potential within a rectangular groove in a conducting block

The configuration of fig. 8.29 is a metal block with a rectangular groove of width $\pi/2$. We take the block to be charged so that all points on the sides and base of the groove are maintained at the potential Φ_0.

To solve the problem of finding the potential at points within the groove, we see that the geometry of the groove is an open three-sided polygon which is obtained from the SC mapping with two branch points (see ex. 8.12 and fig. 8.17).

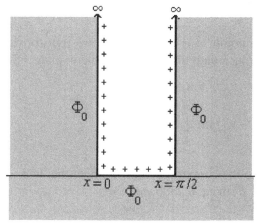

Figure 8.29

A rectangular groove in a metal block

Because the configuration of fig. 8.29 has two vertices, each with angle $\pi/2$, the real axis of the z-plane can be transformed into this geometry by

$$w = \beta + \alpha \int (z - x_1)^{-1/2}(z - x_2)^{-1/2} dz \tag{8.247}$$

By substituting

$$z - x_1 = (x_2 - x_1)\sin^2 \mu \tag{8.248}$$

eq. 8.247 can be written in closed form as

$$w = \beta - 2i\alpha \sin^{-1}\left[\sqrt{\frac{(z - x_1)}{(x_2 - x_1)}}\right] \tag{8.249}$$

This mapping transforms the real axis of the z-plane with branch points at x_1 and x_2 onto the configuration of fig. 8.29 in the w-plane. Therefore, interchanging z and w,

$$z = \beta - 2i\alpha \sin^{-1}\left[\sqrt{\frac{(w - w_1)}{(x_2 - w_1)}}\right] \tag{8.250a}$$

which has branch points at $w = w_1$ and $w = w_2$, transforms the geometry of fig. 8.29 in the z-plane onto the real axis of the w-plane. Solving eq. 8.250a for w we obtain

$$w = w_1 + (w_2 - w_1)\sin^2\left[\frac{i(z - \beta)}{2\alpha}\right] \tag{8.250b}$$

To apply this to the current problem, we note that $w = w_1$ is the image of $z = 0$. Therefore, from eq. 8.250b we have

$$\sin^2\left[\frac{i\beta}{2\alpha}\right] = 0 \tag{8.251a}$$

from which

$$\beta = 0 \tag{8.251b}$$

The image of $z = \pi/2$ is $w = w_2$, so that eq. 8.250b yields

$$\sin^2\left[\frac{i\pi}{4\alpha}\right] = 1 \qquad (8.251c)$$

Therefore,

$$\alpha = \frac{i}{2} \qquad (8.251d)$$

and the mapping of eq. 8.250b becomes

$$w = w_1 + (w_2 - w_1)\sin^2 z \qquad (8.252a)$$

To simplify the arithmetic of this problem, we set $w_1 = 0$ and $w_2 = 1$. Then

$$w = \sin^2 z \qquad (8.252b)$$

maps the configuration of fig. 8.29 in the z-plane onto the real axis of the w-plane.

We verify this by noting that points on the left wall of the groove are given by

$$z = iy \qquad 0 \le y \le \infty \qquad (8.253a)$$

Points along the base of the groove are defined by

$$z = x \qquad 0 \le x \le \pi/2 \qquad (8.253b)$$

and points along the right side wall are described by

$$z = \pi/2 + iy \qquad 0 \le y \le \infty \qquad (8.253c)$$

Therefore, points along the left wall map onto

$$w = \sin^2(iy) = -\sinh^2(y) \qquad (8.254a)$$

which is real, and so describes points $w = u$ with $-\infty \le u \le 0$. The images of points on the base of the groove are given by

$$w = \sin^2 x \qquad (8.254b)$$

which are also real. Because $0 \le x \le \pi/2$, the image of the base of the groove is defined by $0 \le u \le 1$. Points on the right-side wall are mapped onto points given by

$$w = \sin^2\left(\frac{\pi}{2} + iy\right) = \cosh^2 y \qquad (8.254c)$$

These points are also real and are in the range $1 \le u \le \infty$. Therefore, the images of the three edges of the groove comprise the real axis of the w-plane.

As was shown in general, the image of the interior of the polygon groove is the upper half of the w-plane. Therefore, the configuration of fig. 8.29 in the z-plane maps onto the geometry of fig. 8.30.

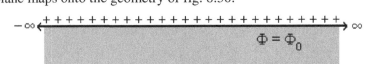

Figure 8.30

Image of the metal block with a rectangular groove of width $\pi/2$
under the mapping of eq. 8.252b

As in ex. 8.18, this geometry has symmetry in the u variable, so the potential in the w-plane only depends on $v = \text{Im}(w)$ the distance above the plate. Following the argument presented in ex. 8.18, the Laplace equation with one Dirichlet boundary condition and the applicable physics results in the potential given by

$$\Phi(v) = \Phi_0 - E_0 v = \Phi_0 - E_0 \text{Im}(w) = \Phi_0 - E_0 \text{Im}(\sin^2 z) \qquad (8.255)$$

Writing

$$\text{Im}\left(\sin^2 z\right) = \text{Im}\left[\left(\sin x \cosh y + i \cos x \sinh y\right)^2\right]$$
$$= \tfrac{1}{2}\sin(2x)\sinh(2y) \qquad (8.256)$$

the expression for the potential in the z-plane becomes

$$\Phi(x, y) = \Phi_0 - \tfrac{1}{2}E_0 \sin(2x)\sinh(2y) \qquad (8.257)$$

Setting the potential to a constant, the equation for the family of equipotential curves is given by

$$\sin(2x)\sinh(2y) = \Lambda = constant \qquad (8.258)$$

□

Example 8.21: Potential outside a pair of semicircular discs separated by an insulator

We consider a circular metal plate of radius R, centered at the origin (the cross-section of an infinitely long cylinder). The two halves of the plate are electrically separated by an insulating strip which, for simplicity, we orient

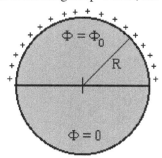

Figure 8.31

Two semicircular plates maintained at different potentials

along the x-axis from $-R$ to R. The upper section is charged to a potential $\Phi = \Phi_0$; the lower section is maintained at a potential $\Phi = 0$.

We first consider the bilinear transformation

$$z' \equiv x' + iy' = i\left(\frac{R-z}{R+z}\right) \qquad (8.259a)$$

Expressing z in polar form, this can be written

$$x' + iy' = i\left(\frac{R-re^{i\theta}}{R+re^{i\theta}}\right) = i\left(\frac{R-re^{i\theta}}{R+re^{i\theta}}\right)\left(\frac{R+re^{-i\theta}}{R+re^{-i\theta}}\right)$$

$$= \frac{2rR\sin\theta}{\left(R^2+r^2+2rR\cos\theta\right)} + i\frac{(R^2-r^2)}{\left(R^2+r^2+2rR\cos\theta\right)} \qquad (8.259b)$$

Points on the insulating strip along the positive x-axis are defined by $\theta = 0$ with $0 \leq r \leq R$. For points on the insulating strip along the negative x-axis, $\theta = \pi$. Because $\sin\theta = 0$ for these angles, the image of points along the insulating strip is given by

$$z' = x' + iy' = i\frac{\left(R^2 - r^2\right)}{\left(R^2 + r^2 + 2rR\cos\theta\right)}\bigg|_{\theta=0 \, or \, \pi} = \begin{cases} i\dfrac{(R-r)}{(R+r)} & \theta = 0 \\[3mm] i\dfrac{(R+r)}{(R-r)} & \theta = \pi \end{cases}$$

(8.260)

Therefore, for points on the insulating strip,

$$x' = 0 \tag{8.261a}$$

and

$$y' = \begin{cases} \dfrac{(R-r)}{(R+r)} & \theta = 0 \\[3mm] \dfrac{(R+r)}{(R-r)} & \theta = \pi \end{cases}$$

(8.261b)

Varying r between 0 and R, we see that the right half of the strip maps onto $0 \leq y' \leq 1$. The image of the left half of the strip is given by $1 \leq y' \leq \infty$. Thus, the insulating strip is mapped onto the positive imaginary axis of the z'-plane.

Referring to eq. 8.259b, we see that because $r < R$, the image of a point inside either semicircle has a positive imaginary part. Thus, in addition to the insulating strip, the interior of the disc maps onto the upper half of the z'-plane. We see from eq. 8.259b that the images of points in the interior of the upper semicircle, defined by $\sin\theta > 0$, have positive real parts and therefore, map to the first quadrant of the z'-plane. The images of points within the lower semicircle, for which $\sin\theta < 0$, have negative real parts and thus map to the second quadrant of the z'-plane.

Setting $r = R$ in eq. 8.259b, we find the image of points on the circumference of the disc to be of the form

$$z' = i\left(\frac{1 - e^{i\theta}}{1 + e^{i\theta}}\right) = -i\left(\frac{e^{-i\theta/2} - e^{i\theta/2}}{e^{-i\theta/2} + e^{i\theta/2}}\right) = \tan\left(\frac{\theta}{2}\right) \tag{8.262}$$

Points on the circumference of the semicircle in the upper half-plane are in the range $0 < \theta < \pi$. Because the lower semicircle is isolated (electrically) from the upper semicircle, points on the circumference of the lower semicircle can be defined by $-\pi < \theta < 0$. With these definitions, we see from eq. 8.262 that the circumference of the upper semicircle maps onto the positive x'-axis and the image of the circumference of the lower semicircle is the negative x'-axis.

Therefore, the image of the configuration of fig. 8.31 under the bilinear mapping of eq. 8.259a is that shown in fig. 8.32.

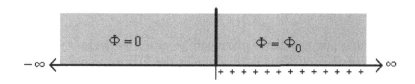

Figure 8.32
Image of the semicircular discs of fig. 8.31
under the bilinear mapping of eq. 8.259a

We note that the geometric configuration shown in fig. 8.32 is very similar to that of fig. 8.27 which is the image of the wedge with an insulating strip, under the mapping

$$w = z^{\pi/\gamma} \tag{8.221}$$

We recall that the configuration of fig. 8.27 was mapped onto a parallel plate capacitor by the logarithm function. Therefore, the configuration of fig. 8.32 will be transformed to a parallel plate capacitor by the mapping

$$w = \ell n(z') \tag{8.263}$$

Points on the edge of plate in the first quadrant are of the form

$$z' = r'e^{i0} = x' \quad x' \geq 0 \tag{8.264a}$$

The images of these points are given by

$$w = \ell n(x') \tag{8.264b}$$

These points map to the entire u-axis.

Because the insulator is along the positive imaginary axis in the z'-plane, points on the edge of the plate in the second quadrant must be accessed by a clockwise rotation by π through the vacuum region so we do not cross the line of discontinuous potential presented by the insulator. Therefore, points on this edge of the plate are given by

$$z' = r'e^{-i\pi} = |x'|e^{-i\pi} \tag{8.265a}$$

The image of this edge is

$$w = \ell n(|x'|) - i\pi \tag{8.265b}$$

This describes points in the w-plane with $-\infty \leq u \leq +\infty$ and $v = -\pi$. Therefore, the image of the split circular disc under the two mappings is the parallel plate capacitor shown in fig. 8.33.

The solution for the potential now proceeds as in the previous examples. The symmetry of the geometry in u allows us to express the potential as a function of v only. As in the previous examples, the solution to Laplace's equation when the potential depends on a single variable is given by

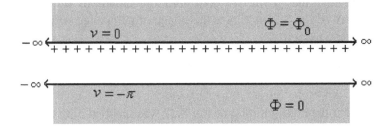

Figure 8.33
Image of the divided semicircular discs under a bilinear mapping
followed by a logarithm mapping

$$\Phi(v) = a_0 + a_1 v \tag{8.230}$$

Applying the Dirichlet boundary conditions

$$\Phi(0) = \Phi_0 \tag{8.266a}$$

and

$$\Phi(-\pi) = 0 \tag{8.266b}$$

we obtain the solution

$$\Phi(v) = \Phi_0\left(1 + \frac{v}{\pi}\right) \tag{8.267}$$

To transform the solution to the z-plane, we note that

$$v = \text{Im}(w) = \text{Im}\left[\ell n(r') + i\theta'\right] = \theta' \tag{8.268a}$$

Using eq. 8.259b, this can be written as

$$\theta' = \tan^{-1}\left(\frac{R^2 - r^2}{2rR\sin\theta}\right) \tag{8.268b}$$

Therefore,

$$\Phi(r,\theta) = \Phi_0\left[1 + \frac{1}{\pi}\tan^{-1}\left(\frac{R^2 - r^2}{2rR\sin\theta}\right)\right] \tag{8.269}$$

Referring to fig 8.31, we note that $\theta = 0$ and $\theta = \pm\pi$ are points on the insulating strip. Therefore, points on the conducting surface do not access these angles. Thus, $\sin\theta \neq 0$ for these points.

On the conducting surface of the upper semicircle, $\sin\theta > 0$. Therefore,

$$\lim_{r \to R}\tan^{-1}\left(\frac{R^2 - r^2}{2rR\sin\theta}\right) = 0 \tag{8.270a}$$

On the conducting surface of the lower semicircle, $\sin\theta < 0$ and

$$\lim_{r \to R}\tan^{-1}\left(\frac{R^2 - r^2}{2rR\sin\theta}\right) = -\pi \tag{8.270b}$$

Thus, the potential given in eq. 8.269 satisfies the specified boundary conditions.

Again, the equation for the family of equipotential curves is found by setting the potential to a constant. For the current example, this equation is

$$\frac{R^2 - r^2}{2rR\sin\theta} = \frac{1}{\Lambda} = constant \tag{8.271a}$$

or equivalently,

$$\sin\theta = \Lambda\left(\frac{R^2 - r^2}{2rR}\right)$$

(8.271b)

□

The examples presented thus far have involved mapping a nonsymmetric geometry, for which Laplace's equation cannot be solved easily, onto a line (the cross-section of a flat plate) or a pair of parallel lines (the cross-section of parallel plates). These images are not the only configurations that exhibit the symmetry needed to make Laplace's equation an easy differential equation to solve.

Example 8.22: Potential in the space between nonconcentric discs

We consider nonconcentric metal cylinders, the cross section of which is the nonconcentric circles shown in fig. 8.34. The large circle has a radius R and is centered at the origin of the z-plane. The small circle, which has a radius r, lies entirely within the large circle. The x-axis is defined by the line joining the centers of the two circles. The large circle is maintained at a potential $\Phi = \Phi_0 > 0$ and the small circle is at a potential $\Phi = 0$.

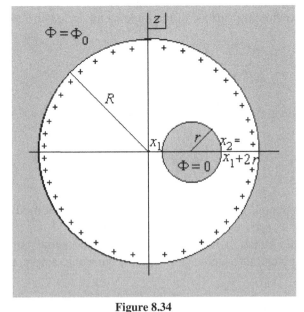

Figure 8.34

Nonconcentric circles

Referring to fig. 14 of Churchill and Brown, 1990, p. 348, a bilinear mapping transforms the circumferences of such nonconcentric circles and the region between them onto the circumferences of, and the region between, two concentric circles. To determine the bilinear mapping that transforms the geometry of fig. 8.34 to two concentric circles we begin with

$$w = \frac{\alpha z + \beta}{\gamma z + \delta} \tag{8.92a}$$

By factoring out α and δ, and defining λ, a, and b by

$$\frac{\alpha}{\delta} \equiv \lambda \tag{8.272a}$$

$$\frac{\gamma}{\delta} = a \tag{8.272b}$$

and

$$\frac{\beta}{\alpha} \equiv b \tag{8.272c}$$

the general bilinear mapping can be written as

$$w \equiv \lambda \left[\frac{z + b}{az + 1} \right] \tag{8.273}$$

The argument of λ is a measure of the overall rotation of the concentric circles in the w-plane. Because nothing is gained by keeping this argument unspecified, we take $\arg(\lambda) = 0$, and thus λ to be real.

We take the points on the circumference of the large circle to map onto the circumference of a circle, centered at the origin of the w-plane, of radius P. Because λ does not generate a rotation, the images of the points $\pm R$ are the points $\pm P$. Therefore,

$$P = \lambda \left[\frac{R + b}{aR + 1} \right] \tag{8.274a}$$

and

$$-P = \lambda \left[\frac{-R + b}{-aR + 1} \right] \tag{8.274b}$$

Combining these, we obtain

$$b = aR^2 \tag{8.275}$$

To demonstrate that the circumference of the large circle maps onto the circumference of a circle of radius P, we substitute $z = Re^{i\theta}$ into eq. 8.273 to obtain

$$w = \lambda \left(\frac{Re^{i\theta} + aR^2}{aRe^{i\theta} + 1} \right) \tag{8.276}$$

With $\phi_a \equiv \arg(a)$, the absolute value of w is given by

$$
\begin{aligned}
|w|^2 &= \lambda^2 \left[\frac{Re^{i\theta} + aR^2}{aRe^{i\theta} + 1} \right] \left[\frac{Re^{-i\theta} + a*R^2}{a*Re^{-i\theta} + 1} \right] \\
&= \lambda^2 R^2 \left[\frac{1 + |a|^2 R^2 + 2|a| R \cos(\theta - \phi_a)}{1 + |a|^2 R^2 + 2|a| R \cos(\theta + \phi_a)} \right]
\end{aligned}
\tag{8.277}
$$

where $\phi_a = \arg(a)$.

In order for these points to lie on the circumference of a circle in the w-plane, $|w|^2$ and therefore the quantity in the square bracket must be constant, independent of θ. This can be achieved by taking $\phi_a = 0$ or $\phi_a = \pi$. Then a is real, the quantity in the square bracket is 1, and therefore $|w| = \text{constant} \equiv P$. Then

$$\lambda = \frac{P}{R} \tag{8.278}$$

To map points on the circumference of the small circle onto a circle of radius p, centered at the origin of the w-plane, we require the image of $z = x_1$ to be $-p$, and the image of $z = x_1 + 2r$ to be p. Then

$$-p = \frac{P}{R} \left[\frac{x_1 + aR^2}{ax_1 + 1} \right] \tag{8.279a}$$

and

$$p = \frac{P}{R} \left[\frac{x_1 + 2r + aR^2}{ax_1 + 2ar + 1} \right] \tag{8.279b}$$

Combining these, we obtain

$$a = \frac{-\left(R^2 + x_1^2 + 2rx_1\right) \pm \sqrt{(R^2 - x_1^2)(R^2 - x_1^2 - 4r^2 - 4rx_1)}}{2R^2(x_1 + r)} \quad (8.280)$$

To see that the circumference of the small circle maps onto the circumference of a circle centered at the origin of the w-plane, we note from fig. 8.34 and eq. 8.273 that the images of $z = x_1 + r + re^{i\theta}$ are given by

$$w = \frac{P}{R}\left[\frac{x_1 + r + aR^2 + re^{i\theta}}{ax_1 + ar + 1 + are^{i\theta}}\right] \quad (8.281)$$

The magnitude of these points satisfies

$$|w|^2 = p^2$$

$$= \frac{P^2}{R^2}\frac{\left(x_1 + r + aR^2\right)}{a\left(ax_1 + ar + 1\right)}\left[\frac{\dfrac{\left(x_1 + r + aR^2\right)^2 + r^2}{\left(x_1 + r + aR^2\right)} + 2r\cos\theta}{\dfrac{\left(ax_1 + ar + 1\right)^2 + a^2r^2}{a\left(ax_1 + ar + 1\right)} + 2r\cos\theta}\right] \quad (8.282)$$

Clearly, for the term in the square bracket to be independent of θ, we must have

$$\frac{\left(x_1 + r + aR^2\right)^2 + r^2}{\left(x_1 + r + aR^2\right)} = \frac{\left(ax_1 + ar + 1\right)^2 + a^2r^2}{a\left(ax_1 + ar + 1\right)} \quad (8.283)$$

It is straightforward to demonstrate that this equality holds when a is given by the expression of eq. 8.280.

Attempting to proceed algebraically becomes a somewhat cumbersome process. We simplify the problem by taking $x_1 = 0$, $R = 5$, and $r = 2$. Then, from eq. 8.280, $a = -2/5$ (so that $aR^2 = -10$) or $a = -1/10$ (and $aR^2 = -5/2$). Taking $a = -2/5$, the mapping is

$$w = P\left(\frac{z - 10}{5 - 2z}\right) \quad (8.284)$$

from which it is straightforward to ascertain that

$$w(R) = w(5) = P \tag{8.285a}$$

$$w(-R) = w(-5) = -P \tag{8.285b}$$

$$w(x_1) = w(0) = -p = -2P \tag{8.285c}$$

and

$$w(2r) = w(4) = p = 2P \tag{8.285d}$$

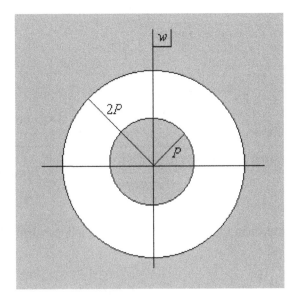

Figure 8.35
Image of the nonconcentric circles of fig. 8.34
under the bilinear mapping of eq. 8.284

Thus, the radius of the image of the small circle is twice as large as the radius of the image of the large circle.

As indicated in fig. 14 on p. 348 of Churchill and Brown, 1990, the interior of the small circle of fig. 8.34 maps onto the exterior of the circle of radius $2P$ and the exterior of the large circle of fig. 8.34 maps onto the interior of the circle of radius P.

To verify this, we note that a point in the exterior of the large circle of fig. 8.34 can be expressed as $z = Qe^{i\theta}$ with $Q > 5$.

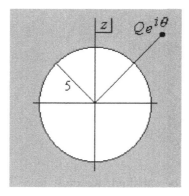

Figure 8.36a

A point in the region outside the large circle of fig. 8.34

The magnitude of the image of such a point is given by

$$|w|^2 = P^2 \left| \frac{Qe^{i\theta} - 10}{5 - 2Qe^{i\theta}} \right|^2 = P^2 \left(\frac{Q^2 + 100 - 20Q\cos\theta}{4Q^2 + 25 - 20Q\cos\theta} \right) \qquad (8.286)$$

If $|w|^2 > P^2$, then

$$Q^2 + 100 - 20Q\cos\theta > 4Q^2 + 25 - 20Q\cos\theta \qquad (8.287)$$

which requires $Q < 5$. Thus, $Q > 5$ requires that $|w| < P$.

Points in the interior of the small circle can be expressed as $z = 2 + qe^{i\theta}$ with $q < 2$. The magnitude of such a point is given by

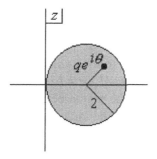

Figure 8.36b

A point in the interior of the small circle of fig. 8.34

$$|w|^2 = P^2 \left[\frac{q^2 + 64 - 16q\cos\theta}{4q^2 + 1 - 4q\cos\theta} \right] \qquad (8.288)$$

If $|w| < 2P$, then

$$\frac{q^2 + 64 - 16q\cos\theta}{4q^2 + 1 - 4q\cos\theta} < 4 \qquad (8.289)$$

For this inequality to be satisfied, we must have $q > 2$. Because $q < 2$, w must satisfy $|w| > 2P$. Therefore, under the bilinear mapping of eq. 8.284, the geometry of fig. 8.34 maps onto the configuration shown in fig. 8.35.

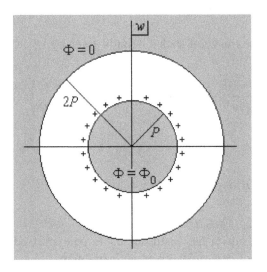

Figure 8.37
Cross-section of a coaxial cylindrical capacitor

Applying this mapping to the cylinders of fig. 8.34, we obtain the cross-section of an infinite *coaxial cylindrical capacitor* shown in fig. 8.37. To describe the potential in the region between the circles, we define the points in the w-plane in terms of the circular coordinates by $w = \rho e^{i\psi}$. It is clear from the symmetry of the geometry of fig. 8.37 that the potential is independent of ψ. Thus,

$$w(\rho, \psi) = w(\rho) \qquad (8.290)$$

Therefore, referring to Cohen, 1992, p. 37, for example, Laplace's equation in circular coordinates becomes

$$\frac{1}{\rho}\frac{d}{d\rho}\left(\rho\frac{d\Phi}{d\rho}\right)=0 \tag{8.291}$$

which has the straightforward solution

$$\Phi(\rho) = A\,\ell n(\rho) + B \tag{8.292}$$

Applying the boundary conditions

$$w(P) = \Phi_0 \tag{8.293a}$$

and

$$w(2P) = 0 \tag{8.293b}$$

we have

$$\Phi(P) = A\,\ell n(P) + B = \Phi_0 \tag{8.294a}$$

and

$$\Phi(2P) = A\,\ell n(2P) + B = 0 \tag{8.294b}$$

from which

$$\Phi(\rho) = \frac{\Phi_0}{\ell n(2)}\ell n\left(\frac{2P}{\rho}\right) \tag{8.295}$$

We transform the potential back to the z-plane by writing

$$\Phi(x,y) = \frac{\Phi_0}{\ell n(2)}\ell n\left(\frac{2P}{|w|}\right) = \frac{\Phi_0}{\ell n(2)}\ell n\left(2\left|\frac{5-2z}{z-10}\right|\right)$$

$$= \frac{\Phi_0}{2\ell n(2)}\ell n\left(4\left[\frac{(5-2x)^2+4y^2}{(x-10)^2+y^2}\right]\right) \tag{8.296}$$

The equation for the family of equipotential curves is obtained by setting the argument of the logarithm to a constant. This equation is

$$\frac{(5-2x)^2 + 4y^2}{(x-10)^2 + y^2} = \Lambda = constant \tag{8.297a}$$

which, for $\Lambda \neq 4$, can be written in the form

$$\left(x - 10\frac{(1-\Lambda)}{(4-\Lambda)}\right)^2 + y^2 = \Lambda\left(\frac{15}{(4-\Lambda)}\right)^2 \tag{8.297b}$$

Therefore, the equipotential curves are circles (infinitely long cylindrical surfaces in three dimensions) centered at $x = 10(1 - \Lambda)/(4 - \Lambda)$, $y = 0$ with radii $R = 15\Lambda^{1/2}/(4 - \Lambda)$. For example, the equation for the equipotential circle with $\Lambda = 1$ is

$$x^2 + y^2 = 25 \tag{8.298}$$

□

Sources containing tables of mappings

The discussions and examples above present the reader with a sample of useful conformal mappings and a guide to the techniques for applying them to problems soluble by Laplace's equation with Dirichlet or zero Neumann boundary conditions. A library search will provide the reader with other sources that contain additional examples and mappings.

Some of this literature includes tables of mappings that the reader should find useful. It would be an overwhelming task to compile a table of all the conformal mappings the author has found in other sources. Instead, this chapter is concluded by providing a short list of such sources that contain useful tables of mappings.

Churchill and Brown, 1990, appendix 2, pp. 345-352

Fisher, 1986, pp. 366–371

Kranz, 1999, pp. 181–194

Marsden and Hoffman, 1999, pp. 340–341

Saff and Snider, 1976, pp. 416–420

Spiegel, 1964, pp. 205–211
    Note: There is an error in this table in item C-4, p. 211. The mapping
        should read

$$w = \ln\left[\cosh\left(z/2\right)\right] \quad \text{instead of} \quad w = \ln\left[\coth\left(z/2\right)\right]$$

Problems

1. Determine if each of the mappings,

    (i) $w = e^z$ (ii) $w = z^2$ (iii) $w = z^3$

    is one-to-one, many-to-one or one-to-many when the region in the z-
plane
    being mapped is

    (a) $-\infty < z = x < \infty$ (b) $z = iy$ with $y > 0$

    (c) $-\infty < z = x < 0$ (d) the circumference of a circle, centered at
                               the origin, of radius R

2. In the mappings below, $|\delta| > |\gamma|$, the integer $N \geq 2$, and μ is an
irrational number. The region being mapped is the interior of the unit circle
in the z-plane. Determine if each mapping below is one-to-one, one-to-
many, or many-to-one. If the mapping is many-to-one or one-to-many,
determine the multiplicity of the mapping.

    (a) $w = \dfrac{1}{\gamma z + \delta}$ (b) $w = \dfrac{1}{\gamma z^N + \delta}$ (c) $w = \dfrac{1}{\gamma z^{1/N} + \delta}$

    (d) $w = \dfrac{1}{\gamma z^\mu + \delta}$ (e) $w = \alpha \ln(z) + \beta$

3. For each mapping and regions specified below, determine whether the
    mapping of the region R of the z-plane to the region S of the w-plane is
    onto or into.

Mapping	Region R	Region S
(a) $w = \sin^2 z$	$-\pi/2 < z = x < \pi/2$	$-1 < w = u < 1$

(b) $w = \sin^3 z$ $-\pi/2 < z = x < \pi/2$ $-1 < w = u < 1$

(c) $w = \sin^2 z$ $-\pi/2 < z = x < 0$ $0 < w = u < 1$

(d) $w = \sin^2 z$ $-\pi/2 < z = x < 0$ $0 \le w = u \le 1$

(e) $w = \sinh^2 z$ $z = iy, 0 < y < \pi/2$ $-1 < w = u < 1$

(f) $w = \sinh^4 z$ $z = iy, 0 \le y \le \pi/2$ $0 \le w = u \le 1$

(g) $w = \sinh^2 z$ $z = x + i\pi/4, 0 < x < \infty$ $w = -\frac{1}{2} + iv, -\infty < v < \infty$

4. Let $w = f(z)$ be a conformal mapping with its inverse defined by $z = g(w)$. Prove that if w_0 is the image of z_0 under this mapping, then eq. 8.59c is satisfied.

5. Prove that for any real number $N \ne 0$, the mapping $w = z^N$ and its inverse $z = w^{1/N}$ satisfy eqs. 8.59b and 8.59c.

6. For the following mappings, identify any finite z at which the mapping is not conformal.

    (a) $w = e^z$ (b) $w = \cosh z$ (c) $w = z^N$ integer $N \ge 2$

    (d) $w = \sin^2 z$ (e) $w = \dfrac{1}{z^2 + 1}$ (f) $w = \tan z$ (g) $w = |z|^2$

7. A straight line in the z-plane connects the points $z = 2 + i$ and $1 + 2i$. Under each of the mappings below
    • Find the equation of the image of the line.
    • Find the images of the endpoints.
    • Find the distance between the images of the endpoints.
    • Sketch a graph that could represent the image of the line.

    (a) $w = z^2$ (b) $w = \ln(z)$ (c) $w = \sqrt{z}$ (d) $w = \sqrt{2}e^{i\pi/4}z + (1 + 3i)$

    (e) $w = \dfrac{2z + 6i}{z - 3i}$ (f) $w = \dfrac{3\sqrt{2}e^{i\pi/4}z + 2(i - 2)}{3iz + (3 - i)}$

8. (a) Find the linear mapping that rotates a rectangle by $\pi/4$, doubles its area and displaces its center from the point $z = 2 + 2i$ to the point $z = 4 + 5i$.

   (b) Determine the invariant point(s) of this mapping.

   (c) Find the inverse mapping.

9. The quantities α, β, γ and δ are complex constants. Find the Jacobian determinant of the transformation of the area element under the mappings

   (a) $w = \alpha z + \beta$ (b) $w = z^2$ (c) $w = \dfrac{\alpha z + \beta}{\gamma z + \delta}$ (d) $w = \ell n \left[\dfrac{e^z + 1}{e^z - 1} \right]$

   (e) $w = \beta + \alpha \int (z - x_1)^{-k_1} (z - x_2)^{-k_2} dz$

10. The *cross-ratio* of four points z_0, z_1, z_2, and z_3 is defined by

$$(z_0, z_1, z_2, z_3) \equiv \frac{(z_0 - z_1)(z_2 - z_3)}{(z_0 - z_3)(z_2 - z_1)}$$

   (a) What values of z satisfy

$$(1, 2, 3, z) = (z, 1, 2, 3)$$

   (b) Show that

$$\left(1, e^{i\theta}, e^{i2\theta}, e^{i3\theta} \right) = -\frac{\sin\left(\dfrac{\theta}{2}\right)}{\sin\left(\dfrac{3\theta}{2}\right)} \qquad \theta \neq \tfrac{2}{3} N\pi \text{ with } N \text{ an integer}$$

   (c) Evaluate

$$\left(1, e^{i2\theta}, e^{i3\theta}, e^{i\theta} \right)$$

   Express the result in the same form as that in part (b).

(d) Find the value of k that satisfies

$$\left(1, e^{i\theta}, e^{i2\theta}, e^{ik\theta}\right) = \frac{\sin(2\theta)}{\sin(3\theta)}$$

11. Find the invariant point(s) of the following mappings:

(a) $w = \dfrac{z}{z-1}$ (b) $w = \dfrac{z^2 - 25}{z + 15}$ (c) $w = \dfrac{z^3 - 1}{z^2 - z + 2}$

12. What is the image of the circumference of the unit circle under each of the following mappings?

(a) $w = z + \dfrac{1}{z}$ (b) $w = z - \dfrac{1}{z}$ (c) $z^N + \dfrac{1}{z^N}$ $N = \text{integer} \geq 2$

13. What is the image of the second quadrant of the z-plane under the mapping

$$w = \frac{z - i}{z + i}$$

14. What is the image of
    (a) The line $x = \pi/2$ under the mapping $w = \sin z$?
    (b) The region $0 \leq x \leq \pi/2$, $-\infty \leq y \leq \infty$ under the mapping $w = \cos z$?
    (c) The region $0 \leq x \leq \pi/4$, $y > 0$ under the mapping $w = \tan z$?
    Sketch each region of the z-plane and its image in the w-plane.

15. Determine the image of a line of slope m that passes through the origin of the z-plane under the inversion

$$w = \frac{1}{z}$$

16. Find the image of the
    (a) upper half   (b) lower half   (c) right half   (d) left half
    of the z-plane under the mapping

$$w = \ell n(z)$$

when the cut associated with the logarithm branch point extends along the real axis to $+\infty$ and z is a point on the principal sheet.

17. The parameter b is real. Find the image of the line
    (a) $z = x + ib$ (b) $z = b + iy$
    under the mapping

    $w = e^z$

18. Find the image of the each of the following geometries.
    (a) The quarter circle of radius R shown in fig. P8.1a under the mapping

    (i) $w = z^2$ (ii) $w = \left(\dfrac{R^2 + z^2}{R^2 - z^2} \right)^2$

    (b) The semicircle of radius R shown in fig. P8.1b under the mapping

    (i) $w = \left(\dfrac{R+z}{R-z} \right)^2$ (ii) $w = z + \dfrac{1}{z}$

    (c) The 1/6 circle of radius R shown in fig. P8.1c under the mapping

    $w = z^3$

    (d) The 1/6 circle of radius R shown in fig. P8.1d under the mapping

    $w = z^3$

    (e) The first quadrant of the z-plane shown in fig. P8.1e under the mapping

    (i) $w = \dfrac{z^2 - i}{z^2 + i}$ (ii) $w = \dfrac{z - 1}{z + 1}$

    (f) The unit circle of fig. P8.1f under the mapping

    $w = \sqrt{\dfrac{iz + 1}{iz - 1}}$

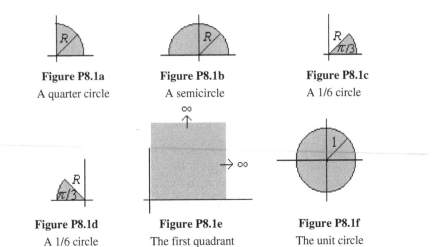

Figure P8.1a

A quarter circle

Figure P8.1b

A semicircle

Figure P8.1c

A 1/6 circle

Figure P8.1d

A 1/6 circle

Figure P8.1e

The first quadrant

Figure P8.1f

The unit circle

19. Deduce the mapping that transforms the $1/2N$ circle of radius R to a semicircle of radius $2R$ as shown in fig. P8.2.

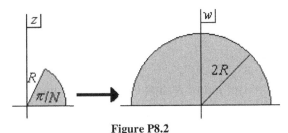

Figure P8.2

A $1/2N$ circle of radius R mapped to a semicircle of radius $2R$

20. Find the image of each region of the z-plane listed below under the mapping

    $w = e^z$

    (a) The strip defined by $-\infty \le x \le \infty$, $\pi \le y \le 2\pi$.
    (b) The boundary and interior of the rectangle with vertices at $z_1 = -1$, $z_2 = 1$, $z_3 = -1 + i\pi$, and $z_4 = 1 + i\pi$
    Sketch each region in the z-plane and its image in the w-plane.

21. A rectangle in the z-plane has sides that extend from $x = 0$ to $x = 1$, $y = 0$ to $y = 2$. Find the image of the perimeter of this rectangle under each mapping below.

    (a) $w = z + (i-1)$ (b) $w = 2ze^{i\pi/3} + (i-1)$

22. Find the linear mapping that transforms a square with sides L centered at the origin of the z-plane to a square in the w-plane with sides $2L$ centered at the point $(1,1)$.

23. Find a linear mapping that transforms the rectangle in the z-plane to the rectangle in the w-plane as shown in fig. P8.3.

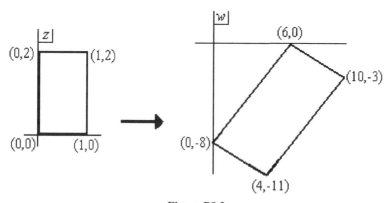

Figure P8.3

The mapping of a rectangle to a rectangle

24. Determine the linear mapping that transforms a circle in the z-plane of radius R centered at z_0 to a circle in the w-plane of radius kR (k is real and positive) centered at w_0. The image of z_0 is w_0 and the image of $z_0 + iR$ is $w_0 + kR$.

25. The equation of an ellipse centered at the origin of the w-plane is

$$\frac{u^2}{a^2} + \frac{v^2}{b^2} = 1$$

where a and b are real nonzero constants and $a \neq b$.

(a) Show that the mapping

$$w = \frac{1}{2}\left(\frac{z}{\beta} + \frac{\beta}{z}\right) \qquad \beta \geq 1$$

transforms the circumference of the unit circle centered at the origin of the z-plane, onto the circumference of an ellipse, centered at the origin of the w-plane.

(b) Determine the constants a and b in terms of β.

(c) What region of the w-plane is the image of the interior of the circle?

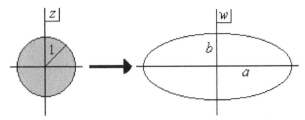

Figure P8.4

Mapping of the unit circle onto an ellipse

26. The functions $F(z)$ and $G(z)$ are defined by

$$F(z) \equiv \frac{z+a}{z+b} \quad \text{and} \quad G(z) \equiv z+c$$

where a, b, and c are nonzero constants with $a \neq b \neq c$. Find

(a) $F(G(z))$ (b) $G(F(z))$ (c) $F(F(z))$ (d) $F^{-1}(G(z))$

Find the value(s) of z for which

(e) $F(G(z)) = G(F(z))$ (f) $G^{-1}(F(z)) = z$

27. A function is given by

$$\Phi(x, y) = \frac{xy}{x^2 + y^2}$$

Find $\Phi(u,v)$, the transformation of this function, under each of the following mappings:

(a) $w = z^2$ (b) $w = e^z$

28. (a) Prove that the positive u-axis is the image of the entire x-axis under the mapping

$$w = e^z$$

(b) Show that

$$\Phi(x, y) = e^{-x} \cos y$$

satisfies Laplace's equation in the z-plane at all finite x and y.

(c) Find $\Phi(u,v)$, the transformation of $\Phi(x, y)$, under the mapping given in part (a).

(d) Show that $\Phi(u,v)$ satisfies Laplace's equation in the w-plane under the mapping given in part (a).

(e) It is shown above that the image of the x-axis is the u-axis under the mapping given in part (a). The normal to the x-axis is y and the normal to the u-axis is v. If $\Phi(x, y)$ satisfies the Neumann boundary condition

$$\left.\frac{\partial \Phi}{\partial n}\right|_{y=0} = \left.\frac{\partial \Phi}{\partial y}\right|_{y=0} = 0$$

at all points on the x-axis, prove that $\Phi(u,v)$ satisfies the boundary condition

$$\left.\frac{\partial \Phi}{\partial n}\right|_{v=0} = \left.\frac{\partial \Phi}{\partial v}\right|_{v=0} = 0$$

in the w-plane at all points along the positive u-axis.

29. Find the linear mapping that transforms the equilateral triangle with sides L to the equilateral triangle with sides $2L$ as shown in fig. P8.5.

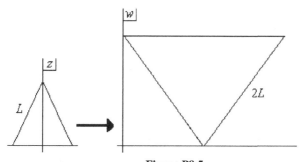

Figure P8.5

Transformation of an equilateral triangle

30. Each pair of points below represents the two invariant points of a bilinear mapping for which the image of $z = 0$ is $w = 3$. Find each bilinear mapping.

(a) $+1$ and -1 (b) $+N$ and $-N$ (N real and positive)   (c) $+1$ and -2

(d) $+2$ and -1

31. (a) Find the bilinear mapping $f(z)$ that satisfies

$$f(1) = 0, \quad f(i) = 1, \quad f(1-i) = i$$

(b) What is the image of the point $(1+i)$ under this mapping?
(c) What are the invariant points of this mapping?

32. A circle of radius R, centered at z_0, is transformed by a bilinear mapping to a circle of radius ρ centered at w_0. The image of z_0 is w_0 and the images of $z = +R$ and $z = -R$ are $w = -\rho$ and $w = +\rho$, respectively. Referring to ex. 8.22,
(a) Determine the bilinear mapping.
(b) How are the radii R and ρ related if the origin is an invariant point of the mapping?

33. Find the bilinear mapping that transforms the y-axis onto the circumference of the unit circle, centered at the origin of the w-plane for which $+i$ and $-i$ are invariant points.

34. (a) Determine the linear mapping under which the image of the point $(1,0)$ in the z-plane is the origin of the w-plane and the image of $(1,1)$ in the z-plane is $(2,1)$ in the w-plane.
(b) What is the image of the line $x = 1$ under this mapping?

35. A bilinear mapping transforms the circumference and interior of the unit circle, centered at the origin of the z-plane onto the circumference and exterior of the unit circle, centered at the origin of the w-plane as shown in fig. P8.6.
(a) Find the mapping if the points $+1$ and -1 are invariant points.
(b) What are the images of $z = +i$ and $z = -i$ under this mapping?

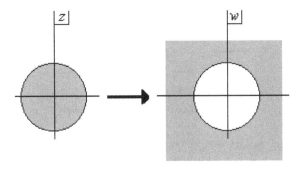

Figure P8.6
Mapping of the interior and circumference of the unit circle onto
the exterior and circumference of the unit circle

36. (a) Show that a general bilinear transformation can be written in the form

$$w = \frac{\alpha z + \beta}{\gamma z + \delta} = w_0 + \frac{z - z_0}{z - z_0^*}$$

Identify α, β, γ and δ in terms of z_0 and w_0.

(b) Referring to ex. 8.9, determine the image of the x-axis and the upper half of the z-plane under a general bilinear mapping.

37. Let w_0, w_1, w_2, and w_3 be the images of z_0, z_1, z_2, and z_3 respectively.

(a) Prove that under a general bilinear mapping

$$(w_0, w_1, w_2, w_3) = (z_0, z_1, z_2, z_3)$$

The quantities (z_0, z_1, z_2, z_3) and (w_0, w_1, w_2, w_3) are the cross-ratios defined in prob. 10 of this chapter.

(b) Prove that if w_m is the image of z_m for $m = 1, 2,$ and 3, and if

$$(w, w_1, w_2, w_3) = (z, z_1, z_2, z_3)$$

then the mapping $w = f(z)$ must be a bilinear mapping

$$w = \frac{\alpha z + \beta}{\gamma z + \delta}$$

Identify α/δ, β/δ, and γ/δ in terms of the fixed points z_m and w_m, for $m = 1, 2,$ and 3.

(c) Find the image of the unit circle and its interior under the bilinear mapping for which $(w, -i, 1, i) = (z, -i, i, 1)$.

38. A SC mapping has one branch point on the x-axis. Repeat the analysis of ex. 8.11 when points in region II are accessed along the bottom of the cut associated with the branch point.

39. Consider a SC mapping that has two branch points on the x-axis. Refer to eqs. 8.121, 8.124, and 8.127 to deduce the angles of the polygon image of the x-axis analogous to those shown in figs. 8.14 when

(a) Points are accessed along the top of the cut(s) in region II and along the bottom of the cut(s) in region III of the z-plane.

(b) Points are accessed along the bottom of the cut(s) in region II and along top of the cut(s) in region III of the z-plane.

Sketch illustrations of the polygon images analogous to those of figs. 8.14 for each of these cases.

40. What is the polygon image of the x-axis under each of the following SC mappings?

(a) $w(x) = \beta + \alpha \int \left(1 - x^2\right)^{-3/4} dx$ (b) $w(x) = \beta + \alpha \int \left(1 - x^2\right)^{-2/3} dx$

(c) $w(x) = \beta + \alpha \int \left[x \left(1 - x^2\right) \right]^{-1/2} dx$

41. If all points are accessed along the top of each cut, determine the SC transformation that maps the x-axis with two branch points to the triangle of fig. P8.7. You may leave your answer in terms of an integral.

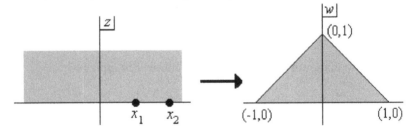

Figure P8.7

The mapping of the upper half of the z-plane, including the x-axis,

to an isosceles triangle in the w-plane

42. (a) Referring to ex. 8.18, show that the mapping

$$w = iz^{\pi/2\gamma}$$

transforms the edges and interior of the wedge in the z-plane to the u-axis and upper half of the w-plane as shown in fig. P8.8.

(b) Using the methods of conformal mapping, determine the potential in the region of the wedge when the conducting part of the wedge is maintained at a potential $\Phi_0 > 0$.

(c) Determine the equation for the family of equipotential curves in the vacuum region of the wedge.

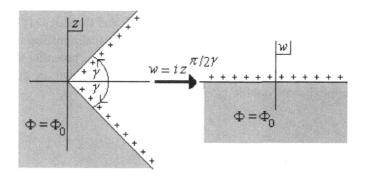

Figure P8.8

Mapping of a wedge to the upper half-plane

43. A groove is cut out of a metal block. The base of the groove extends along the real axis from $x = -\pi/2$ to $x = \pi/2$, and both sides of the groove extend to ∞ as shown in fig. P8.9. The edges of the sides and the base of the groove are maintained at a potential $\Phi_0 > 0$.

   (a) Determine the image of this geometry under the mapping

   $$w = \sin z$$

   (b) Determine the potential in the region within the groove.
   (c) Determine the equation for the family of equipotential curves in the region within the groove.

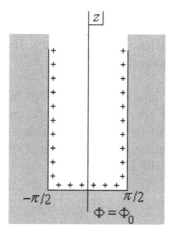

Figure P8.9

A groove cut out of a metal block

44. A groove is cut out of a metal block. The base of the groove extends along the real axis from $x = -\pi/2$ to $x = \pi/2$, and the sides of the groove extend to ∞. The left side of the groove is electrically insulated from the base and right side. The left side is maintained at a potential Φ_0 and the base and right side are at zero potential as shown in fig. P8.10. Determine the potential in the region within the groove.

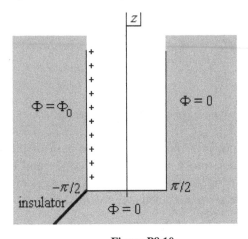

Figure P8.10
A groove in a metal block with one side
electrically isolated from the other two sides

(Hint: It is noted in problem 43 that

$$z' = \sin z$$

is a useful transformation for this geometry. Then refer to ex. 8.19 for an indication as to how to proceed from there.)

45. A groove is cut out of a metal block. The base of the groove extends along the real axis from $x = -\pi/2$ to $x = \pi/2$, and the sides of the groove extend to ∞. The base of the groove is electrically insulated from the two sides. The sides are maintained at a potential Φ_0 and the base is at zero potential as shown in fig. P8.11.

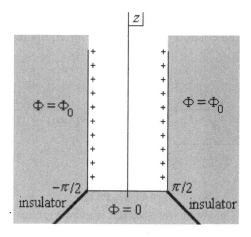

Figure P8.11

A groove in a metal block with sides and base
electrically isolated from one another

Determine the potential in the region within the groove.
(Hint: It is noted in prob. 43 that $z' = \sin z$ is a useful transformation for this
geometry. For the current problem, follow this by the mapping

$$w = \ell n \left(\frac{z' + 1}{z' - 1} \right)$$

46. Referring to ex. 8.12, it was shown in eqs. 8.145 that the SC mapping

$$w = \beta + \alpha \ell n \left[\frac{z + \sqrt{z^2 - z_1^2}}{z_1} \right]$$

maps the upper half of the z-plane, including the real axis, onto the sides
and interior of the three-sided open polygon of fig. P8.12. The origin is
an invariant point of the mapping and the point $z = i\pi$ is the image of u_1,
a point on the real axis of the w-plane.
(a) If the geometry of fig. P8.12 is a configuration in the z-plane, find the
SC mapping that transforms that geometry onto the upper half of the
w-plane, including the u-axis.
(b) If the figure of fig. P8.12 is the cross-section of a metal slab that is
uniformly charged to a potential Φ_0, use conformal mapping

methods to find the potential at any point in the z-plane exterior to the slab.

(c) Determine the equation for the family of equipotential curves for this charged slab.

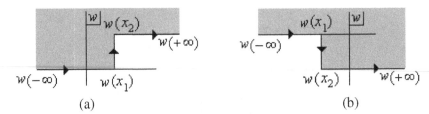

(a)   (b)

Figure P8.12

Cross-section of a charged, semi-infinite metal slab

47. Consider a SC mapping of the x-axis with branch points at x_1 and $x_2 > x_1$. With all points accessed along the tops of the associated cuts, find the SC mapping that transforms the upper half of the z-plane, including the x-axis, to each of the "step" geometries in the w-plane shown in figs. P8.12.

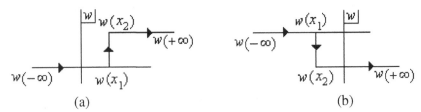

(a)   (b)

Figure P8.13

The mapping of the upper half of the z-plane, including the x-axis, to "steps" of finite height in the w-plane

48. Two identical discs of radius r_0, the cross-sections of two infinitely long metal cylinders, are centered at $\pm i r_0$ in the z-plane as shown in fig. P8.14. A small insulating strip is placed along the axis of contact of the cylinders so they can be maintained at potentials $\pm\Phi_0$.

(a) Show that with α and β real, the bilinear mapping

$$w = \frac{i\alpha z - \beta}{z} = i\alpha - \frac{\beta}{z} = i\alpha - \frac{\beta z^*}{|z|^2}$$

transforms the circumferences of the discs in the z-plane to two parallel plates and the region exterior to the discs maps onto the region between the plates.

(b) If the discs are maintained at potentials $+\Phi_0$ and $-\Phi_0$, what is the potential in the region exterior to the discs?

(c) What is the equation for the family of equipotential curves for the configuration of fig. P8.14?

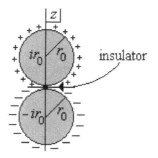

Figure P8.14

Metal discs maintained at different potentials

49. The pair of the semi-infinite parallel metal plates shown in fig. P8.15 is charged so that the upper plate is maintained at a potential $+\Phi_0$ and the lower plate is maintained at $-\Phi_0$, with $\Phi_0 > 0$.

(a) Find the image of this geometry under the sequence of mappings

$$z' = \cosh\left(\frac{\pi z}{2d}\right) \quad \text{followed by} \quad w = \ell n(z')$$

(b) Find the potential at all points in the z-plane.

(c) Determine the equation for the family of equipotential curves for this configuration.

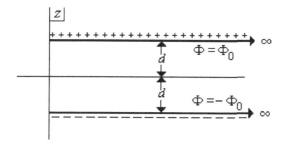

Figure P8.15

Semi-infinite parallel metallic plates

50. A metal disc of radius $r = 1$ is insulated from the wall of a circular cavity in an infinite metal block. The radius of the cavity is $R = 2$. The small disc is maintained at a potential of $-\Phi_0$ and the wall of the cavity is maintained at a potential $+\Phi_0$ with $\Phi_0 > 0$ as shown in fig. P8.16.

    (a) Find the geometry of this configuration under the bilinear mapping

$$w = i\left(\frac{4-z}{z}\right)$$

    (b) Find the potential in the crescent-shaped vacuum region?
    (c) What is the equation for the equipotential curves in the vacuum region?

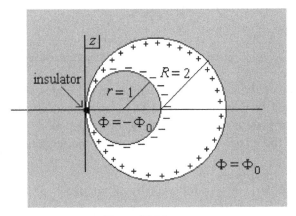

Figure P8.16

Circular metal disc insulated from the wall

of a circular cavity in a metal block

51. A metal block occupies the entire left half (the second and third quadrants) of the z-plane. The edge of the block lies along the y-axis and is maintained at a potential $\Phi = 20$ volts. A metal disc of radius $R = 2$ is positioned with its center at $z = 5/2$. The surface of the disc is maintained at a potential $\Phi = 5$ volts.

    (a) Show that the bilinear mapping under which:

$$z = \pm\infty \to w = +2 \qquad z = 0 \to w = -2 \qquad z = 1/2 \to w = -1$$

    transforms the geometry of a plate and a disc onto two concentric discs centered at the origin of the w-plane of radii $|w| = 1$ and $|w| = 2$ as shown in fig. P8.17.

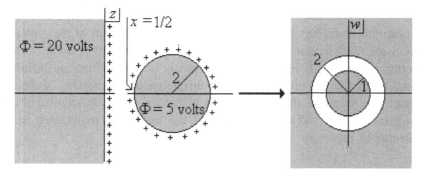

Figure P8.17
Mapping of a metal block and disc onto two concentric discs

(b) Find the potential in the right half of the z-plane outside the disc.
(c) Find the equation for the family of equipotential curves for this geometry.

52. For an integer $m \geq 2$, a conducting wedge of angle π/m is charged to a potential $\Phi_0 > 0$. The wedge is oriented with one side along the x-axis of the z-plane and a circular section of radius 1 is removed from the apex of the wedge as shown in fig. P8.18.

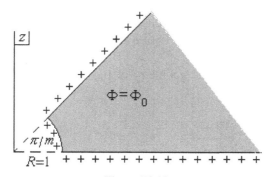

Figure P8.18
A metal wedge of angle π/m with a circular section removed

(a) Find the image of the wedge under the sequence of mappings.

$$z' = z^m \qquad w = \frac{1}{2}\left(z' + \frac{1}{z'}\right)$$

(b) Find the potential at points in the vacuum region outside the conductor.

(c) Find the equation for the equipotential curves in the vacuum region.

53. An infinite metal block covering the entire first quadrant of the z-plane has a quarter circular cavity of radius $R = 1$ removed from it. The section of the block in the first quadrant is insulated from an infinite metal block that occupies the second through fourth quadrants of the z-plane. The edge of the arc of the circular cavity is maintained at a potential $\Phi_0 > 0$ and the flat edges of the cavity are maintained at a potential of zero.

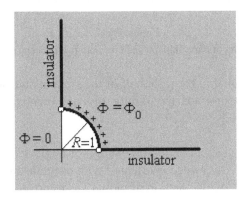

Figure P8.19

A block with a quarter circular section removed

that is electrically isolated from infinite metal blocks

(a) Find the image of the geometry of fig. P8.19 under the sequence of mappings:

$$z' = z^2 \qquad z'' = \frac{1+z'}{1-z'} \qquad z''' = (z'')^2 \qquad w = \ell n(z''')$$

(b) Find the potential at points within the quarter circular cavity.

(c) Find the equation for the equipotential curves within the cavity.

54. Two infinitely long metal cylinders are positioned near each other. The discs of fig. P8.20a are the cross-sections of these cylinders. One disc has a radius R and is centered at the origin of the z-plane. The other has

a radius of βR and is centered at $(1 + \alpha + \beta)R$ (clearly both α and β are positive). The line joining their centers defines the x-axis.

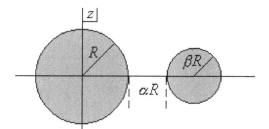

Figure P8.20a
Two metal discs

(a) Consider the bilinear mapping

$$w = \frac{z - aR^2}{az - 1}$$

where, with $\alpha > 0$ and $\beta > 0$,

$$a = \frac{1}{2R}\left[\frac{1 + (1+\alpha)(1+\alpha+2\beta) + \sqrt{\alpha(2+\alpha)(\alpha+2\beta)(2+\alpha+2\beta)}}{(1+\alpha+\beta)}\right]$$

Using ex. 8.22 as a guide, select numerical values for α and β and show that this bilinear mapping transforms the configuration of fig. P8.20a onto the configuration of fig. P8.20b.

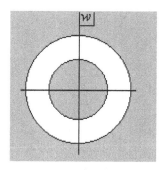

Figure P8.20b
A circular disc in a circular cavity in an infinite conducting medium

(b) The disc of radius R is maintained at a potential $\Phi_0 > 0$ and the disc of radius βR is at a potential $-\Phi_0$. Find the expression for the potential at points in the vacuum region of the z-plane outside the two discs of fig. P8.20a.

(c) Determine the equation for the family of equipotential curves in the vacuum region of the z-plane.

Chapter 9

DISPERSION RELATIONS

Dispersion relations were originally developed by physicists H. Kramers and R. Kronig (Kramers, 1926, p.775 and Kronig, 1926, p. 547). These dispersion relations are a pair of integrals that were developed to relate the real and imaginary parts of a function called the *complex index of refraction*. This index describes how a material medium affects electromagnetic radiation passing through it, and depends on the frequency of the radiation.

The real part of the index of refraction describes how radiation of a given frequency refracts (changes speed) when passing from one medium to another. When a beam of radiation enters a medium along a line that is not perpendicular to the surface of the medium, refraction causes a change in the direction in which the beam is traveling. Beams of different frequencies bend at different angles. This results in the spreading or *dispersion* of the radiation.

The imaginary part of the index describes the absorption of radiation by the medium as a function of frequency. The real part of the index is a function that can be derived from theory. Kramers and Kronig derived dispersion relations to determine the imaginary (absorptive) part of the index from the real (dispersive) part.

The meaning of dispersion relations has been broadened to include various types of integral representations of complex functions. They are tools used to determine a complex function from a knowledge of just the real part or just the imaginary part of that function.

Let $f(z')$ have poles at $z' = z_1, z_2, \ldots, z_N$ inside a closed contour C with residues $R_1, R_2, \ldots, R_N$. If the point z is inside the contour, the function $f(z')/(z' - z)$ has $N + 1$ poles at $z, z_1, \ldots, z_N$. Then, the extension of Cauchy's integral representation of $f(z)$ given in eq. 3.101b becomes

$$\frac{1}{2\pi i}\oint_C \frac{f(z')}{(z'-z)}dz' = f(z) + \sum_{n=1}^{N} Res\left[\frac{f(z')}{(z'-z)}\right]_{z'=z_n} \tag{9.1}$$

where $f(z)$ is the residue of the pole of the integrand at $z' = z$ and the nth residue in the sum arises from the pole of $f(z')$ at z_n. Because $f(z)$ is not completely determined, the residues of poles of $f(z')/(z'-z)$ at z_n cannot be determined. As such, a set of dispersion relations is an integral representation for a function that is analytic everywhere within and on the contour. Then the sum of residues on the right-hand side of eq. 9.1 is zero, and eq. 9.1 becomes Cauchy's integral representation of $f(z)$

$$f(z) = \frac{1}{2\pi i}\oint_C \frac{f(z')}{(z'-z)}dz' \tag{3.101b}$$

9.1 Kramers-Kronig Dispersion Relations Over the Entire Real Axis

In many applications, z represents a physical quantity such as the energy of a particle or, as in the Kramers–Kronig analysis, the frequency of radiation. Because these physical quantities are real, contours appropriate for the integrals in the analysis extend from $-\infty$ to $+\infty$ along the real axis. To apply Cauchy's theorem, the contour must be closed in one of the half-planes as shown in fig. 9.1 and z must be inside the contour.

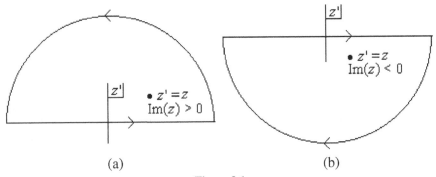

(a)   (b)

Figure 9.1

Contours appropriate for dispersion relations involving real quantities

Of course, in order for Cauchy's integral representation to be defined, $f(z')/(z'-z)$ must vanish on the infinite semicircular segments of the contour.

Because integration is along the real axis from $-\infty$ to $+\infty$, the integral around the contour of fig. 9.1a is taken in the counterclockwise direction. For this contour, eq. 3.101b becomes

$$\frac{1}{2\pi i}\oint_C \frac{f(z')}{(z'-z)}dz' = \begin{cases} f(z) & \text{Im}(z)>0 \\ 0 & \text{Im}(z)<0 \end{cases} \qquad (9.2a)$$

Integrating along the real axis from $-\infty$ to $+\infty$ and closing in the lower half-plane causes the integral to be taken in the clockwise direction. Then

$$\frac{1}{2\pi i}\oint_C \frac{f(z')}{(z'-z)}dz' = \begin{cases} 0 & \text{Im}(z)>0 \\ -f(z) & \text{Im}(z)<0 \end{cases} \qquad (9.2b)$$

Unless there is a reason to close the contour in the lower half-plane, we proceed with the analysis by closing in the upper half-plane so that the contour is traversed in the counterclockwise direction as in fig. 9.1a. With $\text{Im}(z)>0$,

$$f(z) = \frac{1}{2\pi i}\oint_C \frac{f(z')}{(z'-z)}dz' = \frac{1}{2\pi i}\int_{-\infty}^{\infty} \frac{f(x')}{(x'-z)}dx' \qquad (9.3)$$

where the integral along the infinite semicircle is zero and has been omitted.

Because $\text{Im}(z)>0$, when z represents a real quantity, z must approach the real axis from the upper half-plane. This is accomplished by taking

$$z = \lim_{\varepsilon\to 0}(x+i\varepsilon) \qquad (9.4)$$

so that, with $f(x)$ analytic at all x, eq. 9.3 becomes

$$f(x+i\varepsilon) \to f(x) = \frac{1}{2\pi i}\lim_{\varepsilon\to 0}\int_{-\infty}^{\infty} \frac{f(x')}{(x'-x-i\varepsilon)}dx' \qquad (9.5)$$

With

$$\lim_{\varepsilon\to 0}\frac{1}{(x'-x\pm i\varepsilon)} = \frac{1}{(x'-x)_p}\mp i\pi\delta(x'-x) \qquad (5.71)$$

and using the property of the Dirac δ symbol,

$$\int_a^b F(y)\,\delta(x-y)\,dy = \begin{cases} F(x) & a < x < b \\ 0 & x < a \text{ or } x > b \end{cases} \qquad (5.69)$$

eq. 9.5 becomes

$$f(x) = \frac{1}{2\pi i} P\int_{-\infty}^{\infty} \frac{f(x')}{(x'-x)}\,dx' + \frac{1}{2}f(x) \qquad (9.6a)$$

from which

$$f(x) = \frac{1}{\pi i} P\int_{-\infty}^{\infty} \frac{f(x')}{(x'-x)}\,dx' \qquad (9.6b)$$

Substituting

$$f(x) = \text{Re}\big[f(x)\big] + i\,\text{Im}\big[f(x)\big] \qquad (9.7)$$

into eq. 9.6b and equating real and imaginary parts, we obtain

$$\text{Re}\big[f(x)\big] = \frac{1}{\pi} P\int_{-\infty}^{\infty} \frac{\text{Im}\big[f(x')\big]}{(x'-x)}\,dx' \qquad (9.8)$$

and

$$\text{Im}\big[f(x)\big] = -\frac{1}{\pi} P\int_{-\infty}^{\infty} \frac{\text{Re}\big[f(x')\big]}{(x'-x)}\,dx' \qquad (9.9)$$

Thus, if one part of $f(x)$ is known ($\text{Re}[f(x)]$ or $\text{Im}[f(x)]$), eqs. 9.8 or 9.9 can be used to determine the unknown part of $f(x)$ ($\text{Im}[f(x)]$ or $\text{Re}[f(x)]$).

Example 9.1: Dispersion relations for a function defined over the entire real axis

Let $f(z)$ be a function that is analytic in the upper half-plane and has an imaginary part given by

$$\text{Im}\big[f(x)\big] = \frac{1}{\left(x^2 + \alpha^2\right)} \qquad -\infty \le x \le \infty \qquad (9.10)$$

Because $f(z)$ is analytic in the upper half of the z-plane, we find the real part of $f(x)$ from eq. 9.8 to be

$$\text{Re}[f(x)] = \frac{1}{\pi} P \int_{-\infty}^{\infty} \frac{1}{\left(x'^2 + \alpha^2\right)(x' - x)} dx'$$

$$= \frac{1}{\pi} \lim_{\varepsilon \to 0} \text{Re}\left[\int_{-\infty}^{\infty} \frac{1}{\left(x'^2 + \alpha^2\right)(x' - x \pm i\varepsilon)} dx' \right]$$

(9.11)

This integral can be evaluated straightforwardly using Cauchy's theorem. We note that if we take the denominator to contain $+i\varepsilon$, the integrand of the integral has poles at $z' = \pm i\alpha$ and at $z' = x - i\varepsilon$. Therefore, by constructing a contour comprised of the x'-axis and an infinite semicircle in the upper half-plane, only the pole at $z' = + i\alpha$ is inside the contour. Then, because the integrand is zero on the infinite semicircle,

$$\lim_{\varepsilon \to 0} \oint \frac{1}{\left(z'^2 + \alpha^2\right)(z' - x + i\varepsilon)} dz' = \lim_{\varepsilon \to 0} \int_{-\infty}^{\infty} \frac{1}{\left(x'^2 + \alpha^2\right)(x' - x + i\varepsilon)} dx'$$

$$= 2\pi i R(i\alpha) = -\frac{\pi}{\alpha(x - i\alpha)}$$

(9.12)

and eq. 9.11 becomes

$$\text{Re}[f(x)] = -\frac{1}{\alpha} \text{Re}\left[\frac{1}{(x - i\alpha)} \right] = -\frac{x}{\alpha\left(x^2 + \alpha^2\right)}$$

(9.13)

Therefore, on the real axis,

$$f(x) = \text{Re}[f(x)] + i \text{Im}[f(x)] = -\frac{1}{\alpha(x + i\alpha)}$$

(9.14a)

and at any point in the z-plane,

$$f(z) = -\frac{1}{\alpha(z + i\alpha)}$$

(9.14b)

We note that $f(z)$ has one pole at $z = -i\alpha$ which is in the lower half-plane. Therefore, as required, $f(z)$ is analytic in the upper half-plane. □

9.2 Kramers-Kronig Dispersion Relations Over Half The Real Axis

In some cases, $f(x)$ is defined only at positive (or only at negative) values of x (particularly for those problems in which x represents a physical quantity). Then the dispersion relations must be expressed so that the integrals are taken over the appropriate range of x'.

If $f(x')$ is defined only at positive values of x', we write eq. 9.6b as

$$f(x) = \frac{1}{i\pi} P\left[\int_{-\infty}^{0} \frac{f(x')}{(x'-x)} dx' + \int_{0}^{\infty} \frac{f(x')}{(x'-x)} dx' \right] \tag{9.15a}$$

then substitute $x' = -x''$ in the integral from $-\infty$ to 0. We then rename x'' as x' to obtain

$$f(x) = \frac{1}{i\pi} P\left[\int_{0}^{\infty} \frac{f(x')}{(x'-x)} dx' - \int_{0}^{\infty} \frac{f(-x')}{(x'+x)} dx' \right]$$

$$= \frac{1}{i\pi} P\left[\int_{0}^{\infty} \frac{x'(f(x') - f(-x'))}{(x'^2 - x^2)} dx' + x \int_{0}^{\infty} \frac{(f(x') + f(-x'))}{(x'^2 - x^2)} dx' \right]$$

$$\tag{9.15b}$$

To proceed further with this analysis, we must be able to relate $f(x')$ to $f(-x')$.

Reflection symmetry around the imaginary axis

Let $z = x + iy$ be a point in the z-plane. The reflection of the point $z = x + iy$ about the imaginary axis is given by $-z^* = -x + iy$. If

$$f(-z^*) = \pm f^*(z) \tag{9.16}$$

then $f(z)$ is said to exhibit even (+) or odd (−) *reflection symmetry* under such a reflection. A function that has such symmetry can be described by dispersion integrals over half the real axis. This discussion involves integration over the range $0 \le x' \le \infty$. In prob. 7, the reader develops an equivalent description for integration over $-\infty \le x' \le 0$.

For points on the real axis, reflection about the imaginary axis is reflection about the origin. That is, when $z' = x'$, $-z'^* = -x'$. Therefore, a function of x' that exhibits reflection symmetry about the origin satisfies

$$f(-x') = \pm f^*(x') \tag{9.17}$$

Let $f(x')$ be even under this reflection. Then eq. 9.15b can be written

$$
\begin{aligned}
f(x) &= \text{Re}[f(x)] + i\,\text{Im}[f(x)] \\
&= \frac{1}{i\pi} P\left[\int_0^\infty \frac{x'\left(f(x') - f^*(x')\right)}{\left(x'^2 - x^2\right)} dx' + x\int_0^\infty \frac{\left(f(x') + f^*(x')\right)}{\left(x'^2 - x^2\right)} dx' \right] \\
&= \frac{1}{i\pi} P\left[2i\int_0^\infty \frac{x'\,\text{Im}[f(x')]}{\left(x'^2 - x^2\right)} dx' + 2x\int_0^\infty \frac{\text{Re}[f(x')]}{\left(x'^2 - x^2\right)} dx' \right]
\end{aligned}
\tag{9.18}
$$

Equating the real parts we obtain

$$\text{Re}[f(x)] = \frac{2}{\pi} P \int_0^\infty \frac{x'\,\text{Im}[f(x')]}{\left(x'^2 - x^2\right)} dx' \tag{9.19a}$$

and from the equality of the imaginary parts we have

$$\text{Im}[f(x)] = -\frac{2x}{\pi} P \int_0^\infty \frac{\text{Re}[f(x')]}{\left(x'^2 - x^2\right)} dx' \tag{9.19b}$$

These are the dispersion relations for a function that is even under reflection about the origin.

For a function that is odd under reflection, which satisfies

$$f(-x') = -f^*(x') \tag{9.20}$$

it is straightforward to show that eq. 9.15b yields

$$\text{Re}[f(x)] = \frac{2x}{\pi} P \int_0^\infty \frac{\text{Im}[f(x')]}{\left(x'^2 - x^2\right)} dx' \tag{9.21a}$$

and

$$\text{Im}\big[f(x)\big] = -\frac{2}{\pi} P \int_0^\infty \frac{x' \text{Re}\big[f(x')\big]}{\big(x'^2 - x^2\big)} dx' \tag{9.21b}$$

The dispersion relations of eqs. 9.19 and 9.21 are the ones originally developed by Kramers and Kronig to determine the absorptive (imaginary) part of the complex index of refraction from the dispersive (real) part. In their analysis, x represents the (positive) frequency of the radiation.

Example 9.2: Dispersion relations for a function defined over half the real axis

Let us determine the function $f(x)$ that has the following properties:

- $f(x)$ is even under reflection about the origin.
- The real part of $f(x)$ is given in terms of the real positive constant α by

$$\text{Re}\big[f(x)\big] = \frac{1}{\big(x^2 + \alpha^2\big)} \qquad 0 \le x \le \infty \tag{9.22}$$

Because $f(x)$ is even under reflection and $\text{Re}[f(x)]$ is known, we determine $\text{Im}[f(x)]$ from eq. 9.19b. We obtain

$$\text{Im}\big[f(x)\big] = -\frac{2x}{\pi} P \int_0^\infty \frac{1}{\big(x'^2 + \alpha^2\big)\big(x'^2 - x^2\big)} dx'$$

$$= \frac{2}{\pi} \frac{x}{\big(x^2 + \alpha^2\big)} \left[\int_0^\infty \frac{1}{\big(x'^2 + \alpha^2\big)} dx' - P \int_0^\infty \frac{1}{\big(x'^2 - x^2\big)} dx' \right]$$

$$= \frac{2}{\pi} \frac{x}{\big(x^2 + \alpha^2\big)} \left[\frac{\pi}{2\alpha} - P \int_0^\infty \frac{1}{\big(x'^2 - x^2\big)} dx' \right]$$

$$\tag{9.23}$$

Writing

$$\frac{1}{\big(x'^2 - x^2\big)_P} = \lim_{\varepsilon \to 0} \text{Re}\left(\frac{1}{\big(x'^2 - x^2 - i\varepsilon\big)} \right) \tag{9.24}$$

we obtain

$$P\int_0^\infty \frac{1}{\left(x'^2 - x^2\right)} dx' = \lim_{\varepsilon \to 0} \text{Re} \int_0^\infty \frac{1}{\left(x'^2 - x^2 - i\varepsilon\right)} dx'$$

$$= \frac{1}{2x} \lim_{\varepsilon \to 0} \text{Re}\left[ln\left(\frac{x'-x-i\varepsilon}{x'+x+i\varepsilon}\right)\right]_0^\infty \qquad (9.25)$$

Taking the logarithm cuts to extend from the branch points to $+\infty$

$$\lim_{\varepsilon \to 0} ln(x' - x - i\varepsilon) = ln|x' - x| + 2\pi i \qquad (9.26a)$$

and

$$\lim_{\varepsilon \to 0} ln(x' + x + i\varepsilon) = ln|x' + x| \qquad (9.26b)$$

Then eq. 9.25 becomes

$$P\int_0^\infty \frac{1}{\left(x'^2 - x^2\right)} dx' = \frac{1}{2x} \text{Re}\left[ln\left(\left|\frac{x'-x}{x'+x}\right|\right) + 2\pi i\right]_0^\infty = 0 \qquad (9.27)$$

Therefore, eq. 9.23 becomes

$$\text{Im}\left[f(x)\right] = \frac{x}{\alpha\left(x^2 + \alpha^2\right)} \qquad (9.28)$$

from which

$$f(x) = \left(1 + i\frac{x}{\alpha}\right)\frac{1}{\left(x^2 + \alpha^2\right)} = \frac{i}{\alpha(x + i\alpha)} \qquad (9.29)$$
$$\square$$

9.3 Dispersion Relations for a Function With Branch Structure

If $f(z)$ contains branch points, the development of dispersion relations presented above must be modified to account for the branch structure of the function. Let $f(z)$ have no poles and have a single branch point on the real axis at $z = x_0$. We take the associated cut to extend from x_0 to $+\infty$ along the x-axis. Then, the Cauchy's integral representation of $f(z)$

$$f(z) = \frac{1}{2\pi i} \oint \frac{f(z')}{(z'-z)} dz' \tag{3.101b}$$

is valid for the contour shown in fig. 9.2.

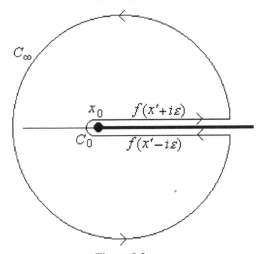

Figure 9.2
Contour and branch structure for the integral representation
of a function with a branch point

Writing the integral as a sum of integrals over the various segments of the contour and taking the contributions from C_∞ and C_0 to be zero, we have

$$f(z) = \frac{1}{2\pi i}\left[\int_{x_0}^\infty \frac{f(x'+i\varepsilon)}{(x'-z)} dx' + \int_\infty^{x_0} \frac{f(x'-i\varepsilon)}{(x'-z)} dx'\right]$$
$$= \frac{1}{2\pi i}\int_{x_0}^\infty \frac{[f(x'+i\varepsilon)-f(x'-i\varepsilon)]}{(x'-z)} dx' = \frac{1}{2\pi i}\int_{x_0}^\infty \frac{\Delta(x')}{(x'-z)} dx' \tag{9.30}$$

where

$$\Delta(x') \equiv f(x'+i\varepsilon) - f(x'-i\varepsilon) \tag{9.31}$$

is the discontinuity across the cut.

Functions that are odd/even under reflection about the real axis

In order to relate the discontinuity across the cut to either the real or imaginary part of $f(x')$, we define a reflection symmetry across the real axis. The reflection of a point $z = x + iy$ around the real axis is $z^* = x - iy$. $f(z)$ is said to be even (+) or odd (−) under reflection around the real axis if

$$f(z^*) = \pm f^*(z) \tag{9.32a}$$

For points on the real axis with $x > x_0$, this reflection symmetry is given by

$$f\left((x + i\varepsilon)^*\right) = f(x - i\varepsilon) = \pm f^*(x + i\varepsilon) \tag{9.32b}$$

If $f(z)$ is even under this reflection,

$$f(x' - i\varepsilon) = +f^*(x' + i\varepsilon) \tag{9.33}$$

and the discontinuity across the cut becomes

$$\Delta(x') = f(x' + i\varepsilon) - f^*(x' + i\varepsilon) = 2i\,\mathrm{Im}\left(f(x' + i\varepsilon)\right) \tag{9.34}$$

Then eq. 9.30 becomes the integral representation

$$f(z) = \frac{1}{\pi} \int_{x_0}^{\infty} \frac{\mathrm{Im}\left[f(x' + i\varepsilon)\right]}{(x' - z)} dx' \tag{9.35}$$

Equation 9.35 is also referred to as a dispersion relation.

Because the integrand involves $\mathrm{Im}[f(x + i\varepsilon)]$, $\mathrm{Im}[f(z)]$ must be known along the top of the cut in order to determine the complete function. For points on the real axis, with $z = x < x_0$, $1/(x' - z) = 1/(x' - x)$ is analytic everywhere. Then, the integral of eq. 9.35 can be evaluated in principle (numerically when necessary). When $x > x_0$, we set $z = x + i\varepsilon$, and eq. 9.35 becomes

$$f(x + i\varepsilon) = \frac{1}{\pi} \int_{x_0}^{\infty} \frac{\mathrm{Im}\left[f(x' + i\varepsilon)\right]}{(x' - x - i\varepsilon)} dx'$$

$$= \frac{1}{\pi}\left[P\int_{x_0}^{\infty} \frac{\mathrm{Im}\left[f(x' + i\varepsilon)\right]}{(x' - x)} dx' + i\pi\,\mathrm{Im}\left[f(x + i\varepsilon)\right]\right] \tag{9.36}$$

Writing $f(x + i\varepsilon)$ as $\text{Re}[f(x + i\varepsilon)] + i\text{Im}[f(x + i\varepsilon)]$, eq. 9.36 becomes

$$\text{Re}\left[f(x+i\varepsilon)\right] = \frac{1}{\pi} P \int_{x_0}^{\infty} \frac{\text{Im}\left[f(x'+i\varepsilon)\right]}{(x'-x)} dx' \qquad (9.37)$$

Because $f(z)$ is even under reflection, we see from eq. 9.33 that

$$\text{Re}\left[f(x-i\varepsilon)\right] = \text{Re}\left[f*(x+i\varepsilon)\right] = \text{Re}\left[f(x+i\varepsilon)\right] \qquad (9.38a)$$

and

$$\text{Im}\left[f(x-i\varepsilon)\right] = \text{Im}\left[f*(x+i\varepsilon)\right] = -\text{Im}\left[f(x+i\varepsilon)\right] \qquad (9.38b)$$

Therefore, we see from eq. 9.37 that, at points along the bottom of the cut

$$\text{Re}\left[f(x-i\varepsilon)\right] = \frac{1}{\pi} P \int_{x_0}^{\infty} \frac{\text{Im}\left[f(x'+i\varepsilon)\right]}{(x'-x)} dx'$$
$$= -\frac{1}{\pi} P \int_{x_0}^{\infty} \frac{\text{Im}\left[f(x'-i\varepsilon)\right]}{(x'-x)} dx' \qquad (9.39)$$

If $f(z)$ is odd under this reflection, then

$$f(x'-i\varepsilon) = -f*(x'+i\varepsilon) \qquad (9.40)$$

Then, as the reader will show in prob. 10

$$f(z) = \frac{1}{i\pi} \int_{x_0}^{\infty} \frac{\text{Re}\left[f(x'+i\varepsilon)\right]}{(x'-z)} dx' \qquad (9.41)$$

from which

$$\text{Im}\left[f(x+i\varepsilon)\right] = -\frac{1}{\pi} P \int_{x_0}^{\infty} \frac{\text{Re}\left[f(x'+i\varepsilon)\right]}{(x'-x)} dx' \qquad (9.42a)$$

and

$$\begin{aligned}
\mathrm{Im}\left[f(x-i\varepsilon)\right] &= -\frac{1}{\pi}P\int_{x_0}^{\infty}\frac{\mathrm{Re}\left[f(x'+i\varepsilon)\right]}{(x'-x)}dx' \\
&= \frac{1}{\pi}P\int_{x_0}^{\infty}\frac{\mathrm{Re}\left[f(x'-i\varepsilon)\right]}{(x'-x)}dx'
\end{aligned}$$

(9.42b)

As stated above, the dispersion relations of eq. 9.35 or eq. 9.41 apply to a function $f(z)$ that has only one singularity, a branch point on the real axis at x_0. The cut associated with x_0 must extend to $+\infty$ along the real axis.

Let x be a point on the real axis with $x < x_0$. We define the region R to be the set of points inside and on the circle of convergence of the Taylor series representation of $f(z)$ expanded about some $x < x_0$.

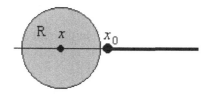

Figure 9.3
Circle of convergence of the Taylor series about $x < x_0$

This Taylor series is

$$f(z) = \sum_{n=0}^{\infty}\frac{f^{(n)}(x)}{n!}(z-x)^n$$

(9.43)

If $f(z)$ is real at $z = x$ then each derivative $f^{(n)}(x)$ is also real for all x in R. Then, referring to eq. 9.43,

$$f^*(z) = \sum_{n=0}^{\infty}\frac{f^{(n)}(x)}{n!}(z^*-x)^n = f(z^*)$$

(9.44)

That is, when $f(x)$ is real for any $x < x_0$, $f(z)$ is even under reflection about the real axis. Such a function is said to satisfy the *Schwarz reflection condition*.

If $f(z)$ is imaginary at $z = x$, then each of its derivatives at x is imaginary, and it is straightforward to see that eq. 9.43 yields

$$f^*(z) = \sum_{n=0}^{\infty} \frac{\left(f^{(n)}(x)\right)^*}{n!} (z^*-x)^n = -\sum_{n=0}^{\infty} \frac{f^{(n)}(x)}{n!} (z^*-x)^n = -f(z^*)$$

$$(9.45)$$

That is, such a function is odd under reflection about the real axis.

Example 9.3: Dispersion relations for a function with one branch point

Let us determine the function $f(z)$ with the following properties:

- $f(z)$ has a branch point at $z = 0$ with an associated cut extending to $+\infty$ along the real axis.
- $f(z) \to 0$ for all $|z| \to \infty$.
- $f(z)$ is even under reflection about the real axis.
- The imaginary part of $f(z)$ on the real axis above the cut is given by

$$\text{Im}\left[f(x+i\varepsilon)\right] = \frac{1}{\left(x^2 + \alpha^2\right)} \qquad x \geq 0 \qquad (9.46)$$

with α real and positive.

Because the imaginary part of the function is specified, eq. 9.35 becomes

$$f(z) = \frac{1}{\pi} \int_0^{\infty} \frac{1}{\left(x'^2 + \alpha^2\right)(x' - z)} \, dx' \qquad (9.47)$$

This integral can be evaluated straightforwardly using, for example, the method of partial fractions. The result is

$$f(z) = -\frac{1}{\left(z^2 + \alpha^2\right)} \left[\frac{z - 2i\alpha}{2\alpha} + \frac{1}{\pi} \ln\left(\frac{z}{\alpha}\right) \right] \qquad (9.48)$$

By taking the cut associated with the branch point of $\ln(z)$ at $z = 0$ to extend to $+\infty$ along the x-axis, $\ln(i) = i\pi/2$ and $\ln(-i) = i3\pi/2$. Then, it is straightforward to show that

$$\lim_{z \to \pm i\alpha} \left[\frac{z - 2i\alpha}{2\alpha} + \frac{1}{\pi} \ell n \left(\frac{z}{\alpha} \right) \right] = 0 \tag{9.49}$$

and therefore, $f(z)$ is analytic at $z = \pm i\alpha$. Using l'Hopital's rule, we obtain

$$f(\pm i\alpha) = \frac{1}{\alpha^2} \left[\frac{1}{\pi} \pm \frac{i}{2} \right] \tag{9.50}$$

$\square$

9.4 Subtracted Dispersion Relations

The dispersion relations developed in the first three sections do not impose a specified value for a function at a given value of x which we designate x_1. The value of a function described by these dispersion representations at x_1 is obtained from the dispersion integral.

Let $f(x)$ be a function that is described by a dispersion relation developed in one of the first three sections and let $g(x)$ be a function that is described by a dispersion relation and has a specified value $g(x_1)$ at x_1. To develop a dispersion relation for a function $g(x)$ with such a constraint we define

$$f(x) \equiv \frac{(g(x) - g(x_1))}{(x - x_1)} \tag{9.51}$$

We note that because

$$f(x_1) = \lim_{x \to x_1} \frac{(g(x) - g(x_1))}{(x - x_1)} = \frac{dg}{dx}\bigg|_{x_1} \tag{9.52}$$

then $f(x)$ is analytic at x_1.

Subtracted Kramers–Kronig dispersion relations
over the entire real axis

Substituting eq. 9.51 into eqs. 9.8, we obtain

$$\frac{\mathrm{Re}\left[g(x) - g(x_1) \right]}{(x - x_1)} = \frac{1}{\pi} P \int_{-\infty}^{\infty} \frac{\mathrm{Im}\left[g(x') - g(x_1) \right]}{(x' - x_1)(x' - x)} dx' \tag{9.53a}$$

and

$$\frac{Im[g(x)-g(x_1)]}{(x-x_1)}=-\frac{1}{\pi}P\int_{-\infty}^{\infty}\frac{Re[g(x')-g(x_1)]}{(x'-x_1)(x'-x)}dx' \qquad (9.53b)$$

These can be expressed as

$$Re[g(x)]=Re[g(x_1)]+\frac{(x-x_1)}{\pi}P\int_{-\infty}^{\infty}\frac{Im[g(x')-g(x_1)]}{(x'-x_1)(x'-x)}dx'$$

$$(9.54a)$$

and

$$Im[g(x)]=Im[g(x_1)]-\frac{(x-x_1)}{\pi}P\int_{-\infty}^{\infty}\frac{Re[g(x')-g(x_1)]}{(x'-x_1)(x'-x)}dx'$$

$$(9.54b)$$

These are called *once-subtracted* or *singly subtracted* dispersion relations over the entire real axis.

Example 9.4: Subtracted dispersion relations for a function defined over the entire real axis

Let us find the function $g(x)$ that has the following properties.

- $g(z) \to 0$ when $|z| \to \infty$.
- $Im[g(x)] = 1/(x^2 + \alpha^2) -\infty \leq x \leq \infty$ with α real and positive.
- $Re[g(1)] = 2$.

Because $Im[g(x)]$ is given, $g(x)$ is found by determining $Re[g(x)]$ from eq. 9.53a. With the above properties, this becomes

$$Re[g(x)]=Re[g(1)]+\frac{(x-1)}{\pi}P\int_{-\infty}^{\infty}\frac{\left[\dfrac{1}{(x'^2+\alpha^2)}-\dfrac{1}{(1+\alpha^2)}\right]}{(x'-1)(x'-x)}dx'$$

$$(9.55)$$

The integral of eq. 9.55 can be evaluated by Cauchy's residue theorem by writing

$$\frac{1}{(x'-x)_P}=Re\left[\frac{1}{x'-x+i\varepsilon}\right] \qquad (5.72a)$$

which excludes the pole at $x' = x$ from a contour closed in the upper half of the z'-plane. Then

$$P\int_{-\infty}^{\infty} \frac{\left[\dfrac{1}{(x'^2+\alpha^2)}-\dfrac{1}{(1+\alpha^2)}\right]}{(x'-1)(x'-x)}dx' = \mathrm{Re}\left[\int_{-\infty}^{\infty} \frac{\left[\dfrac{1}{(x'^2+\alpha^2)}-\dfrac{1}{(1+\alpha^2)}\right]}{(x'-1)(x'-x+i\varepsilon)}dx'\right]$$

$$= \mathrm{Re}\left[2\pi i a_{-1}(i\alpha)\right] = \frac{\pi}{\alpha}\mathrm{Re}\left[\frac{1}{(i\alpha-1)(i\alpha-x)}\right] = \frac{-\pi(x+1)}{(\alpha^2+1)(x^2+\alpha^2)}$$

$$(9.56)$$

Therefore,

$$\mathrm{Re}[g(x)] = 2 - \frac{x^2-1}{(1+\alpha^2)(x^2+\alpha^2)} \tag{9.57a}$$

Then, with $\mathrm{Im}[g(x)]$ given in the statement of the problem,

$$g(x) = 2 - \frac{x^2-1}{(1+\alpha^2)(x^2+\alpha^2)} + i\frac{1}{(x^2+\alpha^2)} \tag{9.57b}$$

$$\square$$

Subtracted Kramers–Kronig dispersion relations over half the real axis

If $g(x)$ is defined only over the interval $0 \le x \le \infty$, we write eqs. 9.53 as

$$\mathrm{Re}[g(x)] = \mathrm{Re}[g(x_1)]$$
$$+\frac{(x-x_1)}{\pi}P\left[\int_{-\infty}^{0}\frac{\mathrm{Im}[g(x')-g(x_1)]}{(x'-x_1)(x'-x)}dx' + \int_{0}^{\infty}\frac{\mathrm{Im}[g(x')-g(x_1)]}{(x'-x_1)(x'-x)}dx'\right]$$
$$= \mathrm{Re}[g(x_1)]$$
$$+\frac{(x-x_1)}{\pi}P\left[\int_{0}^{\infty}\frac{\mathrm{Im}[g(-x')-g(x_1)]}{(x'+x_1)(x'+x)}dx' + \int_{0}^{\infty}\frac{\mathrm{Im}[g(x')-g(x_1)]}{(x'-x_1)(x'-x)}dx'\right]$$

$$(9.58a)$$

and

$$\mathrm{Im}\left[g(x)\right] = \mathrm{Im}\left[g(x_1)\right]$$

$$-\frac{(x-x_1)}{\pi}P\left[\int_{-\infty}^{0}\frac{\mathrm{Re}\left[g(x')-g(x_1)\right]}{(x'-x_1)(x'-x)}dx' + \int_{0}^{\infty}\frac{\mathrm{Re}\left[g(x')-g(x_1)\right]}{(x'-x_1)(x'-x)}dx'\right]$$

$$= \mathrm{Im}\left[g(x_1)\right]$$

$$-\frac{(x-x_1)}{\pi}P\left[\int_{0}^{\infty}\frac{\mathrm{Re}\left[g(-x')-g(x_1)\right]}{(x'+x_1)(x'+x)}dx' + \int_{0}^{\infty}\frac{\mathrm{Re}\left[g(x')-g(x_1)\right]}{(x'-x_1)(x'-x)}dx'\right]$$

$$(9.58b)$$

In section 9.2, we noted that if $g(x)$ is even under a reflection about the imaginary axis, then

$$g(-z^*) = g^*(z) \tag{9.59a}$$

so that

$$g(-x') = g^*(x') \tag{9.59b}$$

This requires that

$$\mathrm{Re}\left[g(-x')\right] = \mathrm{Re}\left[g(x')\right] \tag{9.60a}$$

and

$$\mathrm{Im}\left[g(-x')\right] = -\mathrm{Im}\left[g(x')\right] \tag{9.60b}$$

Then, eqs. 9.58 become

$$\mathrm{Re}\left[g(x)\right] = \mathrm{Re}\left[g(x_1)\right]$$

$$+\frac{2(x-x_1)}{\pi}P\int_{0}^{\infty}\frac{\left(x'(x+x_1)\mathrm{Im}\left[g(x')\right]-(x'^2+xx_1)\mathrm{Im}\left[g(x_1)\right]\right)}{(x'^2-x^2)(x'^2-x_1^2)}dx'$$

$$(9.61a)$$

and

$$Im[g(x)] = Im[g(x_1)]$$
$$-\frac{2(x-x_1)}{\pi}\int_0^\infty \frac{(x'^2 + xx_1)(Re[g(x')] - Re[g(x_1)])}{(x'^2 - x^2)(x'^2 - x_1^2)}dx' \qquad (9.61b)$$

Example 9.5: Subtracted dispersion relations for a function defined over half the real axis

Let us find the function $g(x)$ that has the following properties.

- $g(z) \to 0$ when $|z| \to \infty$.
- $Re[g(x)] = 1/(x^2 + 1)$ $0 \le x \le \infty$.
- $Im[g(0)] = 2$.

Because $Re[g(x)]$ is given, $g(x)$ is found by determining $Im[g(x)]$ from eq. 9.61b. With the above properties, this becomes

$$Im[g(x)] = 2 + \frac{2x}{\pi}P\int_0^\infty \frac{x'^2}{(x'^2 - x^2)(x'^2 + 1)}dx'$$
$$= 2 + \frac{2x}{\pi}\frac{1}{(x^2 + 1)}\int_0^\infty\left[\frac{1}{(x'^2 - x^2)_P} - \frac{1}{(x'^2 + 1)}\right]dx' \qquad (9.62)$$

The principal value integral can be evaluated by writing

$$\int_0^\infty \frac{1}{(x'^2 - x^2)_P}dx' = \frac{1}{2x}\int_0^\infty\left[\frac{1}{(x' - x)_P} - \frac{1}{(x' + x)_P}\right]dx'$$
$$= \frac{1}{2x}Re\left[\int_0^\infty\left[\frac{1}{(x' - x \pm i\varepsilon)} - \frac{1}{(x' + x \pm i\varepsilon)}\right]dx'\right]$$
$$= \frac{1}{2x}\ell n\left|\frac{x' - x}{x' + x}\right|_{x'=0}^{x'=\infty} \qquad (9.63)$$

Then, eq. 9.62 becomes

$$Im[g(x)] = 2 + \frac{2x}{\pi}\frac{1}{(x^2 + 1)}\left[\frac{1}{2x}\ell n\left|\frac{x' - x}{x' + x}\right|_{x'=0}^{x'=\infty} - \int_0^\infty \frac{1}{(x'^2 + 1)}dx'\right] \qquad (9.64)$$

As shown in section 9.5, the integrated term

$$\frac{1}{2x}\ln\left|\frac{x'-x}{x'+x}\right|_{x'=0}^{x'=\infty} = \delta(2x) \neq 0 \tag{9.65a}$$

However,

$$2x\left[\frac{1}{2x}\ln\left|\frac{x'-x}{x'+x}\right|_{x'=0}^{x'=\infty}\right] = \ln\left|\frac{x'-x}{x'+x}\right|_{x'=0}^{x'=\infty} = 0 \tag{9.65b}$$

Therefore, eq. 9.64 becomes

$$\text{Im}\left[g(x)\right] = 2 - \frac{2x}{\pi}\frac{1}{(x^2+1)}\int_0^\infty \frac{1}{(x'^2+1)}dx' = 2 - \frac{x}{(x^2+1)} \tag{9.66}$$

from which

$$g(x) = 2i + \frac{1}{(x^2+1)}(1-ix) = 2i - \frac{i}{(x-i)} \tag{9.67}$$

Derivations of once-subtracted dispersion relations for functions that are defined over the interval $0 \leq x \leq \infty$ and are odd under reflection around the imaginary axis are identical to those above. So too are derivations of once-subtracted dispersion relations for functions defined over the interval $-\infty \leq x \leq 0$. These derivations are not presented here, but are left as an exercise for the reader.

Subtracted dispersion relations for a function with a branch point

If $f(z)$ has a branch point on the x-axis at x_0, then defining $g(z)$ by

$$f(z) \equiv \frac{\left(g(z) - g(x_1)\right)}{(z - x_1)} \tag{9.51}$$

requires $g(z)$ to have a branch point at x_0. Substituting this into the Cauchy integral representation

$$f(z) = \frac{1}{2\pi i}\oint_C \frac{f(z')}{(z'-z)}dz' \tag{3.101b}$$

we have

$$\frac{\left(g(z)-g(x_1)\right)}{(z-x_1)}=\frac{1}{2\pi i}\oint_C \frac{\left(g(z')-g(x_1)\right)}{(z'-x_1)(z'-z)}dz' \tag{9.68}$$

where the contour C is that shown in fig. 9.2. We again take the integrand to be zero on C_∞ and C_0. Then eq. 9.68 becomes

$$\frac{\left(g(z)-g(x_1)\right)}{(z-x_1)}$$
$$=\frac{1}{2\pi i}\left[\int_{x_0}^\infty \frac{g(x'+i\varepsilon)-g(x_1)}{(x'-x_1)(x'-z)}dx' - \int_{x_0}^\infty \frac{g(x'-i\varepsilon)-g(x_1)}{(x'-x_1)(x'-z)}dx'\right]$$
$$=\frac{1}{2\pi i}\int_{x_0}^\infty \frac{g(x'+i\varepsilon)-g(x'-i\varepsilon)}{(x'-x_1)(x'-z)}dx' \equiv \frac{1}{2\pi i}\int_{x_0}^\infty \frac{\Delta(x')}{(x'-x_1)(x'-z)}dx' \tag{9.69a}$$

After a little algebra, this becomes

$$g(z)=g(x_1)+\frac{(z-x_1)}{2\pi i}\int_{x_0}^\infty \frac{\Delta(x')}{(x'-x_1)(x'-z)}dx' \tag{9.69b}$$

Because $g(x_1)$ is uniquely defined, x_1 must be in the region where $g(x)$ is single-valued. Because the cut extends from x_0 to $+\infty$, it is necessary that $x_1 < x_0$ and there is no need to avoid the pole of the integrand at x_1. Therefore, the integral of eq. 9.69b is not a principal value integral.

If $g(z)$ is even under reflection about the real axis, then for $x > x_0$,

$$g(x'-i\varepsilon)=g^*(x+i\varepsilon) \tag{9.70}$$

Then

$$\mathrm{Re}\left[g(x-i\varepsilon)\right]=\mathrm{Re}\left[g(x+i\varepsilon)\right] \tag{9.71a}$$

and

$$\mathrm{Im}\left[g(x-i\varepsilon)\right]=-\mathrm{Im}\left[g(x+i\varepsilon)\right] \tag{9.71b}$$

so the discontinuity in $g(z')$ across the cut is given by

$$\Delta(x') = g(x'+i\varepsilon) - g*(x'+i\varepsilon) = 2i\,\text{Im}\big[g(x'+i\varepsilon)\big] \qquad (9.72)$$

Then eq. 9.69b becomes

$$g(z) = g(x_1) + \frac{(z-x_1)}{\pi} \int_{x_0}^{\infty} \frac{\text{Im}\big[g(x'+i\varepsilon)\big]}{(x'-x_1)(x'-z)} dx' \qquad (9.73)$$

To determine $g(z)$ along the real axis at points above the cut, we set $z = x + i\varepsilon$, express $g(x + i\varepsilon)$ in terms of its real and imaginary parts, and use the identity

$$\frac{1}{(x'-x-i\varepsilon)} = \frac{1}{(x'-x)_P} + i\pi\delta(x'-x) \qquad (5.71)$$

to obtain

$$\text{Re}\big[g(x+i\varepsilon)\big] = g(x_1) + \frac{(x-x_1)}{\pi} \int_{x_0}^{\infty} \frac{\text{Im}\big[g(x'+i\varepsilon)\big]}{(x'-x_1)(x'-x)_P} dx' \qquad (9.74a)$$

All terms in eq. 9.74a are real, therefore we see that $g(x_1)$ must be real. Thus, if $g(z)$ is even under reflection about the real axis, it must satisfy the Schwarz reflection principal.

To find $g(z)$ on the real axis below the cut, we set $z = x - i\varepsilon$. Because $g(z)$ is even under reflection about the real axis, we refer to eqs. 9.73 to obtain

$$\text{Re}\big[g(x-i\varepsilon)\big] = g(x_1) + \frac{(x-x_1)}{\pi} \int_{x_0}^{\infty} \frac{\text{Im}\big[g(x'+i\varepsilon)\big]}{(x'-x_1)(x'-x)_P} dx'$$

$$= g(x_1) - \frac{(x-x_1)}{\pi} \int_{x_0}^{\infty} \frac{\text{Im}\big[g(x'-i\varepsilon)\big]}{(x'-x_1)(x'-x)_P} dx' \qquad (9.74b)$$

As the reader will show in prob. 18, if a function is antisymmetric across the real axis and has a branch point at x_0 with associated cut extending to $+\infty$ it can be found from the dispersion relation

$$g(z) = g(x_1) + \frac{(z-x_1)}{i\pi} \int_{x_0}^{\infty} \frac{\text{Re}\big[g(x'+i\varepsilon)\big]}{(x'-x_1)(x'-z)} dx' \qquad (9.75)$$

from which

$$\mathrm{Im}\big[g(x+i\varepsilon)\big]=-ig(x_1)-\frac{(x-x_1)}{\pi}\int_{x_0}^{\infty}\frac{\mathrm{Re}\big[g(x'+i\varepsilon)\big]}{(x'-x_1)(x'-x)_P}dx' \quad (9.76a)$$

and

$$\mathrm{Im}\big[g(x-i\varepsilon)\big]=-ig(x_1)+\frac{(x-x_1)}{\pi}\int_{x_0}^{\infty}\frac{\mathrm{Re}\big[g(x'-i\varepsilon)\big]}{(x'-x_1)(x'-x)_P}dx' \quad (9.76b)$$

Because $\mathrm{Im}[g(x \pm i\varepsilon)]$ is real, it is evident from eqs. 9.76 that when $g(z)$ is odd under reflection about the real axis, $g(x_1)$ must be imaginary.

Example 9.6: Subtracted dispersion relations for a function with one branch point

Let us determine the function $g(z)$ with the following properties.

- $g(z)$ has a branch point on the real axis at the origin.
- $g(z)$ is analytic at all points on the negative real axis.
- $g(-1) = 1$.
- $g(z)$ satisfies the Schwarz reflection condition (it is even under reflection about the real axis, and, as a consequence, it is real at any $x < 0$).
- $\mathrm{Im}[g(x + i\varepsilon)] = 1/(x + 2)$ for $x \geq 0$, and because $g(z)$ is even under reflection, we see from eq. 9.71b that $\mathrm{Im}[g(x - i\varepsilon)] = -1/(x + 2)$.

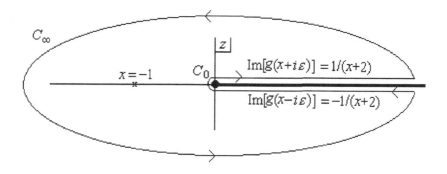

Figure 9.4
Contour and cut structure of $g(z)$

These properties of $g(z)$ and the contour used to determine it are shown in fig. 9.4. The integrals around C_∞ and C_0 are zero, so that $g(z)$ is given by eq. 9.73 to be

$$g(z) = -1 + \frac{(z+1)}{\pi} \int_0^\infty \frac{\text{Im}\left[g(x'+i\varepsilon)\right]}{(x'+1)(x'-z)} dx'$$

$$= -1 + \frac{(z+1)}{\pi} \int_0^\infty \frac{1}{(x'+2)(x'+1)(x'-z)} dx' \tag{9.77}$$

This is evaluated straightforwardly (using partial fractions, for example). We obtain

$$g(z) = -1 - \frac{(z+1)}{\pi(z+2)} \left[\frac{1}{(z+1)} \ell n(-z) + \ell n(2) \right] \tag{9.78}$$

$\text{Re}[g(z)]$ at points on the positive real axis above and below the cut are obtained from eqs. 9.71a and 9.74 to be

$$\text{Re}\left[g(x+i\varepsilon)\right] = \text{Re}\left[g(x-i\varepsilon)\right]$$

$$= -1 + \frac{(x+1)}{\pi} \int_0^\infty \frac{1}{(x'+2)(x'+1)(x'-x)_P} dx' \tag{9.79}$$

$$= -1 - \frac{(x+1)}{\pi} \left[\frac{\ell n(x)}{(x+1)(x+2)} + \frac{\ell n(2)}{(x+2)} \right]$$

9.5 Dispersion Relations and a Representation of the Dirac δ Symbol

The Kramers–Kronig dispersion relations of eqs. 9.8 and 9.9 can be used to develop an integral representation of the Dirac δ symbol. Substituting the expression for $\text{Im}[f(x)]$ given in

$$\text{Im}\left[f(x')\right] = -\frac{1}{\pi} P \int_{-\infty}^\infty \frac{\text{Re}\left[f(x'')\right]}{(x''-x')} dx'' \tag{9.9}$$

into the integrand of

$$Re[f(x)] = \frac{1}{\pi} P \int_{-\infty}^{\infty} \frac{Im[f(x')]}{(x'-x)} dx' \tag{9.8}$$

we obtain

$$Re[f(x)] = -\frac{1}{\pi^2} \int_{-\infty}^{\infty} \frac{1}{(x'-x)_P} \int_{-\infty}^{\infty} \frac{Re[f(x'')]}{(x''-x')_P} dx'' \, dx' \tag{9.80a}$$

Changing the order of integration, and writing $(x''-x')$ as $-(x'-x'')$, this becomes

$$Re[f(x)] = \frac{1}{\pi^2} \int_{-\infty}^{\infty} \left\{ Re[f(x'')] \int_{-\infty}^{\infty} \frac{1}{(x'-x)_P(x'-x'')_P} dx' \right\} dx'' \tag{9.80b}$$

The reader will recall that the definition of the Dirac δ symbol is given by

$$\int_a^b F(y)\delta(x-y)dy = \begin{cases} F(x) & a < x < b \\ 0 & x < a, x > b \end{cases} \tag{5.69}$$

Because eq. 9.80b has been obtained for any $f(x)$, $Re[f(x)]$ is an arbitrary function. Then, comparing eq. 9.80b to the definition of the δ symbol, we see that

$$\delta(x-x'') = \frac{1}{\pi^2} \int_{-\infty}^{\infty} \frac{1}{(x'-x)_P(x'-x'')_P} dx' \tag{9.81}$$

To further convince the reader of the validity of this representation, recall that it was shown in eq. 5.85 and fig. 5.9 that when the δ symbol is viewed as a function, it has the values

$$\delta(x-x'') = \begin{cases} 0 & x \neq x'' \\ \infty & x = x' \end{cases} \tag{5.85}$$

We evaluate the integral of eq. 9.81 by writing

$$\int_{-\infty}^{\infty} \frac{1}{(x'-x)_P (x'-x'')_P} dx' = \frac{1}{(x-x'')} \int_{-\infty}^{\infty} \left[\frac{1}{(x'-x)_P} - \frac{1}{(x'-x'')_P} \right] dx'$$

$$= \frac{1}{(x-x'')} \text{Re} \left\{ \int_{-\infty}^{\infty} \left[\frac{1}{(x'-x-i\varepsilon)_P} - \frac{1}{(x'-x''-i\varepsilon)_P} \right] dx' \right\}$$

$$= \frac{1}{(x-x'')} \ln \left| \frac{x'-x}{x'-x''} \right|_{x'=-\infty}^{x'=\infty}$$

$$\tag{9.82}$$

For all finite x and x'',

$$\ln \left| \frac{x'-x}{x'-x''} \right|_{x'=-\infty}^{x'=\infty} = 0 \tag{9.83}$$

Therefore, if $x \neq x''$

$$\frac{1}{(x-x'')} \ln \left| \frac{x'-x}{x'-x''} \right|_{x'=-\infty}^{x'=\infty} = 0 \tag{9.84a}$$

If $x = x''$, we use l'Hopital's rule, for example, to determine that

$$\lim_{x \to x''} \frac{1}{(x-x'')} \left[\ln \left| \frac{x'-x}{x'-x''} \right|_{x'=-\infty}^{x'=\infty} \right] = \infty \tag{9.84b}$$

Comparing eqs. 9.84 to eq. 5.85 on the previous page, we see that the integral representation given in eq. 9.81 has the same functional values that the δ symbol has when viewed as the function shown in fig. 5.9.

Problems

Unless explicitly stated otherwise, the function referred to in each problem is single-valued.
1. Derive dispersion relations equivalent to eqs. 9.8 and 9.9 for a function that is analytic in the lower half-plane by closing the contour as in fig. 9.1b, taking Im(z) < 0.

2. It was shown in section 5.2 of chapter 5 that certain types of Fourier integrals are evaluated by Cauchy's theorem by closing in the upper half- or lower half-plane. For k real and positive, find the function that satisfies the Kramers–Kronig dispersion relations of eqs. 9.8 and 9.9 when

(a) $\mathrm{Re}\left[f(x)\right]=\sin(kx)$ (b) $\mathrm{Im}\left[f(x)\right]=\cos(kx)$

(Hint: Use the identity)

$$\frac{1}{(x'-x)_P}=\frac{1}{2}\left[\frac{1}{(x'-x-i\varepsilon)}+\frac{1}{(x'-x+i\varepsilon)}\right] \qquad (5.75a)$$

3. Determine the function $f(z)$ that has the following properties.

- $f(z) \to 0$ when $|z| \to \infty$.
- $\mathrm{Re}[f(x)] = x/(x^2 + \alpha^2)$ $-\infty \le x \le \infty$ α real and positive.

4. Determine the function $f(z)$ that has the following properties:

- $f(z) \to 0$ when $|z| \to \infty$.
- $f^*(-z^*) = -f(z)$.
- $\mathrm{Re}[f(x)] = x/(x^2 + \alpha^2)$ $0 \le x \le \infty$ α real and positive.

5. Determine the function $f(z)$ that has the following properties:

- $f(z) \to 0$ when $|z| \to \infty$.
- $f^*(-z^*) = f(z)$.
- $\mathrm{Re}[f(x)] = x/[(x + 1)(x + 2)]$ $0 \le x \le \infty$.

6. (a) Find a function $H(x)$ that satisfies

$$\frac{1}{\pi}P\int_{-\infty}^{\infty}\frac{H(x')}{(x'-x)}dx'=\frac{1}{\left(x^2+\alpha^2\right)^2}$$

(b) Prove that an arbitrary constant can be added to $H(x)$ without affecting its definition given in part (a).

7. (a) Equations 9.19 and 9.21 are the dispersion relations for functions defined over the interval $0 \le x \le \infty$ that are even and odd, respectively, under reflection about the imaginary axis. Derive equivalent

dispersion relations for functions that are even and odd functions under reflection that are defined over the interval $-\infty \leq x \leq 0$.

(b) Determine the function $f(z)$ that has the following properties:

- $f(z) \to 0$ when $|z| \to \infty$.
- $f^*(-z^*) = f(z)$.
- $\text{Im}[f(x)] = 1/(x-\alpha)^2$ $-\infty \leq x \leq 0$, α real and positive, $x \neq -\alpha$.

8. Determine the function $g(z)$ that has the following properties:

- $g(z) \to 0$ when $|z| \to \infty$.
- $g^*(-z^*) = g(z)$.
- $g(1/2) = 2$.
- $\text{Im}[g(x)] = 1/[(x+3)(x+1)]$ $0 \leq x \leq \infty$.

9. Determine the function $g(z)$ that has the following properties:

- $g(z) \to 0$ when $|z| \to \infty$.
- $g(-z^*) = g^*(z)$.
- $\text{Re}[g(1/2)] = 0$.
- $\text{Im}[g(x)] = 1/(x+\alpha)^2$ $0 \leq x \leq \infty$ α real, positive.

10. (a) A function $f(z)$ has a branch point on the real axis at x_0 with the associated cut taken to extend from x_0 to $+\infty$. $f(z)$ is odd under reflection about the real axis. Show that $f(z)$ satisfies the dispersion relation

$$f(z) = \frac{1}{i\pi} \int_{x_0}^{\infty} \frac{\text{Re}[f(x'+i\varepsilon)]}{(x'-z)} dx'$$

(b) Use this result to show that at points on the real axis above the cut

$$\text{Im}[f(x+i\varepsilon)] = -\frac{1}{\pi} P \int_{x_0}^{\infty} \frac{\text{Re}[f(x'+i\varepsilon)]}{(x'-x)} dx'$$

and at points on the real axis below the cut

$$\text{Im}[f(x-i\varepsilon)] = \frac{1}{\pi} P \int_{x_0}^{\infty} \frac{\text{Re}[f(x'-i\varepsilon)]}{(x'-x)} dx'$$

11. Determine the function $f(z)$ that has the following properties:

- $f(z) \to 0$ when $|z| \to \infty$.
- $f(z)$ has a branch point on the real axis at $x = 1$, with an associated cut taken to extend from the branch point to $+\infty$.
- $f(z^*) = f^*(z)$.
- $\mathrm{Im}[f(x + i\varepsilon)] = 1/(x + 2)^2 \quad 1 \le x \le \infty$.

12. Determine the function $f(z)$ that has the following properties:

- $f(z) \to 0$ when $|z| \to \infty$.
- $f(z)$ has a branch point on the real axis at $x = 0$, with an associated cut that is taken to extend from the branch point to $+\infty$.
- $f(z^*) = -f^*(z)$.
- $\mathrm{Re}[f(x + i\varepsilon)] = x/[(x + 1)(x + 2)] \quad 1 \le x \le \infty$.

13. The function $f(z)$ has the following properties:

- $f(z) \to 0$ when $|z| \to \infty$.
- $f(z)$ has two branch points on the real axis at $x = \pm x_0$ (take $x_0 > 0$). The cut associated with $+x_0$ is taken to extend from the branch point to $+\infty$. The cut associated with $-x_0$ extends from that branch point to $-\infty$.
- $f(z^*) = f^*(z)$.

(a) Prove that if $\mathrm{Im}[f(z)]$ is even under reflection about the real axis, then

$$f(z) = \frac{2z}{\pi} \int_{x_0}^{\infty} \frac{\mathrm{Im}[f(x' + i\varepsilon)]}{(x'^2 - z^2)} dx'$$

(b) Prove that if $\mathrm{Im}[f(z)]$ is odd under reflection about the real axis, then

$$f(z) = \frac{2}{\pi} \int_{x_0}^{\infty} x' \frac{\mathrm{Im}[f(x' + i\varepsilon)]}{(x'^2 - z^2)} dx'$$

14. The function $f(z)$ has the following properties:

- $f(z) \to 0$ when $|z| \to \infty$.
- $f(z)$ has two branch points on the real axis at $x = \pm x_0$ (take $x_0 > 0$). The cut associated with $+x_0$ is taken to extend from the branch point to $+\infty$. The cut associated with $-x_0$ extends from that branch point to $-\infty$.

- $f(z^*) = -f^*(z)$.

   (a) Prove that if $\mathrm{Re}[f(z)]$ is even under reflection about the real axis, then

$$f(z) = \frac{2z}{i\pi} \int_{x_0}^{\infty} \frac{\mathrm{Re}\left[f(x'+i\varepsilon)\right]}{\left(x'^2 - z^2\right)} dx'$$

   (b) Prove that if $\mathrm{Re}[f(z)]$ is odd under reflection about the real axis, then

$$f(z) = \frac{2}{i\pi} \int_{x_0}^{\infty} x' \frac{\mathrm{Re}\left[f(x'+i\varepsilon)\right]}{\left(x'^2 - z^2\right)} dx'$$

15. Use the results of problems 13 and 14 to find the function $f(z)$ that has
   (a) $\mathrm{Im}[f(x + i\varepsilon)] = 1/(x^2 + \alpha^2)$ (b) $\mathrm{Re}[f(x + i\varepsilon)] = 1/(x^2 + \alpha^2)$
   with α real and positive, and has the following properties.

- $f(z) \to 0$ when $|z| \to \infty$.
- $f(z)$ has two branch points on the real axis at $x = \pm 1$ with associated cuts that extend from $+1$ to $+\infty$ and from -1 to $-\infty$.
- $f(z^*) = -f^*(z)$.

16. Use the results of prob. 13 and 14 to find the function $f(z)$ that has
   (a) $\mathrm{Im}[f(x + i\varepsilon)] = 1/[x(x + 1/2)]$ (b) $\mathrm{Re}[f(x + i\varepsilon)] = 1/[x(x + 1/2)]$
   and has the following properties:

- $f(z) \to 0$ when $|z| \to \infty$.
- $f(z)$ has two branch points on the real axis at $x = \pm 1$ with associated cuts that extend from $+1$ to $+\infty$ and from -1 to $-\infty$.
- $f(z^*) = f^*(z)$.

17. Use the results of probs. 13 and 14 to find the function $f(z)$ that has
   (a) $\mathrm{Im}[f(x + i\varepsilon)] = 1/(x^2 - \alpha^2)$ (b) $\mathrm{Re}[f(x + i\varepsilon)] = 1/(x^2 + \alpha^2)$
   with α real and $\alpha^2 < 1$ and has the following properties:

- $f(z) \to 0$ when $|z| \to \infty$.
- $f(z)$ has two branch points on the real axis at $x = \pm 1$ with associated cuts that extend from $+1$ to $+\infty$ and from -1 to $-\infty$.
- $f(z^*) = f^*(z)$.

18. If a function $g(z)$

- Has a branch point at x_0 with the associated cut extending to $+\infty$
- Is odd under reflection about the real axis

- Has a specified value $g(x_1)$ at a point $x_1 < x_0$ on the real axis

(a) show that

$$g(z) = g(x_1) + \frac{(z-x_1)}{i\pi} \int_{x_0}^{\infty} \frac{\mathrm{Re}[g(x'+i\varepsilon)]}{(x'-x_1)(x'-z)} dx'$$

(b) Use the result of part (a) to show that at points on the real axis along the top of the cut

$$\mathrm{Im}[g(x+i\varepsilon)] = -ig(x_1) - \frac{(x-x_1)}{\pi} \int_{x_0}^{\infty} \frac{\mathrm{Re}[g(x'+i\varepsilon)]}{(x'-x_1)(x'-x)_P} dx'$$

and at points on the real axis along the bottom of the cut

$$\mathrm{Im}[g(x-i\varepsilon)] = -ig(x_1) + \frac{(x-x_1)}{\pi} \int_{x_0}^{\infty} \frac{\mathrm{Re}[g(x'-i\varepsilon)]}{(x'-x_1)(x'-x)_P} dx'$$

from which we deduce straightforwardly that $g(x_1)$ is imaginary.

19. Determine the function $g(z)$ that has the following properties:
- $g(z) \to 0$ when $|z| \to \infty$.
- $g(z^*) = g^*(z)$.
- $g(z)$ has a branch point on the real axis at $x = 2$ with the associated cut extending from the branch point to $+\infty$.
- $g(1) = 2$.
- $\mathrm{Im}[g(x+i\varepsilon)] = 1/x$ $\qquad x \geq 2$.

20. Determine the function $g(z)$ that has the following properties:
- $g(z) \to 0$ when $|z| \to \infty$.
- $g(z^*) = -g^*(z)$.
- $g(z)$ has a branch point on the real axis at $x = 2$ with the associated cut extending from the branch point to $+\infty$.
- $g(1) = 2$.
- $\mathrm{Re}[g(x+i\varepsilon)] = 1/x^2$ $\qquad x \geq 2$.

21. Determine the function $g(z)$ that has the following properties:
- $g(z) \to 0$ when $|z| \to \infty$.

- $g(z^*) = g^*(z)$.
- $g(z)$ has a branch point at $x = 4$ with the associated cut extending to $+\infty$.
- $g(1) = 2$.
- $\text{Im}[g(x + i\varepsilon)] = 1/[x(x + 1)]$ $x \geq 4$.

22. (a) Using the representation

$$\delta(x - x'') = \frac{1}{\pi^2} \int_{-\infty}^{\infty} \frac{1}{(x' - x)_P (x' - x'')_P} dx' \qquad (9.75)$$

show that

$$\delta(x) = \frac{1}{\pi^2} \int_{-\infty}^{\infty} \frac{1}{(x'^2 - \frac{1}{4}x^2)_P} dx'$$

is an integral representation of $\delta(x)$.

(b) Use the integral representation of $\delta(x)$ given in part (a) to evaluate

$$P\int_0^{\infty} \int_0^{\infty} \frac{e^{-x^2}}{\left(y^2 - \frac{1}{4}(x - \alpha)^2\right)} dy\, dx \qquad \alpha \text{ real and positive}$$

Chapter 10

ANALYTIC CONTINUATION

Analytic continuation is a process of extending the domain of a function $F(z)$ beyond an open region R_1 to an open region R_2. Let R_1 and R_2 be open regions in the complex plane such that a function $F(z)$ is represented by $f_1(z)$ in R_1 and $f_2(z)$ in R_2. Let there be a set of points that is common to both R_1 and R_2 called R_{12}. Then R_1 and R_2 are said to overlap in this region.

Because R_{12} contains points in R_1 and also points in R_2, both $f_1(z)$ and $f_2(z)$ represent $F(z)$ at all z in R_{12}. At points in R_1 and at points in R_2 that are not in R_{12}, $f_1(z)$ and $f_2(z)$ are not the same representations of $F(z)$. They are said to be analytic continuations of each other.

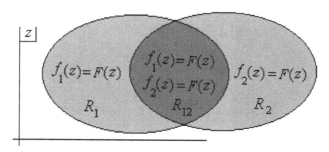

Figure 10.1

Analytic continuations of a function

As discussed in chapter 4, a series does not converge on the boundary of its circle of convergence even if the function it represents within that circle is analytic at points on the boundary. If R_1 is a closed region with a defined boundary, $f_1(z)$ will be singular at every point on that boundary. Then $f_1(z)$ cannot be analytically continued beyond the boundary. Thus, in order to be able to analytically continue $f_1(z)$ to $f_2(z)$, the regions R_1 and R_2 must be open regions, at least in R_{12} where they overlap.

Example 10.1: An analytic continuation of $1/(1-z)$

Using the Cauchy ratio (see appendix 4) it is straightforward to see that the geometric series

$$f_1(z) = \sum_{n=0}^{\infty} z^n \tag{10.1a}$$

converges to $1/(1-z)$ in the region defined by $|z| < 1$. But outside the unit circle

$$\sum_{n=0}^{\infty} z^n \neq \frac{1}{(1-z)} \tag{10.1b}$$

The integral

$$f_2(z) = \int_0^{\infty} e^{-t(1-z)} dt \tag{10.2}$$

converges to $1/(1-z)$ for $x = \text{Re}(z) < 1$.

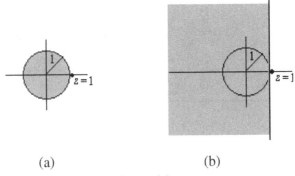

(a)   (b)

Figure 10.2

(a) Circle of convergence of the geometric series $f_1(z)$ and
(b) region of convergence of the integral $f_2(z)$

Because the region in which $f_2(z)$ is defined contains the entire region in which $f_1(z)$ is defined, $f_2(z)$ is the analytic continuation of $f_1(z)$ into the region outside the unit circle, but $f_1(z)$ is not an analytic continuation of $f_2(z)$. □

In the above example, $F(z)$ is known in closed form, $1/(1 - z)$. As such, the analytic structure of $F(z)$ and its value at any z (except $z = 1$) is found straightforwardly. If the closed form of $F(z)$ is known, analytic continuation is not needed to find its value at points outside the region of convergence of a representation. But if this closed form is not known, analytic continuation is necessary if one is to find such a value. For example, we see that $1/(1 - z)\big|_{z=2} = -1$ cannot be determined from either representation of ex. 10.1, but is easily found from the closed form of the function.

10.1 Analytic Continuation by Series

Analytic continuation can be achieved through series representation of a function. If a series representation $f(z)$ is to be analytically continued beyond its circle of convergence to some other series defined by another circle of convergence, the circles must be open circles, particularly in the overlap region.

Let

$$f_1(z) = \sum_{n=n_0}^{\infty} a_n(z_0)(z - z_0)^n \tag{10.3}$$

with

$$a_n(z_0) \equiv \frac{1}{2\pi i} \oint_C \frac{f_1(w)}{(w - z_0)^{n+1}} \, dw \tag{4.25}$$

If $f_1(z)$ converges to a function $F(z)$ that is analytic at z_0, then the series of eq. 10.3 is a Taylor series, for which $n_0 \geq 0$ and

$$a_n(z_0) = \frac{f_1^{(n)}(z_0)}{n!} \tag{10.4}$$

If $f_1(z)$ represents a function $F(z)$ that is singular at z_0, the series is a Laurent series with $n_0 < 0$ and

$$a_n(z_0) \neq \frac{f_1^{(n)}(z_0)}{n!} \tag{10.5}$$

If the singularity at z_0 is a pole of order M, then $n_0 = -M$ (see eq. 4.34). If the singularity is an essential singularity, $n_0 = -\infty$.

If z_1 is the singularity of $F(z)$ closest to z_0, the (Taylor or Laurent) series converges at all z (except z_0 for a Laurent series) within a circle centered at z_0 of radius $| z_1 - z_0 |$.

Referring to fig. 10.3, because z_α is inside the circle of convergence, then from eq. 10.3,

$$f_1(z_\alpha) = \sum_{n=n_0}^{\infty} a_n(z_0)(z_\alpha - z_0)^n \tag{10.6}$$

However, because z_β is outside the circle of convergence, $f_1(z_\beta)$ cannot be obtained from eq. 10.3.

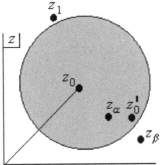

Figure 10.3
The circle of convergence for a series that includes z_α but not z_β

Let $z_0{'}$ be a point of analyticity of $f_1(z)$, and therefore within the circle of convergence of eq. 10.3. Then $F(z)$ can be expanded about $z{'}_0$ with a circle of convergence that includes z_β which is $f_2(z)$, an analytic continuation of $f_1(z)$. The circle of convergence of $f_2(z)$ has a radius defined by the singularity of $F(z)$ closest to $z{'}_0$.

If $f_1(z)$ is a Laurent series, then z_0 is outside the circle of convergence of $f_2(z)$. If $f_1(z)$ is a Taylor series it is possible to include z_0 within the circle of convergence of $f_2(z)$.

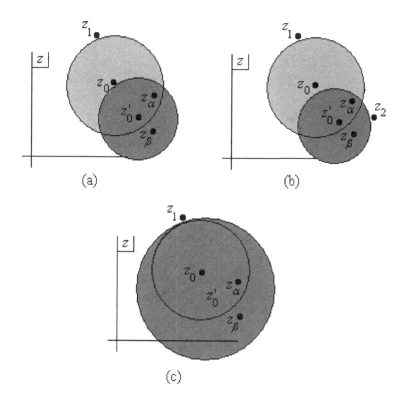

Figure 10.4

Possible circles of convergence of an analytic continuation of the series of eq. 10.3 when that series is

(a) a Laurent series and z_0 is the singularity closest to z_0',

(b) either Laurent or Taylor and z_2 is the singularity closest to z_0', or

(c) a Taylor series and z_1 is the singularity closest to z_0'

Example 10.2: Analytic continuation of a series

From the Cauchy ratio test and the fact that the sum starts at $n = -1$, we see that

$$f_1(z) = \sum_{n=-1}^{\infty} z^n \qquad (10.7)$$

converges at all z within the unit circle except at $z = 0$.

As shown in fig. 10.5a $z = -5/4$ and $z = -5/4 + i/4$ are outside the circle of convergence of $f_1(z)$. Because $z = -3/4$ is within the circle of convergence and is a point of analyticity of $f_1(z)$, we can expand $f_1(z)$ about this point.

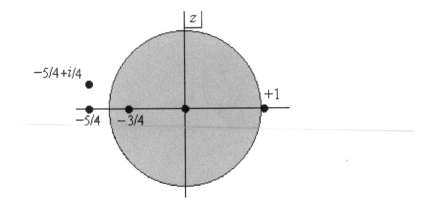

Figure 10.5a

Circle of convergence for the Laurent series of eq. 10.7

We write the series of eq. 10.7 as

$$f_1(z) = \frac{1}{z} + \sum_{n=0}^{\infty} z^n \tag{10.8}$$

Referring to appendix 3, eq. A3.4, we see that for points within the unit circle,

$$f_1(z) = \frac{1}{z} + \frac{1}{1-z} \tag{10.9}$$

Therefore,

$$\frac{f_1^{(k)}(z)}{k!} = \frac{(-1)^k}{z^{k+1}} + \frac{1}{(1-z)^{k+1}} \tag{10.10a}$$

from which

$$\frac{f_1^{(k)}(-\frac{3}{4})}{k!} = -\left(\frac{4}{3}\right)^{k+1} + \left(\frac{4}{7}\right)^{k+1} \tag{10.10b}$$

Therefore, an analytic continuation of the series of eq. 10.7 is

$$f_2(z) = -\sum_{k=0}^{\infty}\left[\left(\frac{4}{3}\right)^{k+1} - \left(\frac{4}{7}\right)^{k+1}\right]\left(z+\tfrac{3}{4}\right)^k \quad \left|z+\tfrac{3}{4}\right| < \tfrac{3}{4} \qquad (10.11)$$

Referring to fig. 10.5b, we see that the points $z = -5/4$ and $z = -5/4 + i/4$ are within the circle of convergence of $f_2(z)$.

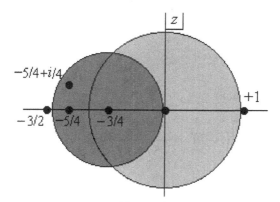

Figure 10.5b
Circle of convergence of the analytic continuation
of the Laurent series of eq. 10.8

Inasmuch as $f_1(z)$ converges to

$$f_1(z) = \frac{1}{z} + \frac{1}{1-z} \qquad (10.8)$$

at points within the unit circle, we see from this closed form that $F(z)$ is analytic at all finite z except at $z = 0$ and 1; then

$$-\sum_{k=0}^{\infty}\left[\left(\frac{4}{3}\right)^{k+1} - \left(\frac{4}{7}\right)^{k+1}\right]\left(z+\tfrac{3}{4}\right)^k = \frac{1}{z} + \frac{1}{1-z} \qquad (10.12)$$

within the circle of convergence defined by $\left|z + 3/4\right| < 3/4$. This circle encloses $-5/4$ and $-5/4 + i/4$.

Setting $z = -5/4$, eq. 10.12 becomes

$$-\sum_{k=0}^{\infty}(-1)^k\left[\left(\frac{4}{3}\right)^{k+1}-\left(\frac{4}{7}\right)^{k+1}\right]\left(\frac{1}{2}\right)^n = -4\sum_{k=0}^{\infty}(-1)^k(2)^k\left[\frac{1}{3^{k+1}}-\frac{1}{7^{k+1}}\right]$$

$$=\frac{1}{-5/4}+\frac{1}{9/4}=-\frac{16}{45}$$

$$(10.13a)$$

from which

$$\sum_{k=0}^{\infty}(-1)^k(2)^k\left[\frac{1}{3^{k+1}}-\frac{1}{7^{k+1}}\right]=\frac{4}{45} \qquad (10.13b)$$

For $z = -5/4 + i/4$, we obtain

$$\sum_{k=0}^{\infty}\frac{1}{4^k}\left[\frac{1}{3^{k+1}}-\frac{1}{7^{k+1}}\right](i-2)^k = \frac{1}{-5/4+i/4}+\frac{1}{9/4-i/4}=\frac{(176+56i)}{533}$$

$$(10.14)$$

Determining the value of a series is referred to as *summing a series*. We see that by knowing the closed form of the function, we have been able to sum each of the series in eqs. 10.13b and 10.14. □

Analytic continuation when the function represented by a series cannot be expressed in closed form

As seen in ex. 10.2, by knowing the closed form of a function that is represented by a series, it is possible to sum that series at various values of z (eqs. 10.13b and 10.14). In addition, knowing $F(z)$ in closed form makes the expansion of an analytic continuation of one series representation to obtain another series representation a relatively straightforward process. The construction of an analytic continuation of a series about any point within the circle of convergence requires $d^k F/dz^k$ to extend the domain of the function beyond the circle of convergence of the original series. If $F(z)$ is known in closed form, its derivatives can be obtained in closed form.

If $F(z)$ is not known in closed form, its series can still be analytically continued. The continuation will be a more unwieldy expression and the series cannot, in general, be summed.

To illustrate, let $f_1(z)$ be a series representation of $F(z)$ within a circle of convergence of radius R_0. That is,

$$f_1(z) = \sum_{n=n_0}^{\infty} a_n(z_0)(z - z_0)^n \qquad |z - z_0| < R_0 \tag{10.15}$$

Let z_1 be a point of analyticity of $F(z)$ within this circle of convergence.

Continuing this series to beyond this circle, to a region where the circle of convergence is centered about z_1 and has a radius R_1, we obtain the Taylor series

$$f_2(z) = \sum_{k=0}^{\infty} \frac{f_1^{(k)}(z_1)}{k!}(z - z_1)^k \qquad |z - z_1| < R_1 \tag{10.16}$$

If $F(z)$ is not known in closed form, the only way to obtain $f_1^{(k)}(z_1)$ is from the series of eq. 10.15. If the series of eq. 10.15 is a Taylor series, then $n_0 = 0$ and

$$f_1^{(k)}(z_1) = \sum_{n=k}^{\infty} \frac{n!}{(n-k)!}a_n(z_0)(z_1 - z_0)^{(n-k)} \tag{10.17}$$

If the original series is a Laurent series, then $n_0 < 0$ and we write eq. 10.15 as

$$f_1(z) = \sum_{n=-|n_0|}^{\infty} a_n(z_0)(z - z_0)^n$$

$$= \sum_{n=-|n_0|}^{-1} a_n(z_0)(z - z_0)^n + \sum_{n=0}^{\infty} a_n(z_0)(z - z_0)^n \tag{10.18}$$

$$= \sum_{n=1}^{|n_0|} a_{-n}(z_0)(z - z_0)^{-n} + \sum_{n=0}^{\infty} a_n(z_0)(z - z_0)^n$$

from which

$$f_1^{(k)}(z_1) = \sum_{n=1}^{|n_0|} (-1)^k \frac{(n+k-1)!}{(n-1)!} a_{-n}(z_0)(z_1-z_0)^{-(n+k)}$$

$$+ \sum_{n=k}^{\infty} \frac{n!}{(n-k)!} a_n(z_0)(z_1-z_0)^{(n-k)}$$

(10.19)

Then, the analytic continuation of the series of eq. 10.15 is obtained in the form given in eq. 10.16, with $f_1^{(k)}(z_1)$ given by either eq. 10.17 or 10.19. We see that when $F(z)$ is not known in closed form, the resulting series $f_2(z)$ is generally a cumbersome expression.

Example 10.3: Analytic continuation of a series that cannot be expressed in closed form

Using the Cauchy ratio test (see appendix 4), it is clear that the series

$$f_1(z) = \sum_{n=1}^{\infty} \frac{z^n}{n^2}$$

(10.20)

converges for $|z| < 1$. However, $F(z)$ cannot be expressed in closed form. To see this, we note that

$$-\int \frac{\ell n(1-z)}{z} dz = \sum_{n=1}^{\infty} \frac{z^n}{n^2} + C$$

(10.21)

Because the integral cannot be expressed in closed form, the function that the series represents cannot be expressed in closed form and the constant of integration cannot be evaluated.

However, we see from the integrand that the function has a singularity at $z = 1$. As such, we can continue the series of eq. 10.20 to a region that includes $z = -3/2$ by an expansion about $z = -1/2$. The continuation of $f_1(z)$ is

$$f_2(z) = \sum_{k=0}^{\infty} \frac{f_1^{(k)}(-\frac{1}{2})}{k!} (z+\frac{1}{2})^k$$

(10.22)

with $f_1^{(k)}(-1/2)$ obtained from eq. 10.20. The result is

$$f_1^{(k)}\left(-\tfrac{1}{2}\right) = \sum_{n=n_0}^{\infty} \frac{(-1)^{n-k}}{n^2} \frac{n!}{(n-k)!}\left(\frac{1}{2}\right)^{n-k} \tag{10.23a}$$

where

$$n_0 = \begin{cases} 1 & k=0 \\ k & k \ge 1 \end{cases} \tag{10.23b}$$

Then the analytic continuation of the series of eq. 10.20 is

$$f_2(z)$$
$$= \sum_{n=1}^{\infty} \frac{(-1)^n}{n^2}\left(\frac{1}{2}\right)^n + \sum_{k=1}^{\infty} \frac{1}{k!}\left[\sum_{n=k}^{\infty}(-1)^{n-k}\frac{n!}{n^2(n-k)!}\left(\frac{1}{2}\right)^{n-k}\right]\left(z+\tfrac{1}{2}\right)^k \tag{10.24a}$$

From this, we obtain

$$F\left(-\tfrac{3}{2}\right) = f_2\left(-\tfrac{3}{2}\right) \tag{10.24b}$$
$$= \sum_{n=1}^{\infty} \frac{(-1)^n}{n^2}\left(\frac{1}{2}\right)^n + \sum_{k=0}^{\infty} \frac{1}{k!}\left[\sum_{n=k}^{\infty}(-1)^n\frac{n!}{n^2(n-k)!}\left(\frac{1}{2}\right)^{n-k}\right] \qquad \square$$

Analytic continuation along an arc

Let $f_0(z)$ be a series representation of $F(z)$ with its circle of convergence centered at z_0, with a radius r_0. Let $F(z)$ have one or more singularities (indicated by small x's in fig. 10.6).

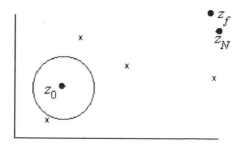

Figure 10.6
Points z_0 and z_N with several singularities of $F(z)$

Referring to fig. 10.6, it may be possible to analytically continue $f_0(z)$ to a representation $f_N(z)$, the circle of convergence of which is centered at z_N and contains the point z_f.

To do so, we choose a point z_1 that is within the circle of convergence of $f_0(z)$. We expand $f_0(z)$ in a Taylor series about z_1 and call this series representation $f_1(z)$. We then choose a point z_2 that is within the circle of convergence of $f_1(z)$ and expand $f_1(z)$ about this point to obtain a representation $f_2(z)$. This process is repeated until we eventually obtain a series $f_N(z)$ about a point z_N that includes z_f. A smooth curve or arc can be drawn connecting the centers of the circles of convergence of the many representations.

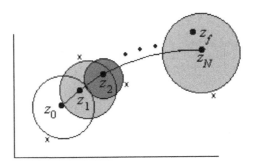

Figure 10.7

Analytic continuation along an arc

Example 10.4: Analytic continuation along an arc

Let $F(z)$ be a function that has poles at $z = 1$, 0, and $-9/4$, and let the Laurent series representation, expanded about the origin be given by

$$f_0(z) = \sum_{n=-1}^{\infty} \left[1 + (-1)^n \left(\tfrac{4}{9} \right)^{n+2} \right] z^n \equiv \frac{5}{9z} + \sum_{n=0}^{\infty} \alpha_n^0 z^n \qquad (10.25)$$

Using the Cauchy ratio and the analytic structure of $F(z)$, we determine that $f_0(z)$ converges within the unit circle and $F(z)$ is represented by $f_0(z)$ within that circle.

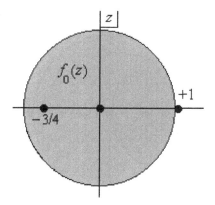

Figure 10.8a

Circle of convergence of $f_0(z)$

Because $z = -3/4$ is within the circle of convergence, we can analytically continue $f_0(z)$ along a segment of the negative real axis by first expanding $f_0(z)$ about $z = -3/4$. To do so, we compute

$$f_0^{(k)}(z) = (-1)^k \frac{5}{9} \frac{k!}{z^{k+1}} + \sum_{n=k}^{\infty} \frac{k!}{(n-k)!} \alpha_n^0 z^{n-k} \qquad (10.26a)$$

from which

$$\frac{f_0^{(k)}\left(-\frac{3}{4}\right)}{k!} = -\frac{5}{9}\left(\frac{4}{3}\right)^{k+1} + \sum_{n=k}^{\infty} \frac{(-1)^k}{(n-k)!} \alpha_n^0 \left(\frac{3}{4}\right)^{n-k} \equiv \alpha_k^1 \qquad (10.26b)$$

Then, the analytic continuation of $f_0(z)$ is

$$f_1(z) = \sum_{k=0}^{\infty} \alpha_k^1 \left(z + \tfrac{3}{4}\right)^k \qquad (10.27)$$

which represents $F(z)$ in the circle centered at $-3/4$, of radius $3/4$ shown below.

Because $z = -5/4$ is within the circle of convergence of $f_1(z)$, we can continue this function farther along the negative real axis by developing a series expanded about this point. From

$$f_1^{(m)}(z) = \sum_{k=m}^{\infty} \alpha_k^1 \frac{m!}{(k-m)!} \left(z + \tfrac{3}{4}\right)^{k-m} \qquad (10.28a)$$

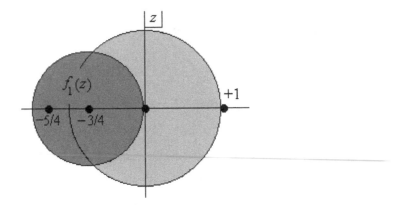

Figure 10.8b

Analytic continuation of $f_0(z)$ along the negative real axis

we have

$$\frac{f_1^{(m)}\left(-\frac{5}{4}\right)}{m!} = \sum_{k=m}^{\infty}(-1)^{k-m}\frac{1}{(k-m)!}\alpha_k^1 \equiv \alpha_m^2 \tag{10.28b}$$

Then

$$f_2(z) = \sum_{m=0}^{\infty}\alpha_m^2\left(z+\tfrac{5}{4}\right)^m \tag{10.29}$$

Because F(z) has a pole at z = –9/4, the circle of convergence of $f_2(z)$ is that shown below in fig. 10.8c.

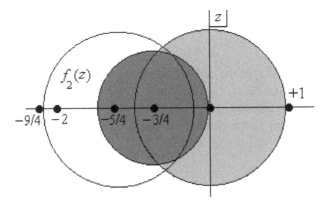

Figure 10.8c

Analytic continuation along the negative real axis

As we can see, the point $z = -2$ is within the circle of convergence of $f_2(z)$. Therefore,

$$F(-2) = f_2(-2) = \sum_{m=0}^{\infty} (-1)^m \alpha_m^2 \left(\frac{3}{8}\right)^m \qquad (10.30)$$

Had we known that

$$F(z) = \frac{1}{z(1-z)(z+\frac{9}{4})} = \frac{4/9}{z} + \frac{4/13}{(1-z)} - \frac{16/117}{(z+\frac{9}{4})} \qquad (10.31a)$$

it would have been unnecessary to use analytic continuation to determine that $F(-2) = -2/3$ and/or it would have been straightforward to obtain

$$\frac{F^{(k)}(z)}{k!} = (-1)^k \frac{4/9}{z^{k+1}} + \frac{4/13}{(1-z)^{k+1}} - (-1)^k \frac{16/117}{(z+\frac{9}{4})^{k+1}} \qquad (10.31b)$$

from which we would have determined that

$$\alpha_k^1 = -\frac{4}{9}\left(\frac{4}{3}\right)^{k+1} + \frac{4}{13}\left(\frac{4}{7}\right)^{k+1} - (-1)^k \frac{16}{117}\left(\frac{2}{3}\right)^{k+1} \qquad (10.32a)$$

$$\alpha_k^2 = -\frac{4}{9}\left(\frac{4}{5}\right)^{k+1} + \frac{4}{13}\left(\frac{4}{9}\right)^{k+1} - (-1)^k \frac{16}{117} \qquad (10.32b)$$
$$\square$$

Analytic continuation along different arcs and the Schwarz reflection

Let $F(z)$ have a finite number of poles in a region between points z_0 and z_N. We can analytically continue $f_0(z)$, a representation of $F(z)$ within the circle centered at z_0 to a representation $f_N(z)$ a representation of $F(z)$ within the circle centered at z_N along different arcs. Of course, these arcs must be chosen to avoid the poles.

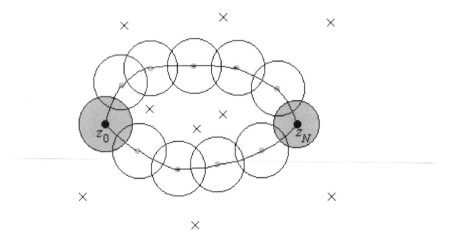

Figure 10.9a
Analytic continuation along different arcs avoiding poles of $F(z)$

Because the regions of these representations all contain points of analyticity of $F(z)$, the analytic continuation from the region containing z_0 to the region containing z_N, is unique, independent of the arc chosen. That is, referring to fig. 10.9a, the representation $f_N(z)$ obtained by continuing along the upper arc is identical to the representation $f_N(z)$ obtained by continuing along the lower arc.

Now let $F(z)$ be a multivalued function with one branch point. For the sake of this discussion, we take the branch point to be on the real axis at a point x_0 and the associated cut to extend along the real axis to $+\infty$. Let $F(z)$ have at most a finite number of poles, and be analytic in a region that includes a segment of the real axis with $x < x_0$.

$F(z)$ can be analytically continued from some point of analyticity in its circle of convergence (which we take to be on the real axis in fig. 10.9b) to a point z_f along an arc as discussed above. Of course, because of the cut, even if the final points are at the same x_f but on opposite sides of the cut, the continuation process will yield different values of $F(z)$.

It was discussed in chapter 9 that if $F(z)$ has a branch point on the real axis at x_0 with an associated cut extending to $+\infty$ along the real axis and the function is real (or imaginary) at a point x_i, then such a function would have to be even or odd under a Schwarz reflection

$$F(z^*) = \pm F^*(z) \tag{9.32a}$$

where $z \rightarrow z^*$ is a reflection about the real axis.

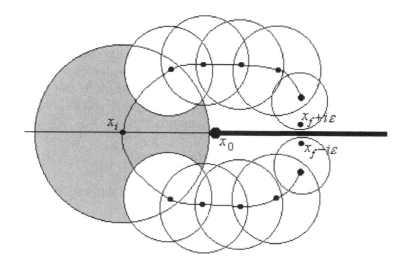

Figure 10.9b
Analytic continuation along arcs to points on two sides of a cut

If $F(z)$ satisfies eq. 9.32, we do not have to perform analytic continuations along two different arcs to find the values of $F(z)$ at points above and below the cut. Once we have found $F(z)$ at a point x_f on one side of the cut by continuing along an arc from x_i to x_f, the value of $F(z)$ at the mirror image of z (i.e., z^*) on the other side of the cut is obtained by Schwarz reflection. Thus, Schwarz reflection provides a method of analytic continuation of a function $F(z)$ with the properties given in eq. 9.32a.

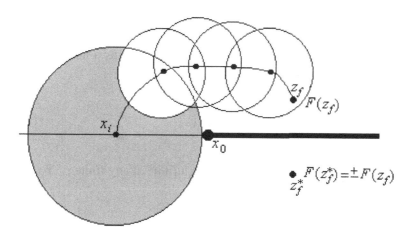

Figure 10.10
Determination of $F(z_f^*)$ from $F(z_f)$ by Schwarz reflection

Natural boundaries

Some series are either singular at every point on the circumference of the circle of convergence or all points of analyticity on the circumference are arbitrarily close to singular points. In such a case, it is not possible to analytically continue beyond the circle of convergence and its circumference forms a *natural boundary* of the representation.

Example 10.5: A natural boundary for a series

Using the Cauchy ratio (see appendix 4), it is straightforward to show that the series

$$f(z) = \sum_{n=0}^{\infty} z^{n!} \tag{10.33}$$

converges for all z within the unit circle. The Cauchy ratio for large n is

$$\rho = \lim_{n \to \infty} \left| \frac{z^{(n+1)!}}{z^{n!}} \right| = \lim_{n \to \infty} \left| z^{(n+1)n!-n!} \right| = \lim_{n \to \infty} \left| z^{n*n!} \right| \tag{10.34a}$$

For the series to converge, z must be constrained by

$$\lim_{N \to \infty} \left| z^{n*n!} \right| < 1 \tag{10.34b}$$

which is satisfied for points within the unit circle.

To see that the circumference of the unit circle forms a natural boundary for this series, we write

$$f(z) = \sum_{n=0}^{N-1} z^{n!} + \sum_{n=N}^{\infty} z^{n!} \equiv \varphi_1(z) + \varphi_2(z) \tag{10.35}$$

Because $\varphi_1(z)$ is a finite sum, it cannot be infinite at any finite z. To analyze $\varphi_2(z)$, we set

$$z = re^{i2\pi k/N} \tag{10.36}$$

where k and N are integers and k/N is an irreducible fraction. Then,

$$\varphi_2(z) = \sum_{n=N}^{\infty} r^{n!} e^{i2\pi kn!/N} \tag{10.37}$$

For all $n = N + k$ with $k \geq 1$,

$$\frac{n!}{N} = \frac{(N+k)(N+k-1)\cdots(N+1)N(N-1)\cdots3*2*1}{N} \tag{10.38}$$
$$= (N+k)(N+k-1)\cdots(N+1)(N-1)\cdots3*2*1$$

That is, for all n in the series $\varphi_2(z)$, $n!/N$ is an integer, and the argument of $z^{n!}$ is an integer multiple of 2π. Therefore, $e^{i2\pi kn!/N} = 1$ and

$$\lim_{r\to1}\varphi_2(z) = \lim_{r\to1}\sum_{n=N}^{\infty} r^{n!} = \infty \tag{10.39}$$

Thus, $\varphi_2(z)$ is singular at every point on the circumference of the unit circle for which the argument can be expressed as $2\pi k/N$.

$\varphi_2(z)$ is not singular at those points on the circumference that cannot be expressed as $z = e^{i2\pi k/N}$, but by an appropriate choice of k and N, any point at which $\varphi_2(z)$ becomes infinite can be made arbitrarily close to any point at which $\varphi_2(z)$ does not become infinite. As such, there is no region on the circumference of the unit circle between a singularity and a point of analyticity over which the series can be continued. Therefore, the series cannot be continued beyond that circumference.

To see that the series

$$g(z) = \sum_{n=1}^{\infty} \frac{z^{n!}}{n!} \tag{10.40}$$

is related to the series $f(z)$ of eq. 10.33, we consider

$$\frac{dg}{dz} = \sum_{n=1}^{\infty} z^{n!-1} = \frac{1}{z}\sum_{n=1}^{\infty} z^{n!} = \frac{1}{z}\left(\sum_{n=0}^{\infty} z^{n!} - 1\right) = \frac{1}{z}(f(z)-1) \tag{10.41}$$

Because the series $f(z)$ converges for $|z| < 1$, and the circumference of the unit circle forms a natural boundary of $f(z)$), the circle of convergence of the series $g(z)$ is the unit circle and its circumference is a natural boundary of $g(z)$. □

10.2   Analytic Continuation of the Factorial

The factorial of a positive nonzero integer is given by

$$N! \equiv N(N-1)(N-2)\cdots 3*2*1 \qquad (10.42)$$

We note that the value of 0! cannot be obtained from this definition.

The gamma function

To see how to obtain this result, we note that we can analytically continue $N!$ using the *gamma function*

$$\Gamma(N+1) \equiv \int_0^\infty x^N e^{-x}dx \qquad (10.43)$$

With N a positive integer, we integrate by parts, taking $u = x^N$ and $dv = e^{-x}dx$. Then

$$\Gamma(N+1) = -x^N e^{-x}\Big|_0^\infty + N\int_0^\infty x^{N-1}e^{-x}dx = N\Gamma(N) \qquad (10.44a)$$

Replacing N by $N-1$ in eq. 10.44a, we obtain

$$\Gamma(N) = (N-1)\Gamma(N-1) \qquad (10.44b)$$

so that

$$\Gamma(N+1) = N(N-1)\Gamma(N-1) \qquad (10.45)$$

Repeating this process N times, we obtain

$$\Gamma(N+1) = N(N-1)(N-2)\cdots 3*2*1*\Gamma(1) \qquad (10.46)$$

With

$$\Gamma(1) = \int_0^\infty e^{-x}dx = 1 \qquad (10.47)$$

this becomes

$$\Gamma(N+1) = N!$$ (10.48)

From this analysis, we see that the gamma function is a representation of $N!$ for integer N. Setting $N = 0$, we obtain

$$0! = \Gamma(1) = 1$$ (10.49)

Replacing N by a complex quantity z, we obtain

$$z! \equiv \Gamma(z+1) = \int_0^\infty x^z e^{-x} dx \quad \mathrm{Re}(z) > -1$$ (10.50)

It is discussed in advanced calculus texts (see, for example, Taylor, 1955, pp. 648–649) that this integral is analytic at all z with $\mathrm{Re}(z) > -1$. Thus, $\Gamma(1 + z)$ is an analytic continuation of the factorial of a postive integer.

Therefore, for a positive integer $N \geq \mathrm{Re}(z)$, $\Gamma(z + N + 1) = (z + N)!$ is analytic. From the iterative property of the gamma function

$$\Gamma(z+N+1) = (z+N)! = (z+N)(z+N-1)\cdots z\Gamma(z)$$
$$= z(z+1)(z+2)\cdots(z+N)\Gamma(z)$$ (10.51a)

or

$$\Gamma(z) = \frac{\Gamma(z+N+1)}{z(z+1)(z+2)\cdots(z+N)}$$ (10.51b)

Because $z + N$ has a nonnegative real part, $\Gamma(z + N + 1)$ is analytic at z. Therefore, $\Gamma(z)$ has simple poles at $z = 0$ and negative integers. Thus, $(-N)!$ is infinite for integer $N = 1, 2, \ldots$.

Half integer factorials

With the gamma function, we can define the factorial of noninteger values of N. For example, for $N = -1/2$, eq. 10.50 is

$$\left(-\tfrac{1}{2}\right)! = \Gamma\left(\tfrac{1}{2}\right) = \int_0^\infty x^{-1/2} e^{-x} dx$$ (10.52)

With the substitution, $y = x^{1/2}$, and referring to eq. A5.9b, we obtain

$$\left(-\tfrac{1}{2}\right)! = 2\int_0^\infty e^{-y^2}\, dy = \sqrt{\pi} \tag{10.53}$$

Then, from the iterative property of the gamma function, we have

$$\left(\tfrac{1}{2}\right)! = \left(\tfrac{1}{2}\right)\left(-\tfrac{1}{2}\right)! = \tfrac{1}{2}\sqrt{\pi} \tag{10.54a}$$

and from

$$\left(-\tfrac{1}{2}\right)! = -\tfrac{1}{2}\left(-\tfrac{3}{2}\right)! = \sqrt{\pi} \tag{10.54b}$$

we obtain

$$\left(-\tfrac{3}{2}\right)! = -2\sqrt{\pi} \tag{10.54c}$$

Using the iterative property of the gamma function, the value of the factorial of any half integer can be determined (see prob. 15).

If the integral representation of $z!$ of eq. 10.50 cannot be evaluated exactly, approximations to the gamma function can be developed.

Approximation of $z!$ for Re(z) positive and large

When Re(z) is a large positive number, we substitute

$$x = yz \tag{10.55}$$

in eq. 10.50 to obtain

$$z! = z^{z+1}\int_0^\infty y^z e^{-zy}\, dy = z^{z+1}\int_0^\infty e^{-z[y - \ell n(y)]}\, dy \tag{10.56}$$

It is straightforward to verify that

$$y - \ell n(y) > 0 \qquad 0 \le y \le \infty \tag{10.57}$$

Thus, when Re(z) is large and positive, the major contribution to

$$e^{-z[y-\ell n(y)]}$$

arises from the values of y in a small region around the minimum of $y - \ell n(y)$ and the function decreases rapidly as y varies from this minimum. The approximation involves adding or ignoring these 1 values of y that are "far" from the minimum as needed. The method of evaluating the integral of eq. 10.56 using this approximation is called *the method of steepest descent.*

Because $y - \ell n(y)$ is a minimum for $y = 1$, we set

$$y = 1 + w \tag{10.58}$$

and expand $(1 + w) - \ell n(1 + w)$ in a MacLaurin series. Because the major contribution to the exponential arises for w near zero, we can approximate this series by the first few terms in w. Thus, we take

$$y - \ell n(y) = 1 + w - \left(w - \frac{w^2}{2} + \frac{w^3}{3} - \cdots\right) \simeq 1 + \frac{w^2}{2} \tag{10.59}$$

and approximate eq. 10.56 by

$$z! \simeq z^{z+1} \int_{-1}^{\infty} e^{-z\left[1 + \frac{w^2}{2}\right]} dw = z^{z+1} e^{-z} \int_{-1}^{\infty} e^{-zw^2/2} dw \tag{10.60}$$

For Re(z) large, the exponential is small for values of w much different from zero. Thus, there is little error made by extending this lower limit, taking

$$\int_{-1}^{\infty} e^{-zw^2/2} dw \simeq \int_{-\infty}^{\infty} e^{-zw^2/2} dw \tag{10.61}$$

Then, eq. 10.60 becomes

$$z! \simeq z^{z+1} e^{-z} \int_{-\infty}^{\infty} e^{-zw^2/2} dw \tag{10.62}$$

We then make the substitution $u = w\sqrt{z/2}$, and recognize that e^{-u^2} is an even function of u, to obtain

$$z! \simeq z^{z+\frac{1}{2}} e^{-z} \sqrt{2} \int_{-\infty}^{\infty} e^{-u^2} du = z^{z+\frac{1}{2}} e^{-z} 2\sqrt{2} \int_{0}^{\infty} e^{-u^2} du \tag{10.63}$$

Referring to appendix 5, eq. A5.9b,

$$\int_0^\infty e^{-u^2} du = \frac{\sqrt{\pi}}{2}$$ (10.64)

and we obtain the *Stirling approximation* of $z!$ for Re(z) large;

$$z! \simeq z^z e^{-z} \sqrt{2\pi z}$$ (10.65a)

Because Re(z) is large, many commercially available calculators do not yield a precise value of z^z. In such cases, it is preferrable to find $z!$ by computing

$$\ell n(z!) \simeq z \, \ell n(z) - z + \frac{1}{2} \ell n(2\pi z)$$ (10.65b)

Example 10.6: The Stirling approximation for $z!$

The exact value of 8! is 40320. By Stirling's approximation, we obtain

$$\ell n(8!) \simeq 8 \, \ell n(8) - 8 + \frac{1}{2} \ell n(16\pi) = 10.5942$$ (10.66)

from which we obtain $8! \cong 39903$, approximately a 1% error.

The Stirling approximation for $z = 8 + 8i$ yields

$$\ell n\big((8+8i)!\big) \simeq 8(1+i)\ell n\big(8(1+i)\big) - 8(1+i) + \frac{1}{2} \ell n\big(16\pi(1+i)\big)$$
                                    (10.67)

Writing this in Cartesian form, we obtain

$$\ell n\big((8+8i)!\big)$$
$$\simeq \left[8\ell n\big(8\sqrt{2}\big) - 2\pi - 8 + \frac{1}{2} \, \ell n\big(16\pi\sqrt{2}\big) \right] + i\left[8\ell n\big(8\sqrt{2}\big) + 2\pi - 8 + \frac{\pi}{8} \right]$$
$$= 7.2569 + 18.0840i$$
                                    (10.68a)

from which

$$(8+8i)! \approx 18.4857e^{1.1892i}$$ (10.68b)

□

Approximation of $z!$ for $|z|$ small

When $|z|$ is small, $z!$ can be approximated by a truncated MacLaurin series. We obtain that MacLaurin series in the form

$$\ell n(z!) = \ell n\left[\Gamma(1+z)\right] \equiv \ell n\left[\Gamma(1)\right] + \sum_{n=1}^{\infty} \frac{\psi^{(n)}(0)}{n!} z^n = \sum_{n=1}^{\infty} \frac{\psi^{(n)}(0)}{n!} z^n$$

(10.69a)

where

$$\psi^{(n)}(0) = \frac{d^n}{dz^n} \ell n(z!)\bigg|_{z=0} = \frac{d^n}{dz^n} \ell n\left(\Gamma(1+z)\right)\bigg|_{z=0}$$ (10.69b)

To determine these coefficients, we substitute the identity

$$e^{-x} = \lim_{m\to\infty}\left(1-\frac{x}{m}\right)^m$$ (10.70)

Then

$$z! = \int_0^{\infty} x^z \lim_{m\to\infty}\left(1-\frac{x}{m}\right)^m dx = \lim_{m\to\infty}\int_0^m x^z\left(1-\frac{x}{m}\right)^m dx$$ (10.71)

Substituting $x = mt$ and referring to the description of the function $\beta(p,q)$ in appendix 8 eq. A8.9, this becomes

$$z! = \lim_{m\to\infty} m^{z+1}\int_0^1 t^z(1-t)^m dt$$

$$= \lim_{m\to\infty} m^{z+1}\beta(z+1,m+1) = \lim_{m\to\infty} m^{z+1}\frac{\Gamma(z+1)\Gamma(m+1)}{\Gamma(z+m+2)}$$ (10.72)

$$= \lim_{m\to\infty} m^{z+1}\frac{m!}{(z+m+1)(z+m)\cdots(z+2)(z+1)}$$

This is called the *Euler limit representation* of $z!$. Using this representation,

$$\ell n(z!) = \lim_{m \to \infty} \left[\ell n(m!) + (1+z) \ell n(m) - \sum_{k=0}^{m} \ell n(1+z+k) \right] \qquad (10.73)$$

from which we obtain the *digamma function*

$$\psi^{(1)}(z) = \frac{d}{dz} \ell n(z!) = \lim_{m \to \infty} \left[\ell n(m) - \sum_{k=0}^{m} \frac{1}{(z+1+k)} \right] \qquad (10.74)$$

Then

$$\psi^{(1)}(0) = \lim_{m \to \infty} \left[\ell n(m) - \sum_{k=0}^{m} \frac{1}{k+1} \right] \qquad (10.75)$$

When $m \to \infty$, both $\ell n(m)$ and the sum are infinite, but their difference is finite. (See Cohen, 1992, p. 146, eq. 3.162 and p. 270, eqs. 6.42–6.45.) The value of $\psi^{(1)}(0)$ is $-0.577216 \equiv -\gamma$. γ is called the *Euler–Mascheroni constant*.

Derivatives of $\psi^{(1)}(z)$, called *polygamma functions*, are easily obtained from eq. 10.74. They are

$$\psi^{(n)}(z) = (-1)^n (n-1)! \sum_{k=0}^{\infty} \frac{1}{(z+1+k)^n} \qquad n \geq 2 \qquad (10.76a)$$

Because this series converges for all $n \geq 2$, the limit $m \to \infty$ is obtained simply by replacing m by ∞. Then from eq. 10.76a, the coefficients of the MacLaurin series are given by

$$\frac{\psi^{(n)}(0)}{n!} = \frac{(-1)^n}{n} \sum_{k=0}^{\infty} \frac{1}{(k+1)^n} \qquad n \geq 2 \qquad (10.76b)$$

The series in this expression is referred to as the *Riemann zeta function* for integer $n \geq 2$, denoted by $\zeta(n)$ and eq. 10.76b can be written

$$\frac{\psi^{(n)}(0)}{n!} = (-1)^n \frac{\zeta(n)}{n} \qquad n \geq 2 \qquad (10.77)$$

Using the Cauchy ratio, the radius of convergence of this series is found from

$$|z| \lim_{n \to \infty} \frac{n}{n+1} \frac{\varsigma(n+1)}{\varsigma(n)} < 1 \tag{10.78}$$

From the definition of the Riemann zeta series, we have

$$\lim_{n \to \infty} \varsigma(n) = \lim_{n \to \infty} \left[1 + \frac{1}{2^n} + \frac{1}{3^n} + \cdots \right] = 1 \tag{10.79}$$

Thus, the circle of convergence of the series of $\ell n(z!)$ is the unit circle, and

$$\ell n(z!) = -\gamma z + \sum_{n=2}^{\infty} \frac{\psi^{(n)}(0)}{n!} z^n \quad |z| < 1 \tag{10.80}$$

We note that all terms in the Riemann zeta series are positive. A series in which all terms are of the same sign is referred to as an *absolute series*. Referring to p. 146, eqs. 3.52, 3.55, and 3.162 of Cohen, 1992, a good approximation of such a series is obtained by writing

$$\sum_{k=0}^{\infty} \frac{1}{(k+1)^n} = \sum_{k=0}^{M-1} \frac{1}{(k+1)^n} + S_2 \tag{10.81a}$$

with

$$S_2 = \sum_{k=M}^{\infty} \frac{1}{(k+1)^n} \simeq \frac{1}{2} \frac{1}{(M+1)^n} + \frac{1}{(n-1)} \frac{1}{(M+1)^{(n-1)}} \tag{10.81b}$$

With this approximation to the Riemann zeta series, the values of coefficients of the series for $\ell n(z!)$ can be obtained. The values of the first ten of these coefficients are given in table 10.1.

$-\gamma$	$\psi^{(2)}(0)/2!$	$\psi^{(3)}(0)/3!$	$\psi^{(4)}(0)/4!$	$\psi^{(5)}(0)/5!$
–0.57722	0.82247	–0.40069	0.27058	–0.27039
$\psi^{(6)}(0)/6!$	$\psi^{(7)}(0)/7!$	$\psi^{(8)}(0)/8!$	$\psi^{(9)}(0)/9!$	$\psi^{(10)}(0)/10!$
0.16956	–0.14405	0.12551	–0.11133	0.10001

Table 10.1

Values of polygamma functions

   As can be seen, the coefficients decrease with increasing n. Therefore, if $|z| < 1$, a reasonable approximation to $z!$ can be obtained from a truncation of the series of eq. 10.80.

Example 10.7: The MacLaurin series for $z!$

   For $z = 0.1$, using the values of coefficients in table 10.1, we obtain

$$\ell n(0.1!) \simeq -0.1\gamma + \sum_{n=2}^{10} \frac{\psi^{(n)}(0)}{n!}(0.1)^n = -0.04987 \qquad (10.82a)$$

from which

$$(0.1)! = \Gamma(1.1) \simeq 0.95135 \qquad (10.82b)$$

   It is also possible to determine the values of the factorials of small numbers using the Stirling approximation for a large number and the iterative property of the gamma function. Assuming, for example, that $(8.1)!$ is well approximated by the Stirling approximation, we have

$$(0.1)! = \frac{(8.1)!}{8.1*7.1*6.1*5.1*4.1*3.1*2.1*1.1} \qquad (10.83)$$

Referring to eq. 10.63b, we have

$$\ell n(8.1!) \simeq 8.1\,\ell n(8.1) - 8.1 + \frac{1}{2}\,\ell n(16.2\pi) = 10.8090 \qquad (10.84)$$

which yields $(8.1)! = 49464$. Then, from eq. 10.83, $(0.1)! = 0.94165$, which is in good agreement with $(0.1)!$ obtained from the truncated MacLaurin series. □

   With the Stirling and truncated MacLaurin series approximations, plus the exact values of the gamma function at integer and half integer values of its argument, we can compute the gamma function of any value of x to produce the graph below.

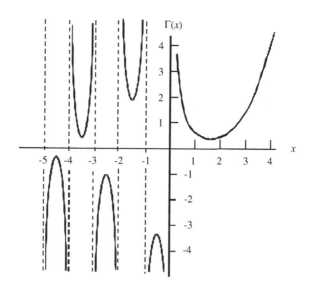

Figure 10.11
The gamma function

Problems

1. Determine the region in which the representation

$$f_3(z) = -\int_0^\infty e^{t(1-z)} dt$$

is the analytic continuation of the geometric series

$$f_1(z) = \sum_{n=0}^\infty z^n$$

2. It is straightforward to see that

$$f_1(z) = \sum_{n=0}^\infty z^{2n}$$

converges to $1/(1 - z^2)$ for all z within the unit circle. The integral representation

$$f_2(z) = \int_0^\infty e^{-t(1-z^2)} dt$$

extends this domain. What is the equation or inequality that describes that extended domain?

3. (a) Determine the radius of convergence of the series

$$f_1(z) = \sum_{n=1}^{\infty} \frac{z^{2n}}{(n+1)\sqrt{n}}$$

(b) Find the series representation $f_2(z)$, expanded about $z = i/2$ that is an analytic continuation of the MacLaurin series given in part (a). What is the radius of convergence of $f_2(z)$?

4. (a) Find a series $f_2(z)$ that is an analytic continuation of

$$f_1(z) = \sum_{n=1}^{\infty} \frac{z^{n+2}}{2^n}$$

and is analytic at $z = -3$.

(b) What is the value of $f_2(-3)$?

5. (a) What is the radius of convergence of the Laurent series representation

$$f_1(z) = \sum_{n=0}^{\infty} (-1)^n \frac{z^{n-1}}{2^{n+1}}$$

(b) Find the Taylor series about $z = 3i/2$ that is an analytic continuation of $f_1(z)$. Show that $z = 5i/2$ is within the circle of convergence of this series.

6. Within the unit circle

$$f_1(z) = \sum_{n=0}^{\infty} z^n$$

converges to $1/(1 - z)$. Show that

$$f_2(z) = \frac{1}{4} \sum_{n=0}^{\infty} \frac{(z+3)^n}{2^{2n}}$$

and

$$f_3(z) = \sum_{n=0}^{\infty} (-1)^{n+1} \frac{(z+1-i)^n}{(2-i)^{n+1}}$$

are analytic continuations of $f_1(z)$. Specify the positions of the centers and the radii of the circles of convergence of $f_2(z)$ and $f_3(z)$.

7. The MacLaurin series

$$f_1(z) = \sum_{n=0}^{\infty} z^{2n}$$

converges to $1/(1 - z^2)$ at all z within the unit circle.
(a) Determine the analytic continuation of this series that converges within a circle centered at $z = 3i/4$. What is the radius of this circle? Prove that $z = 7i/4$ is within this circle of convergence.
(b) Use the fact that $f_1(z)$ converges to $1/(1 - z^2)$ at all z within the unit circle to show that

$$\sum_{m=0}^{\infty} (-1)^m \left(\frac{4}{5}\right)^{2m+1} \left[\begin{array}{c} \cos\left\{(2m+1)\tan^{-1}\left(\frac{3}{4}\right)\right\} \\ -\frac{4}{5}\sin\left\{(2m+1)\tan^{-1}\left(\frac{3}{4}\right)\right\} \end{array} \right] = \frac{16}{65}$$

(Hint: For any number N, one can write

$$\left(1 + \frac{3i}{4}\right)^N = \left(\frac{5}{4}\right)^N e^{iN\tan^{-1}\left(\frac{3}{4}\right)}$$

8. The circle of convergence of the series

$$f_1(z) = \sum_{n=0}^{\infty} (-1)^n z^n$$

is the unit circle. Determine the series $f_2(z)$, expanded about $z = 3$, that is an analytic continuation of $f_1(z)$. What is the radius of convergence of the series $f_2(z)$?

9. The circle of convergence of

$$f_1(z) = \sum_{n=1}^{\infty} \frac{z^n}{n}$$

is the unit circle.

(a) Determine a series $f_2(z)$ that is an analytic continuation of $f_1(z)$, expanded about $z = -1/2$ for which $z = -1$ is contained within the circle of convergence. What is the radius of convergence of this series?

(a) At all z within the unit circle, $f_1(z)$ converges to $-\ell n(1-z)$. Use this fact to sum the series

$$\sum_{n=1}^{\infty} (-1)^n \frac{1}{n} \left(\frac{3}{2}\right)^n$$

10. The series

$$f_1(z) = \sum_{n=0}^{\infty} (-1)^n z^n$$

converges to $1/(1 + z)$ at all z within the unit circle. Find the series that is an analytic continuation of $f_1(z)$ such that its circle of convergence is centered on the real axis, encloses the point $z = (2 + i)$, and has the smallest possible radius.

11. Analytically continue the MacLaurin series

$$f_1(z) = \sum_{n=0}^{\infty} (-1)^n z^{2n}$$

along the real axis. First find a series $f_2(z)$, expanded about $z = 3/4$. Then continue $f_2(z)$ to a series $f_3(z)$ for which the circle of convergence is centered at $z = 7/4$. Evaluate $f_3(z)$ at $z = 15/4$ and use that fact that $f_1(z)$ converges to $1/(1 + z^2)$ to sum the series

$$\sum_{n=0}^{\infty} (-1)^n \sin\left[(n+1)\tan^{-1}\left(\tfrac{4}{7}\right)\right]2^n$$

12. Within its circle of convergence, the Laurent series

$$f_1(z) = \sum_{n=0}^{\infty} b_n z^{n-1}$$

converges to a function with a simple pole at the origin and poles at $z = N \pm i\alpha$, where α is real and positive, and the integer N satisfies $3\alpha > N > 2\alpha$.

(a) What is the radius of the circle of convergence of this series?

(b) In terms of sums over the coefficients b_n, analytically continue this series along the real axis to obtain a Taylor series that has a circle of convergence that includes the point $z = x = 5\alpha/2$.

13. (a) Prove that the series

$$F(z) = \sum_{n=0}^{\infty} z^{2^n}$$

converges for all $|z| < 1$.

(b) Show that the circumference of the unit circle is a natural boundary for this series.

Hint: Let

$$z = re^{i\pi k/2^N}$$

and take the remainder series to be

$$S_2(z) = \sum_{n=N+1}^{\infty} z^{2^n}$$

14. (a) Prove that the circle of convergence of the series

$$F(z) = \sum_{n=1}^{\infty} \frac{z^{n!}}{n}$$

is the unit circle.

(b) Prove that the circumference of the unit circle is a natural boundary of this series.

(Hint: See ex. 10.5.)

15. Determine the values of

   (a) $\left(-\frac{5}{2}\right)!$ (b) $\left(\frac{5}{2}\right)!$

16. (a) Assume that $z!$ is well approximated by the Stirling approximation
      for $\mathrm{Re}(z) > 8$. Obtain a Stirling approximation to $\pi!$.
   (b) Using the iterative property of the gamma function (the factorial),
      estimate the value of $\pi!$ by a sum of the first three terms in the
      MacLaurin series for $(\pi - 3)!$.

17. Estimate the value of

   (a) $(-8.3)!$ (b) $(e)!$ (where e is the base of the natural logarithm)

   using the assumption that the Stirling approximation is satisfactory for
   $\mathrm{Re}(z) > 8$ or that the sum of the first three terms of MacLaurin series for
   $z!$ provides a satisfactory approximation for $|z| < 0.5$.

18. Assume that the sum of the first three terms of its MacLaurin series
   provides an acceptable approximation of $z!$ for $|z| < 0.1$. Use this
   truncated series to estimate the value of $(3.95)!$.

19. Estimate the value of $[(1 + i)/10]!$ by the sum of the first three terms of
   the MacLaurin series for $z!$.

20. Estimate the value of $(10 + 2i)!$ using the Stirling approximation.

21. The *Legendre duplication formula* for the gamma function is given by

$$\frac{\Gamma(1+z)\Gamma(\tfrac{1}{2}+z)}{\Gamma(1+2z)} = \frac{z!\,(z-\tfrac{1}{2})!}{(2z)!} = \frac{\sqrt{\pi}}{2^{2z-1}}$$

   (See, for example, Cohen, 1992, pp. 268–269.)
   Use the Legendre duplication formula to derive a
   (a) Stirling approximation to $(z-1/2)!$ for $\mathrm{Re}(z)$ large.
   (b) MacLaurin series for $(z-1/2)!$ in terms of the Euler–Mascheroni
      constant and the polygamma functions that can be used to
      approximate $(z-1/2)!$ for $|z|$ small. What is the circle of
      convergence of this MacLaurin series?

APPENDIX 1

AC CIRCUITS AND COMPLEX NUMBERS

Using complex numbers, a circuit containing an AC generator, resistors, capacitors, and inductance coils can be analyzed the same way one analyzes a DC circuit containing a battery and resistors.

The need for complex numbers for an AC circuit arises from the fact that there is a *phase difference* between the current through a device and the voltage across the device. AC voltage and current are said to be *sinusoidal* which means that both quantities vary with time according to

$$V(t) = V_p \sin(2\pi ft + \phi_V) \tag{A1.1a}$$

and

$$I(t) = I_p \sin(2\pi ft + \phi_I) \tag{A1.1b}$$

where f is the frequency of the AC generator, and ϕ_V and ϕ_I are called *phase angles*. The phase difference between voltage and current is $\phi_V - \phi_I$.

In an AC circuit, I_R, the current through a resistor, is in phase with V_R, the voltage across the resistor so that $\phi_V - \phi_I = 0$. This means that when one of these quantities (for example, the current) is a maximum, the other (the voltage) is a maximum and when one quantity is zero, the other is zero. The current and voltage are $90°$ *out of phase* with each other for a capacitor and for an inductor. This means that for either device, at the instant when the current is a maximum, the voltage is zero. When the current is zero, the voltage is a maximum. For a capacitor, the voltage is $90°$ behind the current at every instant. That is, $\phi_V - \phi_I = -90°$ for a capacitor. For an inductor the voltage is $90°$ ahead of the current at every instant ($\phi_V - \phi_I = 90°$). These phase relations are represented in figs. A1.1.

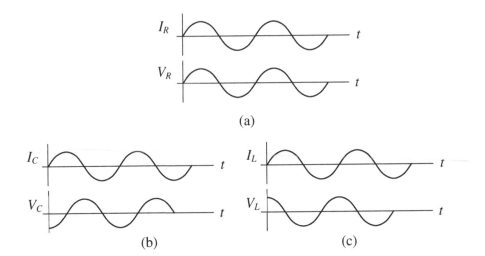

Figure A1.1

Voltage is (a) In phase with the current through a resistor (b) 90° behind
the current through a capacitor and (c) 90° ahead of the current through an inductor

Explanations of how the phase relations between the current and voltage arise are presented in detail in an introductory course in physics or electrical engineering. (See, for example, Serway and Jewett, 2004, pp. 1034–1042.)

An antiquated terminology for voltage that is still occasionally used is the *electromotive force* or *emf.* With E used to represent voltage (emf), the mnemonics below can help the reader remember the phase relations between current and voltage (emf) for the three devices. They are:

ELI ⇒ voltage (E) across an inductor (L) is 90° ahead of current (I) through it.

ICE ⇒ current (I) through a capacitor (C) is 90° ahead of voltage (E) across it.

EIR ⇒ voltage (E) across a resistor (R) and current (I) through it are together (are in phase).

Referring to figs. A1.1, we see that there is as much positive voltage or current as there is negative over one cycle. Therefore, the average of a sinu-- soidal voltage or current over one period is zero. As such, average voltage and current do not describe the maximum or peak values, V_P and I_P. Instead V_P and I_P are found by taking the averages of the squares of the voltage and current over one cycle. These are called the *root mean squared* or *rms* values of the quantities. The rms value of the voltage, for example, is given by

$$V_{rms} \equiv \sqrt{\frac{1}{T}\int_0^T [V(t)]^2\, dt} = \sqrt{\frac{1}{T}V_P^2 \int_0^T \sin^2(2\pi ft)\, dt} = \frac{V_P}{2\sqrt{2}} \qquad \text{(A1.2)}$$

where $T = 1/f$, is the inverse of the frequency of the AC voltage and current. It is the time for the oscillations of these quantities to complete one cycle.

The relation between the rms (or peak) voltage across a resistor and the rms (or peak) current through the resistor is given by *Ohm's law*,

$$V_R = I_R R \tag{A1.3}$$

where R is the resistance in ohms.

Ohm's law relating the rms voltage and current for a capacitor in an AC circuit is

$$V_C = I_C X_C \tag{A1.4}$$

where X_C is a resistance-like quantity called the *capacitive reactance,* defined by

$$X_C = \frac{1}{2\pi f C} \tag{A1.5}$$

The device number C is the *capacitance* of the capacitor, a measure of the maximum charge that can be stored on the device. The value of C is expressed in farads. The capacitance of a typical capacitor is of the order of microfarads (μfd).

Ohm's law for an inductor in an AC circuit is

$$V_L = I_L X_L \tag{A1.6}$$

X_L, the *inductive reactance*, is defined as

$$X_L = 2\pi f L \tag{A1.7}$$

L, the *inductance* of the inductor, a measure of the coil's resistance to a change in the current through it, is measured in henries. The inductance of a typical coil is of the order of millihenries (mH).

The phase difference between the voltage across and the current through these devices can be described graphically on a two-dimensional *phase diagram.*

Series AC circuit

An AC circuit containing a resistor, a capacitor, and inductor in series connected to an AC generator is shown in fig. A1.2.

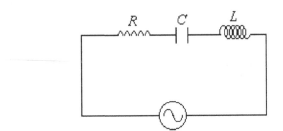

Figure A1.2
An *RCL* AC series circuit

In such a circuit, the same current flows through each device and the sum of the potential differences across the devices is the potential difference across the generator.

Because the current is common to all three devices, the line representing current is taken to be along the positive horizontal axis. The voltages are represented by lines drawn at appropriate phase angles relative to the current. Referring to the mnemonics above, the voltage across the resistor is in phase with the current through the three devices. Therefore, V_R is also along the positive horizontal axis. The voltage across the capacitor is $90°$ behind the current through it, so V_C is along the negative vertical axis. Because V_L leads the current by $90°$, V_L is along the positive vertical axis. The lengths of the lines represent the respective rms or peak values of V_R, V_L, and V_C. This is shown in fig. A1.3a.

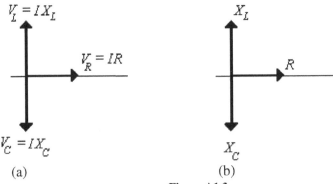

(a)   (b)
Figure A1.3
Phase diagram for an AC series circuit in terms of (a) V_R, V_L, V_C and (b) R, X_L and X_C

When the phase diagram of fig. A1.3a is viewed as a complex plane with complex numbers to represent the current and voltages, the current, being along the real axis, is a real number. Therefore, the voltages become

$$V_R \rightarrow V_R \quad \text{(a real number)} \tag{A1.8a}$$

$$V_C \rightarrow -iV_C \quad \text{(a negative imaginary number)} \tag{A1.8b}$$

and

$$V_L \rightarrow iV_L \quad \text{(a positive imaginary number)} \tag{A1.8c}$$

Each of these reactances acts as a resistor in a DC circuit in that the reactance is a measure of how much the device "impedes" the flow of current. From eqs. A1.3, A1.4, and A1.6 we see that for a given voltage across a device, the larger the reactance of the device, the smaller the current through it. As such, for AC circuits, the complex number that represents the reactance of a device is called its *complex impedence*.

Each voltage can be written as real current multiplied by either a real or imaginary reactance, therefore the phase of each voltage can be associated with the reactances. Thus, the impedence of a resistor is real, the capacitive impedence is negative imaginary, and the inductive impedence is positive imaginary. Then, the phase diagram can be described in terms of these *complex impedences* with the lengths of the lines representing the magnitudes of the reactances of the devices, as shown in fig. A1.3b.

Adding the voltages across the devices in the series circuit, we obtain

$$V_{net} = V_R + iV_L - iV_C \tag{A1.9a}$$

Expressing the voltages in terms of the common current and various impedences, this becomes

$$IZ_{net} = IR + I\left(iX_L - iX_C\right) \tag{A1.9b}$$

from which

$$Z_{net} = R + iX_L + \left(-iX_C\right) = R + i\left(X_L - iX_C\right) \tag{A1.10}$$

Referring to eq. 2.77a, this is the way the equivalent resistance of resistors in series in a DC circuit is calculated. That is, with

$$Z_R = R \qquad\qquad\qquad\qquad\qquad \text{(A1.11a)}$$

$$Z_C = -iX_C \qquad\qquad\qquad\qquad \text{(A1.11b)}$$

and

$$Z_L = iX_L \qquad\qquad\qquad\qquad\qquad \text{(A1.11c)}$$

the net impedence of an AC series circuit is obtained by adding the complex impedences directly.

Parallel AC circuit

For an AC circuit containing a resistor, capacitor, and inductor in parallel, the potential differences across the three devices are the same, and the sum of the currents through the devices adds to the current drawn from the generator.

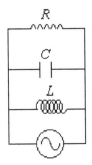

Figure A1.4
An *RCL* AC parllel circuit

Because the voltage across all three devices is the same, the line representing this common voltage is taken to be along the positive horizontal axis and the currents are represented by lines drawn at appropriate phase angles relative to the voltage. Referring to the mnemonics above, the current through the resistor is in phase with the voltage across it. Therefore, I_R is along the positive horizontal axis. The current through the capacitor is 90° ahead of the voltage across it, so I_C is along the positive vertical axis. Because I_L lags behind V_L by 90°, I_L is along the negative vertical axis. The lengths of the lines represent the respective rms or peak values of I_R, I_L, and I_C as shown in fig. A1.5a.

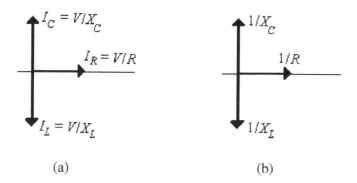

(a)   (b)

Figure A1.5

Phase diagram for an AC series circuit in terms of (a) V_R, V_L, V_C and (b) R, X_L and X_C

When the phase diagram of fig. A1.5a is viewed as a complex plane, the voltage is a real number and the currents carry the phase angles. Therefore,

$$I_R \rightarrow I_R \quad \text{(a real number)} \tag{A1.12a}$$

$$I_C \rightarrow iI_C \quad \text{(a positive imaginary number)} \tag{A1.12b}$$

and

$$I_L \rightarrow -iI_L \quad \text{(a negative imaginary number)} \tag{A1.12c}$$

Then the current drawn from the generator is

$$I_{net} = I_R + i(I_C - I_L) \tag{A1.13a}$$

from which

$$\frac{V}{Z_{net}} = \frac{V}{R} + i\left(\frac{V}{X_C} - \frac{V}{X_L}\right) = \frac{V}{R} + \frac{V}{iX_L} - \frac{V}{iX_C} \tag{A1.13b}$$

Cancelling the common factor V we obtain the phase diagram of fig. A1.5b, and eq. A1.13b yields

$$\frac{1}{Z_{net}} = \frac{1}{R} + \frac{1}{iX_L} + \frac{1}{-iX_C} \tag{A1.14}$$

Referring to eq. 2.77b, this is analogous to the computation one would perform to obtain the equivalent of resistors in a parallel circuit. That is, with

$$Z_R = R \tag{A1.11a}$$

$$Z_C = -iX_C \tag{A1.11b}$$

and

$$Z_L = iX_L \tag{A1.11c}$$

the net impedence of an AC parallel circuit is obtained by adding the inverses of the complex impedences.

Therefore, for both series and parallel combinations, an AC circuit can be treated as a DC resistive circuit, with resistances replaced by the complex impedences given in eqs. A1.11.

APPENDIX 2

DERIVATION OF GREEN'S THEOREM

Let R be a region in the complex plane within a closed contour C. We define an integral around this closed contour (in either the clockwise or counterclockwise direction) as

$$I_C \equiv \oint_C \left[G_x(x, y)\,dx + G_y(x, y)\,dy \right] \tag{A2.1}$$

where $G_x(x, y)$ and $G_y(x, y)$ are two real functions of x and y.

It is expected that the reader has had some basic experience with vector algebra, so is familiar with a dot product, understands that a vector in two dimensions is defined in terms of its components as

$$\vec{G}(x, y) = \left[G_x(x, y), G_y(x, y) \right] \tag{A2.2}$$

and that a two-dimensional vector line element along the contour C is an infinitesimal segment of C pointing in the direction of traversal. Its components are defined by

$$d\vec{\ell} = \left[dx, dy \right] \tag{A2.3}$$

With these descriptions, the integral of eq. A2.1 can be written

$$\oint_C \left[G_x(x, y)\,dx + G_y(x, y)\,dy \right] = \oint_C \vec{G} \cdot d\vec{\ell} \tag{A2.4}$$

Let us segment the region R into a large (ultimately infinite) number of rectangles of small (ultimately zero) area as shown in fig. A2.1.

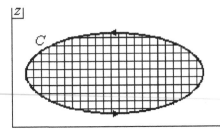

Figure A2.1

The region inside a closed contour segmented into a large
number of small rectangles

We integrate around each rectangle in the counterclockwise direction (the direction of traversal of the contour) and add the integrals around all the rectangles.

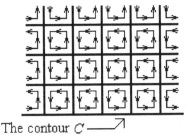

The contour C ——↗

Figure A2.2

Integration around the rectangles that cover the region R

We see in fig. A2.2 that the edge of a rectangle that is not along the contour C is adjacent to the edge of another rectangle, such that these two adjacent edges are traversed in the opposite directions. As such, the integrals along all pairs of interior edges cancel. The edge of a rectangle that comprises a small segment of the contour is not paired with another edge traversed in the opposite direction. Therefore, there is no cancellation of the integral along such an exterior edge. Thus, when the integrals around all rectangles are added together, contributions from all interior sides cancel, and contributions from exterior sides combine to yield the integral around the closed contour. That is,

$$\oint_C \left[G_x(x,y)dx + G_y(x,y)dy \right] = \sum_{rectangles} \oint_{\substack{one \\ rectangle}} \vec{G} \cdot d\vec{\ell} \qquad (A2.5)$$

Consider the integral around a small rectangle with sides of length Δx and Δy, the lower left-hand corner of which is at the point (x,y) as shown in fig. A2.3.

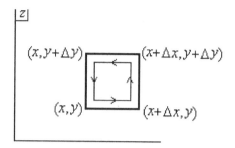

Figure A2.3

The direction of traversal around a small rectangle

The integral around this rectangle is given by

$$\oint_{\substack{one \\ rectangle}} \vec{G} \cdot d\vec{\ell} = \int_{x}^{x+\Delta x} G_x(x, y)\, dx + \int_{y}^{y+\Delta y} G_y(x+\Delta x, y)\, dy$$

$$+ \int_{x+\Delta x}^{x} G_x(x, y+\Delta y)\, dx + \int_{y+\Delta y}^{y} G_y(x, y)\, dx \qquad (A2.6)$$

$$= \int_{x}^{x+\Delta x} \left[G_x(x, y) - G_x(x, y+\Delta y) \right] dx$$

$$+ \int_{y}^{y+\Delta y} \left[G_y(x+\Delta x, y) - G_y(x, y) \right] dy$$

We now take the lengths of the sides of the rectangle to zero to obtain

$$\lim_{\Delta x \to 0} \left[G_y(x+\Delta x, y) - G_y(x, y) \right] = \lim_{\Delta x \to 0} \frac{\partial G_y}{\partial x} \Delta x \qquad (A2.7a)$$

and

$$\lim_{\Delta y \to 0} \left[G_x(x, y+\Delta y) - G_x(x, y) \right] = \lim_{\Delta y \to 0} \frac{\partial G_x}{\partial y} \Delta y \qquad (A2.7b)$$

Therefore,

$$\lim_{\substack{\Delta x \to 0 \\ \Delta y \to 0}} \left\{ \int_x^{x+\Delta x} \left[G_x(x, y) - G_x(x, y + \Delta y) \right] dx \right.$$

$$\left. + \int_y^{y+\Delta y} \left[G_y(x + \Delta x, y) - G_y(x, y) \right] dy \right\} \qquad \text{(A2.8)}$$

$$= \lim_{\substack{\Delta x \to 0 \\ \Delta y \to 0}} \left\{ \Delta x \int_y^{y+\Delta y} \frac{\partial G_y}{\partial x} dy - \Delta y \int_x^{x+\Delta x} \frac{\partial G_x}{\partial y} dx \right\}$$

To determine

$$\lim_{\Delta x \to 0} \int_x^{x+\Delta x} \frac{\partial G_x}{\partial y} dx \quad \text{and} \quad \lim_{\Delta y \to 0} \int_y^{y+\Delta y} \frac{\partial G_y}{\partial x} dy$$

for each of these integrals, we define an indefinite integral to be a function $H(x, y)$. That is,

$$H(x, y) \equiv \int \frac{\partial G_x}{\partial y} dx \qquad \text{(A2.9)}$$

Then

$$\lim_{\Delta x \to 0} \int_x^{x+\Delta x} \frac{\partial G_x}{\partial y} dx = \lim_{\Delta x \to 0} \left[H(x + \Delta x, y) - H(x, y) \right]$$

$$\qquad \text{(A2.10a)}$$

$$= \lim_{\Delta x \to 0} \frac{\partial H}{\partial x} \Delta x = \lim_{\Delta x \to 0} \frac{\partial G_x}{\partial y} \Delta x$$

By identical analysis

$$\lim_{\Delta y \to 0} \int_y^{y+\Delta y} \frac{\partial G_y}{\partial x} dy = \lim_{\Delta y \to 0} \frac{\partial G_y}{\partial x} \Delta y \qquad \text{(A2.10b)}$$

Therefore, the integral of eq. A2.6 around one infinitesimal rectangle is

$$\oint_{\substack{one \\ rectangle}} \vec{G} \cdot d\vec{\ell} = \lim_{\substack{\Delta x \to 0 \\ \Delta y \to 0}} \left[\frac{\partial G_y}{\partial x} - \frac{\partial G_x}{\partial y} \right] \Delta x \, \Delta y \qquad \text{(A2.11)}$$

Summing over all rectangles that comprise the region R, eq. A2.5 becomes

$$\oint_C \left[G_x(x, y)\,dx + G_y(x, y)\,dy \right] = \lim_{\substack{\Delta x \to 0 \\ \Delta y \to 0}} \sum_{rectangles} \left[\frac{\partial G_y}{\partial x} - \frac{\partial G_x}{\partial y} \right] \Delta x\,\Delta y$$

$$= \iint_R \left[\frac{\partial G_y}{\partial x} - \frac{\partial G_x}{\partial y} \right] dx\,dy$$

$$(A2.12)$$

Eq. A2.12 is Green's theorem, which is Stokes' theorem in two dimensions.

APPENDIX 3

DERIVATION OF THE GEOMETRIC SERIES

Let u and t be complex quantities. Starting with the identity

$$\left(u^N - t^N\right) = (u-t)\left(u^{N-1} + tu^{N-2} + t^2 u^{N-3} + \cdots + t^{N-2}u + t^{N-1}\right) \quad \text{(A3.1)}$$

we set $u = 1$ to obtain

$$\left(1 - t^N\right) = (1-t)\left(1 + t + t^2 + \cdots + t^{N-2} + t^{N-1}\right) \quad \text{(A3.2a)}$$

$$\frac{\left(1 - t^N\right)}{(1-t)} = \left(1 + t + t^2 + \cdots + t^{N-2} + t^{N-1}\right) = \sum_{\ell=0}^{N-1} t^\ell \quad \text{(A3.2b)}$$

When $|t| < 1$,

$$\lim_{N \to \infty} t^N = 0 \quad \text{(A3.3)}$$

and eq. A3.2b becomes

$$\lim_{N \to \infty} \frac{\left(1 - t^N\right)}{(1-t)} = \frac{1}{(1-t)} = 1 + t + t^2 + \cdots = \sum_{\ell=0}^{\infty} t^\ell \quad \text{(A3.4)}$$

APPENDIX 4

CAUCHY RATIO TEST FOR CONVERGENCE OF A SERIES

An infinite series is defined as

$$S(z) = \sum_{n=0}^{\infty} \sigma_n(z) \qquad \text{(A4.1)}$$

The Nth *partial sum* for this series is defined by

$$S_N(z) \equiv \sum_{n=0}^{N} \sigma_n(z) \qquad \text{(A4.2)}$$

If the infinite series converges (is finite at z), then

$$\lim_{N \to \infty} S_N(z) = S(z) \qquad \text{(A4.3a)}$$

which can be written as

$$\lim_{N \to \infty} \left[S(z) - S_N(z) \right] = 0 \qquad \text{(A4.3b)}$$

Writing $S(z)$ as

$$S(z) = \sum_{n=0}^{N} \sigma_n(z) + \sum_{n=N+1}^{\infty} \sigma_n(z) \qquad \text{(A4.4)}$$

449

eq. 4.3b becomes

$$\lim_{N\to\infty} \sum_{n=N+1}^{\infty} \sigma_n(z) = 0 \qquad\qquad (A4.5a)$$

Thus, as N approaches ∞, the individual terms must approach zero. Therefore, a condition for a series to converge is that

$$\lim_{N\to\infty} \sigma_N(z) = 0 \qquad\qquad (A4.5b)$$

This condition is necessary but is not sufficient for the convergence of $S(z)$. The Cauchy *ratio test*[†] provides a sufficiency condition as to whether a series converges.

Let two successive terms in $S(z)$ be $\sigma_N(z)$ and $\sigma_{N+1}(z)$ and let N be large enough that we can be assured that $|\sigma_{N+1}(z)| < |\sigma_N(z)|$ and that $\rho_N(z)$ is arbitrarily close to its limiting value

$$\rho(z) = \lim_{N\to\infty} \rho_N(z) \qquad\qquad (A4.6)$$

The Cauchy ratio is defined by

$$\rho_N(z) \equiv \left| \frac{\sigma_{N+1}(z)}{\sigma_N(z)} \right| \simeq \rho(z) \qquad\qquad (A4.7)$$

from which

$$|\sigma_{N+1}(z)| = \rho(z)|\sigma_N(z)| \qquad\qquad (A4.8a)$$

Then,

$$|\sigma_{N+2}(z)| = \rho(z)|\sigma_{N+1}(z)| = \rho^2(z)|\sigma_N(z)| \qquad\qquad (A4.8b)$$

$$\vdots$$

$$|\sigma_{N+n}(z)| = \rho^n(z)|\sigma_N(z)| \qquad\qquad (A4.8c)$$

[†] An earlier version of the ratio test was published some 50 years earlier by D'Lambert. However, Cauchy's name is more commonly associated with the test than is D'Lambert's.

Therefore,

$$\sum_{n=N}^{\infty} |\sigma_n(z)| = |\sigma_N(z)| \sum_{n=0}^{\infty} \rho^n(z) \tag{A4.9}$$

As can be seen from appendix 3, eq. A3.4,

$$\sum_{n=0}^{\infty} \rho^n(z) \text{ is } \begin{cases} \text{finite } \rho(z) < 1 \\ \text{indeterminant } \rho(z) = 1 \\ \text{inifinite } \rho(z) > 1 \end{cases}$$

Therefore, from eq. A4.8, we have

$$\sum_{n=N}^{\infty} \sigma_n(z) \text{ is } \begin{cases} \text{finite } \rho(z) < 1 \\ \text{indeterminant } \rho(z) = 1 \\ \text{inifinite } \rho(z) > 1 \end{cases}$$

From this we determine that

$$\begin{cases} S(z) \text{ converges if } \rho(z) < 1 \\ S(z) \text{ may or may not converge if } \rho(z) = 1 \\ S(z) \text{ diverges if } \rho(z) > 1 \end{cases}$$

Thus, a series converges for those values of z for which

$$\lim_{N \to \infty} \left| \frac{\sigma_{N+1}(z)}{\sigma_N(z)} \right| < 1 \tag{A4.10}$$

APPENDIX 5

EVALUATION OF AN INTEGRAL

Let

$$I \equiv \int_0^\infty e^{-\alpha u^2} du \qquad (A5.1)$$

with α real and positive. Substituting

$$u = \frac{x}{\sqrt{\alpha}} \qquad (A5.2)$$

the integral becomes

$$I = \frac{1}{\sqrt{\alpha}} \int_0^\infty e^{-x^2} dx \qquad (A5.3a)$$

which is unchanged if x is replaced by y. Thus,

$$I = \frac{1}{\sqrt{\alpha}} \int_0^\infty e^{-y^2} dy \qquad (A5.3b)$$

The product of the integrals in eqs. A5.3a and A5.3b is

$$I^2 = \frac{1}{\alpha} \int_0^\infty \int_0^\infty e^{-(x^2+y^2)} dx\, dy \qquad (A5.4)$$

It is evident that the region of integration of the integral of eq. A5.4 is the first quadrant of the x–y plane. As can be seen in fig. A5.1, the quantity *dxdy* is the infinitesimal area element dA_{xy} in Cartesian coordinates.

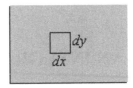

Figure A5.1

Integration over the first quadrant of the x–y plane and an infinitesimal area element in Cartesian coordinates

Transforming to circular coordinates *r* and *θ*, we write

$$r^2 = x^2 + y^2 \qquad\qquad\qquad\qquad\qquad (A5.5a)$$

and

$$\theta = \tan^{-1}\left(\frac{y}{x}\right) \qquad\qquad\qquad\qquad\qquad (A5.5b)$$

Referring to fig. A5.2,

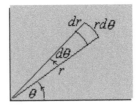

Figure A5.2

Area element in circular coordinates

the infinitesimal area element in circular coordinates is given by

$$dA_{r\theta} = (dr)(rd\theta) = rd\,rd\theta \qquad\qquad\qquad\qquad (A5.6)$$

and the first quadrant is described by

$$0 \le r \le \infty, \quad 0 \le \theta \le \frac{\pi}{2} \qquad\qquad\qquad\qquad (A5.7)$$

Therefore,

$$I^2 = \frac{1}{\alpha} \int_0^{\pi/2} d\theta \int_0^\infty e^{-r^2} r \, dr \tag{A5.8}$$

Eq. A5.8 can be integrated easily to obtain

$$I^2 = \frac{\pi}{4\alpha} \tag{A5.9a}$$

Therefore,

$$I = \int_0^\infty e^{-\alpha u^2} du = \frac{1}{2}\sqrt{\frac{\pi}{\alpha}} \tag{A5.9b}$$

APPENDIX 6

TRANSFORMATION OF LAPLACE'S EQUATION

It was shown in chapter 3 that if $\Phi(x, y)$ satisfies Laplace's equation at some point z, it is analytic in a region containing z. Let w be the image of z under a conformal mapping. Because the mapping is analytic at all points in the region containing z, the derivatives $\partial u/\partial x$, $\partial u/\partial y$, $\partial v/\partial x$, and $\partial v/\partial y$ are defined at z.

We consider

$$\frac{\partial \Phi(u, v)}{\partial x} = \frac{\partial \Phi}{\partial u} \frac{\partial u}{\partial x} + \frac{\partial \Phi}{\partial v} \frac{\partial v}{\partial x} \tag{A6.1}$$

Then

$$\begin{aligned}
\frac{\partial^2 \Phi}{\partial x^2} &= \frac{\partial}{\partial x} \left[\frac{\partial \Phi}{\partial u} \frac{\partial u}{\partial x} + \frac{\partial \Phi}{\partial v} \frac{\partial v}{\partial x} \right] \\
&= \frac{\partial u}{\partial x} \frac{\partial}{\partial x} \left(\frac{\partial \Phi}{\partial u} \right) + \frac{\partial \Phi}{\partial u} \frac{\partial^2 u}{\partial x^2} + \frac{\partial v}{\partial x} \frac{\partial}{\partial x} \left(\frac{\partial \Phi}{\partial v} \right) + \frac{\partial \Phi}{\partial v} \frac{\partial^2 v}{\partial x^2}
\end{aligned} \tag{A6.2}$$

With

$$\begin{aligned}
\frac{\partial}{\partial x} \left(\frac{\partial \Phi}{\partial u} \right) &= \frac{\partial}{\partial u} \left(\frac{\partial \Phi}{\partial u} \right) \frac{\partial u}{\partial x} + \frac{\partial}{\partial v} \left(\frac{\partial \Phi}{\partial u} \right) \frac{\partial v}{\partial x} \\
&= \frac{\partial^2 \Phi}{\partial u^2} \frac{\partial u}{\partial x} + \frac{\partial^2 \Phi}{\partial u \partial v} \frac{\partial v}{\partial x}
\end{aligned} \tag{A6.3a}$$

and

$$\frac{\partial}{\partial x}\left(\frac{\partial\Phi}{\partial v}\right) = \frac{\partial}{\partial u}\left(\frac{\partial\Phi}{\partial v}\right)\frac{\partial u}{\partial x} + \frac{\partial}{\partial v}\left(\frac{\partial\Phi}{\partial v}\right)\frac{\partial v}{\partial x}$$

$$= \frac{\partial^2\Phi}{\partial u\,\partial v}\frac{\partial u}{\partial x} + \frac{\partial^2\Phi}{\partial v^2}\frac{\partial v}{\partial x} \tag{A6.3b}$$

eq. A6.2 becomes

$$\frac{\partial^2\Phi}{\partial x^2} = \frac{\partial\Phi}{\partial u}\frac{\partial^2 u}{\partial x^2} + \frac{\partial\Phi}{\partial v}\frac{\partial^2 v}{\partial x^2}$$

$$+ \left(\frac{\partial u}{\partial x}\right)^2\left(\frac{\partial^2\Phi}{\partial u^2}\right) + \left(\frac{\partial v}{\partial x}\right)^2\left(\frac{\partial^2\Phi}{\partial v^2}\right) + 2\frac{\partial u}{\partial x}\frac{\partial v}{\partial x}\left(\frac{\partial^2\Phi}{\partial u\,\partial v}\right) \tag{A6.4a}$$

By replacing x by y, it is evident that

$$\frac{\partial^2\Phi}{\partial y^2} = \frac{\partial\Phi}{\partial u}\frac{\partial^2 u}{\partial y^2} + \frac{\partial\Phi}{\partial v}\frac{\partial^2 v}{\partial y^2}$$

$$+ \left(\frac{\partial u}{\partial y}\right)^2\left(\frac{\partial^2\Phi}{\partial u^2}\right) + \left(\frac{\partial v}{\partial y}\right)^2\left(\frac{\partial^2\Phi}{\partial v^2}\right) + 2\frac{\partial u}{\partial y}\frac{\partial v}{\partial y}\left(\frac{\partial^2\Phi}{\partial u\,\partial v}\right) \tag{A6.4b}$$

Therefore, Laplace's equation becomes

$$\frac{\partial^2\Phi}{\partial x^2} + \frac{\partial^2\Phi}{\partial y^2} = 0 = \frac{\partial\Phi}{\partial u}\left(\frac{\partial^2 u}{\partial x^2} + \frac{\partial^2 u}{\partial y^2}\right) + \frac{\partial\Phi}{\partial v}\left(\frac{\partial^2 v}{\partial x^2} + \frac{\partial^2 v}{\partial y^2}\right)$$

$$+ \frac{\partial^2\Phi}{\partial u^2}\left[\left(\frac{\partial u}{\partial x}\right)^2 + \left(\frac{\partial u}{\partial y}\right)^2\right] + \frac{\partial^2\Phi}{\partial v^2}\left[\left(\frac{\partial v}{\partial x}\right)^2 + \left(\frac{\partial v}{\partial y}\right)^2\right] \tag{A6.5}$$

$$+ 2\frac{\partial^2\Phi}{\partial u\,\partial v}\left[\frac{\partial u}{\partial x}\frac{\partial v}{\partial x} + \frac{\partial u}{\partial y}\frac{\partial v}{\partial y}\right]$$

Because $f(z)$ is analytic, u and v satisfy Laplace's equation. Therefore,

$$\left(\frac{\partial^2 u}{\partial x^2} + \frac{\partial^2 u}{\partial y^2}\right) = \left(\frac{\partial^2 v}{\partial x^2} + \frac{\partial^2 v}{\partial y^2}\right) = 0 \qquad \text{(A6.6a)}$$

From the CR conditions,

$$\left[\frac{\partial u}{\partial x}\frac{\partial v}{\partial x} + \frac{\partial u}{\partial y}\frac{\partial v}{\partial y}\right] = 0 \qquad \text{(A6.6b)}$$

and

$$\left[\left(\frac{\partial u}{\partial x}\right)^2 + \left(\frac{\partial u}{\partial y}\right)^2\right] = \left[\left(\frac{\partial v}{\partial x}\right)^2 + \left(\frac{\partial v}{\partial y}\right)^2\right] \qquad \text{(A6.6c)}$$

Substituting eqs. A6.6 into eq. A6.5, we obtain

$$\begin{aligned}
&\left[\frac{\partial^2 \Phi}{\partial u^2} + \frac{\partial^2 \Phi}{\partial v^2}\right]\left[\left(\frac{\partial u}{\partial x}\right)^2 + \left(\frac{\partial u}{\partial y}\right)^2\right] \\
&= \left[\frac{\partial^2 \Phi}{\partial u^2} + \frac{\partial^2 \Phi}{\partial v^2}\right]\left[\left(\frac{\partial v}{\partial x}\right)^2 + \left(\frac{\partial v}{\partial y}\right)^2\right] = 0
\end{aligned} \qquad \text{(A6.7)}$$

Because u and v are not, in general, constants

$$\left[\left(\frac{\partial u}{\partial x}\right)^2 + \left(\frac{\partial u}{\partial y}\right)^2\right] = \left[\left(\frac{\partial v}{\partial x}\right)^2 + \left(\frac{\partial v}{\partial y}\right)^2\right] \neq 0 \qquad \text{(A6.8)}$$

Therefore, eq. A6.7 requires that

$$\left[\frac{\partial^2 \Phi}{\partial u^2} + \frac{\partial^2 \Phi}{\partial v^2}\right] = \nabla^2_{u,v}\Phi = 0 \qquad \text{(A6.9)}$$

APPENDIX 7

TRANSFORMATION OF BOUNDARY CONDITIONS

It is expected that the reader understands that points in the two-dimensional plane can be located by vectors as well as by complex numbers. Let C be a curve in the z-plane (e.g., a curve that defines the boundary of a region R). Such a curve is defined by an equation of the form

$$\Phi(z) = \lambda \tag{A7.1}$$

Let $z = (x, y)$ be a point on C and let the point $z + dz = (x + dx, y + dy)$ be differentially distant from z. We define

$$\vec{e}_x \text{ and } \vec{e}_y$$

to be the x and y unit basis vectors directed along the positive x and positive y axes. (These are often called

$$\vec{i} \text{ and } \vec{j}$$

in the literature.) In the vector description, a point is described by

$$\vec{z}(x, y) = x\vec{e}_x + y\vec{e}_y \tag{A7.2a}$$

Because in Cartesian coordinates the basis vectors are constants,

$$d\vec{z} = (dx)\vec{e}_x + (dy)\vec{e}_y \qquad\qquad (A7.2b)$$

The change in $\Phi(z)$ is given by

$$d\Phi(z) = \Phi(x+dx, y+dy) - \Phi(x, y) = \frac{\partial\Phi}{\partial x}dx + \frac{\partial\Phi}{\partial y}dy \qquad (A7.3)$$

From the definition of a dot product, this can be expressed as

$$d\Phi(z) = \left(\frac{\partial\Phi}{\partial x}\vec{e}_x + \frac{\partial\Phi}{\partial y}\vec{e}_y\right) \cdot \left((dx)\vec{e}_x + (dy)\vec{e}_y\right) = \left(\nabla_{xy}\Phi\right) \cdot d\vec{z} \quad (A7.4)$$

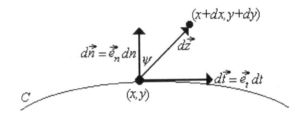

Figure A7.1

Vectors defined in a normal–tangent coordinate system

$d\Phi$ can also be described in a coordinate system defined by one axis along the tangent to C at (x, y) and a second axis along the normal to that tangent, as shown in fig. A7.1. Defining the unit basis vectors in this coordinate system to be

$$\vec{e}_t \quad \text{and} \quad \vec{e}_n$$

the change in $\Phi(z)$ can be written

$$d\Phi(z) = \frac{\partial\Phi}{\partial n}dn + \frac{\partial\Phi}{\partial t}dt \qquad\qquad (A7.5)$$

where

$$\frac{\partial\Phi}{\partial n} \quad \text{and} \quad \frac{\partial\Phi}{\partial t}$$

are the changes in Φ in the normal and tangential directions, and dn and dt are infinitesimal displacements along the normal and tangential basis vectors, respectively. But because the curve C is defined by $\Phi = \lambda =$ constant, Φ does not vary in the tangential direction. That is,

$$\frac{\partial \Phi}{\partial t} = 0 \tag{A7.6}$$

and Φ only depends on the normal coordinate. Therefore, eq. A7.4 becomes

$$d\Phi = \frac{\partial \Phi}{\partial n} dn = \frac{d\Phi}{dn} dn \tag{A7.7}$$

Referring to fig. A7.1, we see that because the basis vector in the direction of the normal is of magnitude 1,

$$dn = |d\vec{z}| \cos \psi = |\vec{e}_n||d\vec{z}| \cos \psi = \vec{e}_n \cdot d\vec{z} \tag{A7.8}$$

Then eq. A7.7 can be written

$$d\Phi = \frac{\partial \Phi}{\partial n} \vec{e}_n \cdot d\vec{z} \tag{A7.9}$$

Comparing the descriptions of $d\Phi$ given in eqs. A7.4 and A7.9, we see that on a curve defined by eq. A7.1,

$$\left[\frac{\partial \Phi}{\partial n} \Big|_{z_{on C}} \right] \vec{e}_n = \left(\nabla_{xy} \Phi \right)_{z_{on C}} = \left(\frac{\partial \Phi}{\partial x} \vec{e}_x + \frac{\partial \Phi}{\partial y} \vec{e}_y \right)_{z_{on C}} \tag{A7.10}$$

This means that when a function is described in terms of normal–tangent coordinates, its gradient in the x–y plane, evaluated at points on a boundary curve, is the derivative of the function with respect to the normal to the curve, and is along that normal.

When quantities are described in the complex plane instead of as vectors, eq. A7.10 becomes

$$\frac{\partial \Phi}{\partial n} \Big|_{z_{on C}} = \left(\nabla_{xy} \Phi \right)_{z_{on C}} = \left(\frac{\partial \Phi}{\partial x} + i \frac{\partial \Phi}{\partial y} \right)_{z_{on C}} \tag{A7.11}$$

Under a conformal transformation $w = f(z)$,

$$\Phi(z) = \Phi(x, y) \rightarrow \Phi(w) = \Phi(u, v) \tag{A7.12}$$

and the gradient transforms as

$$\nabla_{xy}\Phi = \left(\frac{\partial\Phi}{\partial x} + i\frac{\partial\Phi}{\partial y}\right) = \left(\frac{\partial\Phi}{\partial u}\frac{\partial u}{\partial x} + \frac{\partial\Phi}{\partial v}\frac{\partial v}{\partial x}\right) + i\left(\frac{\partial\Phi}{\partial u}\frac{\partial u}{\partial y} + \frac{\partial\Phi}{\partial v}\frac{\partial v}{\partial y}\right)$$

$$= \frac{\partial\Phi}{\partial u}\left(\frac{\partial u}{\partial x} + i\frac{\partial u}{\partial y}\right) + \frac{\partial\Phi}{\partial v}\left(\frac{\partial v}{\partial x} + i\frac{\partial v}{\partial y}\right)$$

$$\tag{A7.13}$$

Because the mapping is conformal, $f(z)$ is analytic everywhere in a region containing the point (x, y), and the CR conditions

$$\frac{\partial u}{\partial x} = \frac{\partial v}{\partial y} \tag{3.7a}$$

and

$$\frac{\partial u}{\partial y} = -\frac{\partial v}{\partial x} \tag{3.7b}$$

are valid. Using them to substitute for $\partial u/\partial x$ and $\partial v/\partial x$, eq. A7.13 becomes

$$\nabla_{xy}\Phi = \left(\frac{\partial\Phi}{\partial u} + i\frac{\partial\Phi}{\partial v}\right)\left(\frac{\partial v}{\partial y} + i\frac{\partial u}{\partial y}\right) = (\nabla_{uv}\Phi)\left(\frac{\partial v}{\partial y} + i\frac{\partial u}{\partial y}\right) \tag{A6.14a}$$

Substituting for $\partial u/\partial y$ and $\partial v/\partial y$, eq. A7.13 becomes

$$\nabla_{xy}\Phi = \left(\frac{\partial\Phi}{\partial u} + i\frac{\partial\Phi}{\partial v}\right)\left(\frac{\partial u}{\partial x} - i\frac{\partial v}{\partial x}\right) = (\nabla_{uv}\Phi)\left(\frac{\partial u}{\partial x} - i\frac{\partial v}{\partial x}\right) \tag{A7.14b}$$

Equations A7.14 describe the transformation of the gradient from the w-plane to the z-plane. The transformation from the z-plane to the w-plane is obtained easily by interchanging u and x and interchanging v and y. The results are

$$\nabla_{uv}\Phi = \left(\nabla_{xy}\Phi\right)\left(\frac{\partial y}{\partial v} + i\frac{\partial x}{\partial v}\right) \tag{A7.15a}$$

and

$$\nabla_{uv}\Phi = \left(\nabla_{xy}\Phi\right)\left(\frac{\partial x}{\partial u} - i\frac{\partial y}{\partial u}\right) \tag{A7.15b}$$

Appendix 8

THE BETA FUNCTION

From the definition of the gamma function

$$(z-1)! = \Gamma(z) = \int_0^\infty u^{(z-1)} e^{-u} du \tag{10.47}$$

we consider

$$\Gamma(z)\Gamma(w) = \int_0^\infty \int_0^\infty u^{(z-1)} v^{(w-1)} e^{-(u+v)} du\, dv \tag{A8.1}$$

We make the substitutions $u = x^2$ and $v = y^2$ to obtain

$$\Gamma(z)\Gamma(w) = 4\int_0^\infty \int_0^\infty x^{(2z-1)} y^{(2w-1)} e^{-(x^2+y^2)} dx\, dy \tag{A8.2}$$

We take x and y represent the Cartesian coordinates and transform to polar coordinates with

$$x = r\cos\theta \tag{2.21a}$$

$$y = r\sin\theta \tag{2.21b}$$

and, as shown in appendix 5,

$$dx\, dy \to rd\, rd\theta \tag{A5.6}$$

Then, eq. A8.2 can be written as

$$\Gamma(z)\Gamma(w) = 4\int_0^{\pi/2} \cos^{(2z-1)}\theta \sin^{(2w-1)}\theta d\theta \int_0^\infty r^{(2z+2w-1)}e^{-r^2}dr \quad (A8.3)$$

Setting $t = r^2$ we obtain

$$\int_0^\infty r^{(2z+2w-1)}e^{-r^2}dr = \frac{1}{2}\int_0^\infty t^{(z+w-1)}e^{-t}dt = \frac{1}{2}\Gamma(z+w) \quad (A8.4)$$

from which

$$\frac{\Gamma(z)\Gamma(w)}{\Gamma(z+w)} = 2\int_0^{\pi/2}\cos^{(2z-1)}\theta\sin^{(2w-1)}\theta d\theta \equiv \beta(z,w) \quad (A8.5)$$

It is trivial to see from its relation to the combination of gamma functions that

$$\beta(z,w) = \beta(w,z) \quad (A8.6)$$

It is straightforward to demonstrate this for the integral representation by making the substitution

$$\phi = \frac{\pi}{2} - \theta \quad (A8.7)$$

Another representation of $\beta(z,w)$ is obtained by substituting

$$\cos^2\theta = \mu \quad (A8.8)$$

Then eq. A8.5 becomes

$$\beta(z,w) = \int_0^1 \mu^{(z-1)}(1-\mu)^{(w-1)}d\mu \quad (A8.9)$$

And by substituting

$$x = \frac{\mu}{1-\mu} \quad (A8.10)$$

we obtain the representation

$$\beta(z,w) = \int_0^\infty \frac{x^{(z-1)}}{(1+x)^{(z+w)}}dx \quad (A8.11)$$

References

•

Abbot, D. (editor), 1985, *Mathematicians,* Peter Bedrick, New York.

Ahlfors, L. , 1976, *Complex Analysis,* 3rd ed., McGraw-Hill, New York.

Bell, E. T., 1937, *Men of Mathematics*, Simon & Shuster, New York.

Bremermann, H., 1965, *Distributions, Complex Variables and Fourier Transforms,* Addison-Wesley, Reading, MA.

Buck, E. and Buck, R., 1978, *Advanced Calculus,* McGraw-Hill, New York.

Carrier, G., Krook, M. and Person, C., 1966, *Functions of a Complex Variable,* McGraw-Hill, New York.

Christoffel, E., 1870, Uber die Abbildung einer N-Blattrigen Einfach Zusammenhangernder Ebenen Fache auf einen Kreise, *Gottingen Nachrichten*: 359

Churchill, R. and Brown, J. 1990, *Complex Variables and Applications,* 5th ed., McGraw-Hill, NewYork.

Cohen, H., 1992, *Mathematics for Scientists and Engineers,* Prentice-Hall, Englewood Cliffs, NJ.

Contantinescu, F., 1980, *Distributions and Their Applications in Physics,* Translated by W. Jones, Pergamon Press, New York.

Eden, R., Landshoff, P., Olive, D. and Polkinghorne, J., 1966, *The Analytic S-Matrix,* Cambridge University Press, Cambridge, MA.

Fisher, S., 1986, *Complex Variables,* Wadsworth, Belmont, CA.

Hadamard, J., 1898, Theorem on Entire Series (in French), *Acta Mathematica* **22**: 55

Kramers, H., 1926, The Theory of Absorption and Refraction of X-rays, *Nature,* **117**: 775

Kranz, S. , 1999, *Handbook of Complex Variables,* Birkhauser, Boston.

Kronig, R., 1926, On the Theory of Dispersion of X-rays, *Jour. of the Optical Soc. of America* **12**: 547

Marion, J., 1965, *Principles of Vector Analysis,* Academic Press, New York.

Marsden, J. and Hoffman, M., 1999, *Basic Complex Analysis,* 3rd ed., W. H. Freeman and Co., New York.

Saff, E. and Snider, A., 1976, *Fundamentals of Complex Analysis for Mathematics, Science and Engineering,* Prentice-Hall, Englewood Cliffs, NJ.

Schwarz, H., 1869, Conforme Abbildung der Oberflache eines Tetraeders auf die Oberflache einer Kugel, *Jour. Reine Ange. Math.,* **70**: 121

Serway, R. and Jewitt, J., 2004, *Physics for Scientists and Engineers,* 6th ed., Brooks/Cole-Thompson Learning, Belmont, CA.

Spiegel, M., 1964, *Theory and Problems of Complex Variables,* Schaum's Outline Series, McGraw-Hill, New York.

Taylor, A., 1955, *Advanced Calculus,* Ginn, Boston.

Zemanian, A., 1987, *Distribution Theory and Transform Analysis; an Introduction to Generalized Functions with Applications,* Dover, New York.

Index